Evolutionary Computation with Biogeography-based Optimization

Metaheuristics Set

coordinated by
Nicolas Monmarché and Patrick Siarry

Volume 8

Evolutionary Computation with Biogeography-based Optimization

Haiping Ma
Dan Simon

iSTE

WILEY

First published 2017 in Great Britain and the United States by ISTE Ltd and John Wiley & Sons, Inc.

ISTE Ltd
27-37 St George's Road
London SW19 4EU
UK

www.iste.co.uk

John Wiley & Sons, Inc.
111 River Street
Hoboken, NJ 07030
USA

www.wiley.com

Library of Congress Control Number: 2016954823

British Library Cataloguing-in-Publication Data
A CIP record for this book is available from the British Library
ISBN 978-1-84821-807-9

Contents

1

The Science of Biogeography

Biogeography is the science studying the distribution of species and ecosystems in geographic space and through time. It is usually considered a subset of physical geography because it often is related to the study of the physical environment, and how it affects species and shapes their distribution across space. It is concerned not only with habitation patterns, but also with the factors responsible for variations in distribution. It aims to analyze where species live, and in what abundance. Biogeography has strong ties to biology, ecology, evolution, climatology and soil science.

Overview of the chapter

This chapter provides the basic notations and ideas that form the foundation of biogeography-based optimization (BBO). This chapter first gives an introduction to natural biogeography in section 1.1, and then focuses on island biogeography in section 1.2. Some interesting factors that influence biogeography and that inspire BBO algorithmic features are described in section 1.3.

1.1. Introduction

The science of biogeography can be traced to the work of 19th Century naturalists, most notably Alfred Wallace [WAL 06] and Charles Darwin [KEY 01] (see Figure 1.1). Wallace is usually considered the father of biogeography, although Darwin is much better known because of his preeminence in publishing the theory of evolution. Science views the distribution of species in the world as a result of continuous evolution. Some species evolve locally.

Figure 1.1. *Photographs of Charles Darwin (left) and Alfred Wallace (right)*

The science of biogeography answers many varied questions. As writer David Quammen put it [QUA 96], "...biogeography does more than ask Which species? and Where. It also asks Why? and, sometimes more crucially, Why not?" Biogeography developed in an attempt to answer some of these questions, such as why there are so many kinds of animals and plants in the world. It seeks to explain why some of these animals and plants are rare while others are common. It seeks to explain why some animals and plants are widely dispersed while others are confined to a limited area. It seeks to explain why some parts of this world are richer in species than others. The study of biogeography helps us to answer these types of questions.

Modern biogeography is the study of the geographical distribution of animals and plants while taking into account species counts, present and past, the habitats in which they are found, and ecological relationships. By observing the geographic distribution of species, we can see that the following factors are associated with biogeography: air pressure, physiography, ocean currents, latitude, temperature, amount of sun light, precipitation and wind. Biogeography combines information and ideas from many fields, ranging from the physiological and ecological constraints on species dispersal, to geological and climatological phenomena that operate at global spatial scales and evolutionary time frames. The short-term interactions within a habitat and between species comprise the ecological application of biogeography. Historical biogeography deals with the long-term, evolutionary periods of time, and broader classifications of species.

There are two important theories in biogeography that have been developed to address the distribution of biological species in the world: the distance-decay theory

[NEK 99] and the island biogeography theory [MAC 67]. The distance-decay theory asserts that the correlation and similarity between species in any two geographical locations will continue decreasing as the distance between the two increases. Island biogeography asserts that those islands that are closely spaced will support more biological species than islands that are far apart. It is this second theory that explains that species on closely spaced islands are rarely threatened by extinction, compared to tiny isolated islands. Geographic information systems scientists say that the above two theories were developed in order to explain the distribution of species, but not the distribution or even the movement of humans. That is, the purpose of these theories is to understand the factors affecting species distribution, to predict future trends in species distribution, and to solve ecological problems that have a spatial aspect.

Because of the focus of this book, we do not emphasize one theory more than the other, and we do not further discuss the essence of the science of biogeography. We focus instead on using island biogeography to inspire an evolutionary algorithm to solve optimization problems: biogeography-based optimization (BBO).

1.2. Island biogeography

In the early 1960s, Robert MacArthur and Edward Wilson began working on mathematical models of island biogeography, culminating in their classic 1967 book *The Theory of Island Biogeography* [MAC 67]. They were mostly interested in the distribution of species between neighboring islands, and mathematical models of the extinction and migration of species. Since MacArthur and Wilson's work, biogeography has become a major subset of biology [HAN 97]. Figure 1.2 shows photographs of Robert MacArthur and Edward Wilson.

Figure 1.2. *Photographs of Robert MacArthur (left) and Edward Wilson (right)*

Biogeography is most keenly focused on islands. Islands are often manageable areas of study because they are more condensed than larger ecosystems on the mainland. Islands are also attractive locations for study because they allow scientists to look at habitats that new invasive species have only recently colonized, and to observe how they disperse throughout the island and change it. Scientists can then apply their understanding to similar but more complex mainland habitats. Islands are very diverse in their biomes, ranging from tropical to arctic climates. This diversity allows for a wide range of species studies in different parts of the world.

Mathematical models of island biogeography describe speciation (the evolution of new species), the migration of species between islands, and the extinction of species. The term island here is descriptive rather than literal. An island is considered any habitat that is geographically isolated from other habitats. In the classic sense of the term, an island is isolated from other habitats by water. But islands can also be habitats that are isolated by stretches of desert, rivers, mountain ranges, predators, man-made artifacts or other obstacles. For example, an island could consist of a riverbank that supports herbs, or a pond that supports insects [HAN 97].

Geographical areas that are friendly to life are said to have a high habitat suitability index (HSI) [WES 87]. Features that correlate with HSI include factors such as rainfall, vegetative diversity, topographic diversity, land area and air temperature. These features that characterize habitability are called suitability index variables (SIVs). In terms of habitability, SIVs are the independent variables of the habitat, and HSI is the dependent variable.

Islands with a high HSI tend to support many species, and islands with a low HSI can support only a few species. Islands with a high HSI have many species that emigrate to nearby habitats, simply by virtue of the large number of species that they host. Emigration from an island with a high HSI does not occur because species want to leave their home; after all, the home island is an attractive place to live. The reason that emigration occurs from these islands is due to the accumulation of random effects on a large number of species with large populations. Emigration occurs as animals ride flotsam, swim, fly or ride the wind to neighboring islands. When a species emigrates from an island, the species does not completely disappear from the island; only a few representatives emigrate, so an emigrating species remains present on its home island while at the same time migrating to a neighboring island.

Islands with a high HSI not only have a high emigration rate, but they have a low immigration rate because they already support many species. The species that arrive at such islands will tend not to survive, even though the HSI is high, because there is too much competition for resources.

Islands with a low HSI have a high immigration rate because of their low populations. Again, this is not because species want to immigrate to such islands; after all, these islands are undesirable places to live. The reason that immigration occurs on these islands is because there is a lot of geographical room for additional species. Whether or not the immigrating species can survive in its new home, and for how long, is another question. However, species diversity is correlated with HSI, so more species arriving at a low HSI island will result in a greater chance that the island's HSI will increase [WES 87].

Figure 1.3 depicts species migration between islands, and Figure 1.4 illustrates a model of species abundance on a single island [MAC 67]. The immigration and emigration rates are functions of the number of species on the island. We have depicted the migration curves as straight lines, but in general they might be more complicated curves, as we will discuss later.

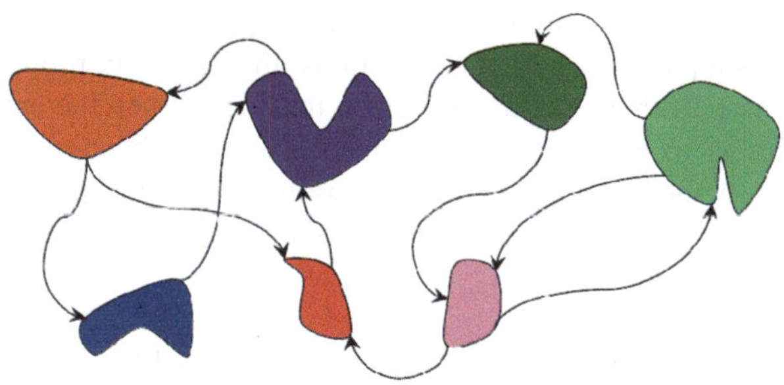

Figure 1.3. *Species migrate between islands via flotsam, wind, flying, swimming and other methods. For a color version of this figure, see www.iste.co.uk/ma-simon/evolutionary.zip*

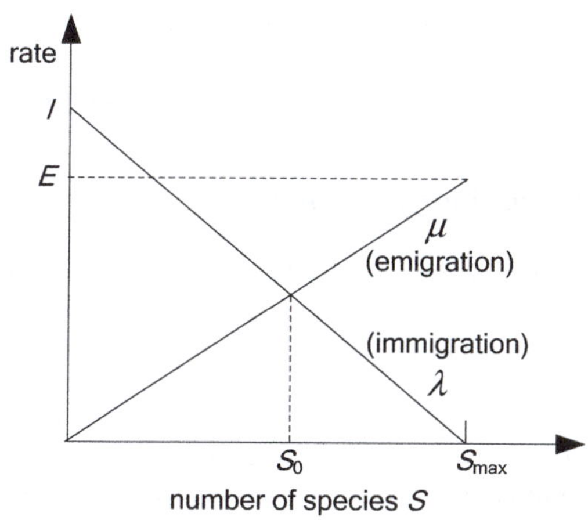

Figure 1.4. *Species migration model of an island, based on [MAC 67], where S_0 is the equilibrium species count*

Consider the immigration curve. The immigration rate λ decreases monotonically as the number of species on the island increases, since as S increases, there is less room on the island for new species. The maximum possible immigration rate to the habitat is I, which occurs when there are zero species on the island. As the number of species increases, the island becomes more crowded, fewer species are able to survive immigration, and the immigration rate decreases. The largest possible number of species that the habitat can support is S_{max}, at which point the immigration rate is zero.

Now consider the emigration curve. The emigration rate μ should be, from reasoning parallel to that above, a monotonically increasing function of S. If area is proportional to population sizes, and emigrations are the chance result of demographic stochasticity, then as the number of species increases, the number of species that are subject to chance emigrations increases in proportion, and the relationship is linear. If there are no species on the island, then the emigration rate is zero. As the number of species on the island increases, it becomes more crowded, more species leave the island, and the emigration rate increases. The maximum emigration rate is E, which occurs when the island contains the largest number of species that it can support.

Finally, consider the equilibrium species count. The migration model predicts that there is a value S_0 at which the immigration rate and the emigration rate balance; there is a dynamic equilibrium. At that point, species on the island immigrate at a rate equal to disappearances due to emigration. This is a stable equilibrium since, if the number of species on the island is perturbed, the imbalance between the immigration and emigration rates at the new S would tend to return island diversity toward its equilibrium value. Below S_0, additional species accumulate and the immigration rate is larger than the emigration rate. Above S_0, the reverse is true, emigrations exceed immigrations, and the number of species declines to S_0.

1.3. Influence factors for biogeography

We have discussed the basic theory of island biogeography, which is the study of the geographical distribution of biological species on islands. There are many interesting influence factors closely associated with biogeography, including the following.

Nonlinear migration: Classic island biogeography theory assumes that the immigration and emigration rates are linear with respect to the number of species, as shown in Figure 1.4. However, there is no reason to suppose that the migration curves are linear. Empirical data suggest that biological migration rates are probably nonlinear functions of the number of species [MAC 67]. Pioneer species are likely to rapidly colonize an island that has a higher immigration rate, and later less robust species follow. They will not only immigrate later, but the rate at which they immigrate will be lower because they have less intrinsic ability for colonization. The rate at which species accumulate on islands is therefore initially rapid and then slower. Also, among poor species, the successful immigration of any one species has less effect on the immigration rates of other species than does the earlier immigration of pioneer species. Therefore, this part of the immigration curve should be flatter; that is, the rate of immigration should be less affected by the arrival of poor species. The result is an immigration rate curve which is nonlinear. For the purpose of simplicity in evaluating the basic implications of the migration model, the emigration curve can be viewed as a mirror image of the immigration curve, and it would also be nonlinear in this case.

Habitat similarity: In island biogeography, immigration rate is correlated with island isolation [ADL 94]. Islands that are isolated are relatively well buffered from immigration. This intuitive idea is called the distance effect [WU 95]. It also stands to reason that emigration rates are correlated with island isolation. The environmental uniqueness of an island is related to island isolation because environmental conditions vary predictably with geographical distance [LOM 00a]. Similar islands could be viewed as clustered together, and belong to the same archipelago. However, dissimilar islands are not part of an archipelago. This tends to increase the immigration and emigration rates between similar islands, and decrease those rates between dissimilar islands. On the other hand, islands in different archipelagos can interact with each other, just as species can migrate across archipelagos. However, migration across archipelagos is less likely than migration within archipelagos. A quantitative way to determine the effect of island isolation on migration rates is given in Hanski [HAN 99]. Figure 1.5 shows two well-known archipelagos: the Fernando de Noronha archipelago in Brazil and the Ksamil archipelago in Albania.

Initial immigration: Classic island biogeography theory indicates that the immigration rate decreases as the number of species increase, as shown in Figure 1.4, which corresponds to a monotonic decrease in immigration rate with the number of species. However, recent advances in biogeography indicate that a monotonic immigration rate curve may be overly simplistic. For some pioneer species, an initial increase in species count results in an initial increase in the immigration rate [WU 95]. This is because these early immigrants modify the island to make it more hospitable to other species. That is, the positive effect of increased diversity due to initial immigration overcomes the negative effect of increased population size, which corresponds to an initial increase in immigration rate as species count increases. This phenomenon can be viewed as a temporary positive feedback mechanism in biogeography. That is, an island with a low HSI accepts species from other islands, increasing its HSI, which subsequently increases its likelihood of accepting even more species from other islands.

Figure 1.5. *The Fernando de Noronha archipelago in Brazil (left) and the Ksamil archipelago in Albania (right)*

Species mobility: Classic island biogeography theory assumes that all species are equal in their migratory ability. In reality, some species are more mobile than others, and some species are better dispersers than others. For example, insect and bird species are generally more mobile than mammals and therefore are more likely to migrate. Figure 1.6 shows that bird species have fast migratory ability and elephant species have slow migratory ability. Efforts have been made in biogeography to incorporate species-specific characteristics into island biogeography theory [LOM 00b]. The species migration model in Figure 1.4 assumes that all species are equally mobile. But the species migration model would be more accurate if species mobility were considered. That is, each individual species would have its own migration curves for each individual island.

Population size: In island biogeography, an island not only has a certain number of species, but each species also has a population size. Those species that are well adapted to their environment tend to increase in population, while those that are not well adapted have a lower equilibrium population. That is, the correlation between a particular species and an island's HSI could be used to determine the equilibrium population of each species. Species with a high HSI contribution would have high equilibrium populations, and those with a low HSI contribution would have low equilibrium populations. We could assume that each species approaches its equilibrium population exponentially [CAS 89]. This approach would result in more flexibility for the species migration model. In addition, those species with a large population would have a greater likelihood of immigrating to neighboring islands. This discussion of population size is similar to species mobility as discussed above.

Figure 1.6. *Bird species with fast migratory ability (left) and elephant species with slow migratory ability (right)*

Species age: In biology, species age influences extinction rate and mobility [GRO 05]. Just as individual mortality is high at a young age, low at middle age and high again at old age, species mortality follows the same pattern. Young species tend to be unstable and susceptible to extinction. Middle-aged species are well established but still mobile. Old species are stagnant and less likely to adapt. That is, species of different ages have different emigration and immigration rates. Species that have been recently introduced to an island have a higher extinction rate and a lower emigration rate, middle-aged species have a lower extinction rate and a higher emigration rate, and older species revert to the pattern of high extinction rate and low emigration rate.

Predator/prey relationships: In biology, certain species have adversarial relationships. These relationships do not necessarily harm the prey species. For instance, prey may respond to predators by reducing the exploitation of their resources, thus benefiting themselves in the long term [HAN 97]. However, the more common scenario is one in which predators reduce prey to such an extent that one or both populations face extinction. Predator/prey relationships can be inferred from a population by examining islands and noting which pairs of species have a low probability of coexisting. Those species can then be modeled as a predator/prey pair. Combining this information with the HSI contribution of each species would result in defining the predator species as the adversary that is positively correlated with island HSI, and the prey species as the adversary that is negatively correlated with island HSI. The predator/prey relationship might lead to a non-zero equilibrium population, or it might lead to the extinction of one or both populations [GOT 08, HAN 97]. Most predator/prey models in biology are for two-species systems, but a more complete description would be obtained if existing predator/prey models could be extended to multi-species systems.

Resource competition: In contrast to the predator/prey relationship described above, we note that similar species compete for similar resources. Therefore, it is unlikely that many similar species occupy the same island, especially if they have large populations [TIL 94]. This fact means that it is unlikely that species immigrate to islands that already have large populations that are similar to them. Alternatively, it could mean that emigration rate is not affected, but survival likelihood is lower following emigration. Resource competition also means that if two species have an equal probability of extinction, then the species that is the most similar to other species is more likely to become extinct. This is a different type of interaction than the predator/prey relationship described above. However, both models are plausible, and competition is generally viewed in biology as a more significant driver of community composition than predator/prey interactions.

Time correlation: In island biogeography, if a species migrates to an island in a given geographical direction, it is likely to continue moving in the same direction to the next island. This feature is due to the fact that migration is influenced by prevailing winds and currents, and those winds and currents have a positive time correlation. This is described by biodiffusion theory, the telegraph equation and the equation of diffusion [OKU 01]. If a species migrates from island A to island B, it is likely to continue in the same direction to the next island in the chain at the next time step. This means that if a species migrates from one island to the next, it is likely to continue migrating in that direction.

Other factors: The influence factors described above are not an exhaustive list of all of the aspects of biogeography; they comprise only a subset which may affect the development of BBO in the next chapter. Other aspects of biogeography could inspire other variations of BBO. The biogeography literature is so rich that there are many possibilities in this direction.

2

Biogeography and Biological Optimization

Bio-inspired metaheuristics are computer systems that are motivated by ideas from the natural world. In addition, computer science research can be used to model and explore biological systems. These two approaches interact to advance the development of artificial intelligence. The approach to artificial intelligence taken by bio-inspired metaheuristics constructs simple systems that are able to evolve into more complex ones. Biological systems have many advantages over computer systems, such as less energy consumption, the ability to survive faults and even the ability to heal. Many of the ideas taken from biological optimization processes have been applied to design bio-inspired metaheuristic algorithms, leading to new developments in artificial intelligence. Biogeography is considered in this chapter as a natural optimization mechanism that can motivate the development of BBO.

Overview of the chapter

This chapter gives an overview of a mathematical model of biogeography in section 2.1 and discusses its interpretation as a naturally occurring optimization process in section 2.2. We discuss some bio-inspired metaheuristic algorithms other than BBO that are motivated by biological optimization processes in section 2.3.

2.1. A mathematical model of biogeography

In the previous chapter, we mentioned a mathematical model of island biogeography which is based on the idea that the number of species in an undisturbed habitat is mostly determined by immigration and emigration. Immigration is the arrival of new species into a habitat or population, while emigration the departure of species. There are also other important factors that influence migration between habitats, including the distance to the nearest neighboring habitat, the size of the habitats, habitat similarity, species mobility and age, predator/prey relationships, resource competition and human activity. These factors make immigration and emigration curves more complicated than the linear model shown in Figure 1.4.

In this section, we discuss the generalized mathematical formulations of island biogeography. We use the term "species count" to refer to the number of species in a given habitat. Biogeography models are based on differential equations for species count probabilities. Consider a model of species count in a single habitat, whose state at any time is represented by the species count. Suppose that the largest possible number of species that the habitat can support is n. Whenever there are k species in the habitat, new arrivals enter the habitat at an immigration rate λ_k, and species leave the habitat at an emigration rate μ_k. Note that as the number of the species increases, the habitat gets more crowded, so the immigration rate decreases and the emigration rate should increase. If there are n species in the habitat, then the immigration rate is zero. On the other hand, if there are no species in the habitat, then the emigration rate is zero. So the immigration and emigration rates are constrained by $\lambda_0 \geq \cdots \geq \lambda_k \geq \cdots \geq \lambda_n = 0$ and $0 = \mu_0 \leq \cdots \leq \mu_k \leq \cdots \leq \mu_n$, respectively, for $k = 0, 1, \cdots, n$.

DEFINITION 2.1.– *The equilibrium species count* k_0 *is the point at which the immigration and emigration rates are equal; that is,* $\lambda_{k_0} = \mu_{k_0}$.

Now consider the probability P_k that the habitat contains exactly k species. P_k changes from time t to time $(t + \Delta t)$ as follows:

$$P_k(t + \Delta t) = P_k(t)(1 - \lambda_k \Delta t - \mu_k \Delta t) + P_{k-1}\lambda_{k-1}\Delta t + P_{k+1}\mu_{k+1}\Delta t \qquad [2.1]$$

This equation holds because in order to have k species at time $(t + \Delta t)$, one of the following conditions must hold:

1) There were k species at time t, and no immigration or emigration occurred between t and $(t + \Delta t)$; or

2) There were $(k-1)$ species at time t, and one species immigrated; or

3) There were $(k+1)$ species at time t, and one species emigrated.

We assume that Δt is small enough so that the probability of more than one immigration or emigration during that time period can be ignored. Taking the limit of equation [2.1] as $\Delta t \to 0$ gives:

$$\dot{P}_k = \begin{cases} -\lambda_0 P_0 + \mu_1 P_1, & k=0 \\ -(\lambda_k + \mu_k) P_k + \lambda_{k-1} P_{k-1} + \mu_{k+1} P_{k+1}, & 1 \le k \le n-1 \\ -\mu_n P_n + \lambda_{n-1} P_{n-1}, & k=n \end{cases} \quad [2.2]$$

It is noted that equation [2.2] is valid for $k = 0, \cdots n$, and $\mu_0 = 0$ and $\lambda_n = 0$. Define $P = [P_0 \quad \cdots \quad P_n]^T$ for notational simplicity. We can then arrange equation [2.2] into the single matrix equation

$$\dot{P} = AP \quad [2.3]$$

where the matrix A is given as:

$$A = \begin{bmatrix} -\lambda_0 & \mu_1 & 0 & \cdots & 0 \\ \lambda_0 & -(\lambda_1 + \mu_1) & \mu_2 & \ddots & \vdots \\ \vdots & \ddots & \ddots & \ddots & \vdots \\ \vdots & \ddots & \lambda_{n-2} & -(\lambda_{n-1} + \mu_{n-1}) & \mu_n \\ 0 & \cdots & 0 & \lambda_{n-1} & -\mu_n \end{bmatrix} \quad [2.4]$$

Now consider the linear model of island biogeography shown in Figure 1.4, where the migration rates are straight lines, and define $n = S_{max}$, $k = S$ and $k_0 = S_0$. We then have:

$$\begin{aligned} \mu_k &= Ek/n \\ \lambda_k &= I(1-k/n) \end{aligned} \quad [2.5]$$

For the special case when the maximum immigration rate and emigration rate are given as $I = E = 1$, we have:

$$\lambda_k + \mu_k = E = I = 1 \quad \text{for all } k \in [0, n] \quad [2.6]$$

and

$$A=\begin{bmatrix} -1 & 1/n & 0 & \cdots & 0 \\ n/n & -1 & 2/n & \ddots & \vdots \\ \vdots & \ddots & \ddots & \ddots & \vdots \\ \vdots & \ddots & 2/n & -1 & n/n \\ 0 & \cdots & 0 & 1/n & -1 \end{bmatrix} \quad [2.7]$$

THEOREM 2.1.– *The steady-state value for the probability of the number of each species is given by:*

$$P_k=\begin{cases} P_0=\dfrac{1}{1+\sum_{i=1}^{n}\dfrac{\lambda_0\lambda_1\cdots\lambda_{i-1}}{\mu_1\mu_2\cdots\mu_i}}, & k=0 \\ P_k=\dfrac{\lambda_0\lambda_1\cdots\lambda_{k-1}}{\mu_1\mu_2\cdots\mu_k\left(1+\sum_{i=1}^{n}\dfrac{\lambda_0\lambda_1\cdots\lambda_{i-1}}{\mu_1\mu_2\cdots\mu_i}\right)}, & 1\le k\le n \end{cases} \quad [2.8]$$

The above theorem was proven in [MA 09]. Sufficient and necessary conditions for these limiting probabilities to exist are that $\mu_k \neq 0$ for all k greater than 0.

Note that Theorem 2.1 above is similar to Theorem 1 in [SIM 08], which was proven by singular value decomposition (SVD). But there are three differences between the two theorems. First, the above theorem is more general than that given in [SIM 08], which only considered the special case that $\lambda_k = 1-k/n$ and $\mu_k = k/n$. Second, we have obtained the probability for any migration rates and species counts, which the theorem in [SIM 08] did not do. This theorem can help us to study how any migration model, including nonlinear models, impact the steady-state value of the probability of species counts.

EXAMPLE 2.1.–

Consider the linear model of island biogeography described in Figure 1.4. Suppose an island can support a maximum of four species, namely, n = 4. The maximum immigration rate and emigration rate $I = E = 1$. Based on equations [2.5] and [2.7], we have:

$$A=\begin{bmatrix} -1 & 1/4 & 0 & 0 & 0 \\ 1 & -1 & 2/4 & 0 & 0 \\ 0 & 3/4 & -1 & 3/4 & 0 \\ 0 & 0 & 2/4 & -1 & 1 \\ 0 & 0 & 0 & 1/4 & -1 \end{bmatrix} \quad [2.9]$$

Theorem 2.1 tells us that the steady-state probability for the number of each species number is:

$$\begin{aligned} &\Pr(S=0)=P_0=1/16 \\ &\Pr(S=1)=P_1=4/16 \\ &\Pr(S=2)=P_2=6/16 \\ &\Pr(S=3)=P_3=4/16 \\ &\Pr(S=4)=P_4=1/16 \end{aligned} \quad [2.10]$$

These results are equivalent to those obtained by Theorem 1 in [SIM 08].

Next, we continue to study the characteristics of the equilibrium species counts in the mathematical model of island biogeography.

THEOREM 2.2.– *If the immigration and emigration rates satisfy* $\lambda_0 \geq \cdots \geq \lambda_k \geq \cdots \geq \lambda_n = 0$ *and* $0 = \mu_0 \leq \cdots \leq \mu_k \leq \cdots \leq \mu_n$, *respectively, for* $k = 0, 1, \cdots, n$, *then the steady-state probability of the number of each species satisfies* $P_0 \leq P_1 \leq \cdots \leq P_{k_0-1} \leq P_{k_0} \geq P_{k_0+1} \geq \cdots \geq P_n$, *where* k_0 *denotes the equilibrium species number.*

It is clear that this theorem is implied by Theorem 2.1. Another interesting characteristic of the steady-state probability of species count is stated in the following theorem.

THEOREM 2.3.– *If the immigration and emigration rates satisfy the conditions that* $\mu_k = \lambda_{n-k}$ *for every* $k = 0, 1, \cdots, n$, *then* $P_k = P_{n-k}$ *for all* $k = 0, 1, \cdots, n$.

Examples of this theorem are shown in equation [2.8] and example 2.1. It implies that the steady-state probability of species count is symmetrical, and the equilibrium point is half of the maximum species count n if $\mu_k = \lambda_{n-k}$ for $k = 0, 1, \cdots, n$. We note that with the special condition $\mu_k = \lambda_{n-k} = k/n$ for $k = 0, 1, \cdots, n$, this theorem reduces to Observation 1 in [SIM 08]. Theorem 2.3 is therefore a generalization of Observation 1 in [SIM 08].

There are several remarks that should be made about Theorem 2.1, Theorem 2.2 and Theorem 2.3. First, the equilibrium species count is given as the index at which the immigration and emigration rates are equal. Second, as shown in Theorem 2.1, the steady-state probabilities P_k are related to λ_k and μ_k; that is, P_k depends on the migration rates. Third, as shown in Theorems 2.2 and 2.3, the steady-state probabilities P_k have the characteristic that the probabilities at the end points are smaller than those near the equilibrium point. Finally, under the condition $\mu_k = \lambda_{n-k}$, the steady-state probabilities P_k are symmetric with respect to the equilibrium point.

2.2. Biogeography as an optimization process

We know that nature includes many processes that optimize [ALE 96]. In fact, this premise is the foundational principle of most EAs. However, is biogeography an optimization process? At first glance, it seems that biogeography simply maintains species count equilibria in habitats, and that it is not necessarily optimal. This section discusses biogeography from the viewpoint of optimality.

Biogeography is nature's way of distributing species, and it has often been studied as a process that maintains equilibrium in habitats. Species equilibrium in a biological habitat occurs when the combined speciation and immigration rates equal the emigration rate. Equilibrium can be seen at the point S_0 shown in Figure 1.4 where the immigration and emigration curves intersect. One reason that biogeography has been viewed from the equilibrium perspective is that this viewpoint was the first to place biogeography on a firm mathematical footing [MAC 67]. We have also discussed this perspective in the previous section. However, since then the equilibrium perspective has been increasingly questioned, or rather expanded, by biogeographers.

Engineers often view stability and optimality as competing objectives; for example, a simple system is typically easier to stabilize than a complex system, while an optimal system is typically more complex and less stable than a simpler system [KEE 97]. However, in biogeography, stability and optimality are two perspectives of the same phenomenon. Optimality in biogeography involves biologically diverse, complex communities that are highly adaptable to their environment. Stability in biogeography involves the persistence of existing populations. Field observations show that complex communities are more adaptable and stable than simple communities [HAR 06, p. 82], and this observation has also been supported by simulation [ELT 58, MAC 55].

Although the complementary nature of optimality and stability in biogeography has been challenged [MAY 73], those challenges have been adequately answered and the idea is generally accepted today [MCC 00, KON 06]. The equilibrium versus optimality debate in biogeography thus becomes a matter of semantics, because equilibrium and optimality are simply two different perspectives on the same phenomenon in biogeography.

A dramatic example of the optimality of biogeography is Krakatoa, a volcanic island in the Indian Ocean which erupted in August 1883 [WIN 08]. The eruption was heard from thousands of miles away and resulted in the death of over 36,000 people, mostly from tidal waves whose remnants were recorded as far away as England. The eruption threw dust particles 30 miles high which remained aloft for months and were visible all around the world. Rogier Verbeek, a geologist and mining engineer, was the first visitor to Krakatoa 6 weeks after the eruption, but the surface of the island was too hot to touch and showed no evidence of life; the island was completely sterilized [WHI 93]. The first animal life (a spider) was discovered on Krakatoa in May 1884, 9 months after the eruption. By 1887, dense fields of grass were discovered on the island. By 1906, plant and animal life was abundant. Although volcanic activity continues today on Krakatoa, by 1983 (one century after its desolation), there were 88 species of trees and 53 species of shrubs, and the species number continues to increase linearly with time [WHI 93]. Life immigrates to Krakatoa, and immigration makes the island more habitable, which in turn makes the island more friendly to additional immigration. Figure 2.1 shows the evolution of Krakatoa over one hundred years.

Biogeography is thus a positive feedback phenomenon – at least to a certain point. When a habitat is highly populated, it has many species and thus is likely to emigrate many species to nearby habitats, while few species immigrate to it because of the lack of additional resources for immigrating species. In the same way, when a habitat is sparsely populated, it has few species and thus is likely to receive many immigrants, while only a few species emigrate because of their sparse populations. The issue of whether or not immigrants can survive after migration is another question, but the immigration of new species can raise the biological diversity of a habitat and thereby improve the habitat's suitability for additional species.

Figure 2.1. *Krakatoa island. The left picture shows the eruption of Krakatoa in 1883, and the right picture shows how it evolved into a habitable island after 100 years*

This is similar to natural selection, or survival of the fittest. As species become more fit, they are more likely to survive. As they thrive, they disperse and become better able to adapt to their environment. Natural selection, like biogeography, entails positive feedback. However, the time scale of biogeography is much shorter than that of natural selection, which hints at the possibility of improved optimization performance by using biogeography rather than natural selection as a motivating paradigm for optimization algorithms. So we regard this positive feedback phenomenon of biogeography as an optimization process. This view of the environment as an optimizing system was first suggested in the 1990s [VOL 97]. Biogeographers claim that "biogeography based on optimizing environmental conditions for biotic activity seems more appropriate than a definition based on homeostasis" [KLE 04].

Another example of biogeography as an optimization process is the Amazon rainforest, which is a typical case of a mutually optimizing life/environment system [HAR 06]. The rainforest has a large capacity to recycle moisture, which decreases aridity and increases evaporation. This leads to cooler and wetter surfaces, which are more amenable to life. This suggests that a view of biogeography "based on *optimizing* environmental conditions for biotic activity seems more appropriate than a definition based on homeostasis" [KLE 04] (emphasis added). This view of the environment as a life-optimizing system was suggested as early as 1997 [VOL 97]. There are many other examples of the optimality of biogeography, such as Earth's temperature [HAR 06], Earth's atmospheric composition [LEN 98] and the ocean's mineral content [LOV 90, LOV 95].

This is not to say that biogeography is optimal for any particular species. For example, investigations of the Bikini Atoll show that the high level of radioactivity resulting from nuclear tests had little effect on its natural ecology, but mammals were seriously affected (Lovelock, page 37). This and similar studies indicate that the Earth "will take care of itself and environmental excesses will be ameliorated, but it's likely that such restoration of the environment will occur in a world devoid of people" [MAR 96]. Interestingly, amid the current warnings about ozone depletion, it is easy to overlook the fact that for the first two billion years of life, Earth had no ozone at all [LOV 95, p. 109]. Life flourishes and evolves regardless of our opinions about Earth's ecology, but not in a human-centric way. Although global warming or an ice age might be disastrous for humans and many other mammals, it would be a minor event in the overall history of biogeography on our planet.

In summary, although the natural phenomenon of biogeography as an optimization process has been challenged, adequate answers have been put forth to answer these challenges. The premise that biogeography is an optimization process has motivated the development of BBO as a metaheuristic algorithm, which we discuss in the next chapter.

2.3. Biological optimization

This section gives an overview of some biological optimization paradigms that motivate bio-inspired metaheuristic techniques, including classical methods such as genetic algorithms (GAs), evolution strategies (ESs) and also some newer algorithms, such as particle swarm optimization (PSO) and artificial bee colony (ABC) optimization [BOU 13, MA 13a, MA 13b, MA 16c]. These algorithms are widely used to solve optimization problems and have attracted increasing attention in recent years.

2.3.1. *Genetic algorithms*

GAs are the earliest, most well-known and most widely used biologically motivated optimization algorithms. GAs were first introduced as a computational analogy of adaptive biological systems by John Holland in his book *Adaptation in Natural and Artificial Systems* [HOL 75]. GAs have proven to be an enormously powerful and successful problem-solving strategy, dramatically demonstrating the power of biologically motivated optimization.

In nature, we have a population of individuals. Some individuals are good, and some are not so good. The good individuals have a relatively high chance of reproducing, while the poor individuals have a relatively low chance of reproducing. Parents beget children, and then the parents drop out of the population to make way for their offspring. As generations come and go, the population as a whole becomes more fit.

GAs are modeled on this natural selection in biological systems. Given an optimization problem, GAs use a set of candidate solutions as a population, and use fitness functions to evaluate these candidate solutions. During the optimization process, the candidate solutions improve through selection, recombination (crossover) and mutation, and then pass on the candidate solutions with the best fitness to the next generation. In GAs, selection is the first step, in which individuals are chosen from the population for breeding. Crossover is the second step, which is used to combine individuals from one generation to create individuals for the next generation. Mutation is the final step, in which individuals are randomly modified, and its purpose is to increase diversity among the population.

2.3.2. *Evolution strategies*

Evolution strategy (ES) was created in the early 1960s and was developed further in the 1970s and later by Ingo Rechenberg, Hans-Paul Schwefel and Peter Bienert [REC 68]. ESs are biologically motivated optimization techniques based on the ideas of adaptation and evolution.

ES uses natural problem-dependent representations, and depends primarily on mutation and selection as search operators. During the process of evolution, mutation is performed by adding a normally distributed random value to each individual (that is, each candidate solution). The step size or mutation strength is often governed by self-adaptation. The selection operator in ES is deterministic and is based on fitness rankings. The simplest ES operates on a population of size two: the current point, called the parent individual, and the result of its mutation. If the mutant's fitness is at least as good as the parent, it becomes the parent individual of the next generation; otherwise, the mutant is discarded. This particular ES is called a (1+1)-ES.

The most popular ES is the (μ, λ)-ES, in which μ parent individuals produce λ offspring individuals using mutation. Each of the λ offspring individuals is assigned a fitness value depending on its quality. The best μ offspring individuals become the next generation's parent individuals. This means that λ must be greater than or equal to μ. Note that the μ and λ that are used in the (μ, λ)-ES notation are not related to the μ and λ that are used in the mathematical model of island biogeography.

2.3.3. *Particle swarm optimization*

Particle swarm optimization (PSO) is a metaheuristic developed by Kennedy and Eberhart in 1995. It is inspired by the social behavior of bird flocking or fish schooling [KEN 95], which is shown in Figure 2.2. During the past several years, PSO has been successfully applied to many research and application areas. It has been demonstrated that PSO achieves better optimization results with a faster, cheaper method compared to many other methods.

Figure 2.2. *Social behavior of bird flocking (left) and fish schooling (right), which inspired the particle swarm optimization algorithm*

PSO shares many similarities with other metaheuristics such as GAs. PSO is initialized with a population of random individuals, and searches for an optimum by updating the population one iteration at a time. However, unlike GAs, PSO does not have evolution operators such as crossover and mutation. In PSO, the individual is called a particle, and it moves through the problem search space by following the best particles in the population.

Each particle keeps track of the coordinates in the problem space which are associated with the best location that it has achieved so far during the optimization process; this location is called P_{best}. Another best location is determined by the globally best position of the entire swarm at the current iteration; this location is called G_{best}. The PSO concept consists of changing the velocity of each particle toward the P_{best} and $G_{best.}$ locations. The particle's velocity is updated based on its current velocity, its previous best location P_{best} and the global best location G_{best} at the current iteration.

2.3.4. *Artificial bee colony algorithm*

The artificial bee colony (ABC) algorithm is another metaheuristic, which was first published in Karaboga and Basturk [KAR 07]. ABC is based on different types of bees and their behaviors, which are shown in Figure 2.3.

First, forager bees, also called employed bees, travel back and forth between a food source and their hive. Each forager is associated with a specific location, and remembers that location as it travels back and forth between the hive. When a forager takes nectar to the hive, it returns to its food source, but it also engages in local exploration as it searches in the nearby vicinity for a better source.

Second, onlooker bees are not associated with any particular food source, but they observe the behavior of the foragers. Onlookers observe the amount of nectar that is returned to the hive by the foragers, and use that information to decide where to search for nectar. The onlookers' search location is decided probabilistically based on their observations of the foragers.

Third, scout bees are explorers and, like onlookers, are not associated with any particular food source. If a scout sees that a forager has stagnated and is not progressively increasing the amount of nectar that it returns to the hive, then the scout randomly searches for a new nectar source in the search space. Stagnation is indicated when the explorer fails to increase the amount of nectar it brings to the hive after a certain number of trips.

These ideas lead to the ABC algorithm, which simulates foraging, onlooking and scouting behaviors to search for an optimal food source. The location of a food source is analogous to a location in the search space of an optimization problem. The amount of nectar at a location is analogous to the fitness of an individual. Each forager randomly modifies its position in the search space. If the random modification results in an improvement, then the forager moves to the new position. The onlooker bees also randomly modify the position of a forager, where the forager that is modified is randomly chosen using roulette-wheel selection. Again, if the random modification improves the forager, then the forager moves to the new position. Finally, a scout replaces a forager if the forager has not improved after a preset number of random modifications.

Figure 2.3. *Behaviors and types of bees, which inspired the artificial bee colony algorithm*

2.4. Conclusion

We have discussed several bio-inspired metaheuristics, including GAs, ES, PSO and ABC, the last two of which are sometimes classified as swarm intelligence rather than EAs. But all these algorithms have certain features in common, and they all adapt biological optimization processes to implement optimization algorithms. It seems that virtually any natural or biological optimization process can be interpreted as an optimization algorithm [ALE 96]. It is therefore difficult to know where one algorithm ends, and another begins. When does a new metaheuristic belong to its own class, and when should it instead be classified as a variation of an existing metaheuristic? One of the challenges for the research community is to encourage new research based on the abundance of biological optimization processes.

There are many other metaheuristics that we have not had time to discuss, including the artificial fish swarm algorithm (AFSA) [LI 03], the shuffled frog leaping algorithm (SFLA) [EUS 03], the firefly algorithm (FA) [YAN 08, Chapter 8], the bacterial foraging optimization algorithm (BFOA) [PAS 02], and so on. These metaheuristics could doubtless provide a lifetime of productivity to the interested student and researcher.

3

A Basic BBO Algorithm

Just as the science of genetics gave rise to genetic algorithms (GAs), and the study of animal swarms gave rise to particle swarm optimization (PSO), and the behaviors of bees gave rise to artificial bee colony optimization, so the science of biogeography has given rise to biogeography-based optimization (BBO).

Overview of the chapter

This chapter shows in section 3.1 how the biogeography theory of the previous chapter can be applied to optimization problems to build a basic BBO algorithm. Section 3.2 discusses the differences between BBO and other bio-inspired optimization algorithms. Section 3.3 demonstrates the performance of basic BBO on a set of standard benchmarks.

3.1. BBO definitions and algorithm

Biogeography is nature's way of distributing species and optimizing environments for life, and operates according to underlying mathematical optimization rules. Suppose that we have an optimization problem and a population of candidate solutions that can be represented as vectors of independent variables; candidate solutions can be referred to as individuals, or solutions. Each independent variable in a solution is considered to be a suitability index variable (SIV) of biogeography. Further suppose that we have some way of assessing the goodness of the solutions. Those solutions that are good are considered to be habitats with a high habitat suitability index (HSI), and those that are poor are considered to be habitats with a low HSI. HSI is analogous to fitness in other bio-inspired optimization algorithms (GAs, for example). Good solutions resist change more than poor solutions, just like habitats with a high HSI have lower immigration rates than habitats with a low HSI. By the same token, good solutions tend to share their SIVs with poor solutions, just like habitats with a high HSI have high emigration rates. Poor solutions are likely to accept new SIVs from good solutions, just like habitats with a low HSI are likely to receive many immigrants from habitats with a high HSI. The addition of new SIVs to poor solutions may raise the quality of those solutions. The bio-inspired optimization algorithm that is based on this approach is called BBO.

3.1.1. *Migration*

High HSI solutions represent habitats with many species, and low HSI solutions represent habitats with few species. We assume that each solution (habitat) has an identical species count curve with $E = I$ for simplicity. Figure 3.1 illustrates the migration rates for a BBO algorithm with these assumptions. The S value represented by the solution depends on its HSI. S_1 in Figure 3.1 represents a low HSI solution due to only a few species in a habitat, while S_2 represents a high HSI solution due to many species in a habitat. The immigration rate λ_1 for S_1 will therefore be higher than the immigration rate λ_2 for S_2. The emigration rate μ_1 for S_1 will be lower than the emigration rate μ_2 for S_2. Figure 3.1 is called a linear migration model since the λ and μ values are linear functions of fitness.

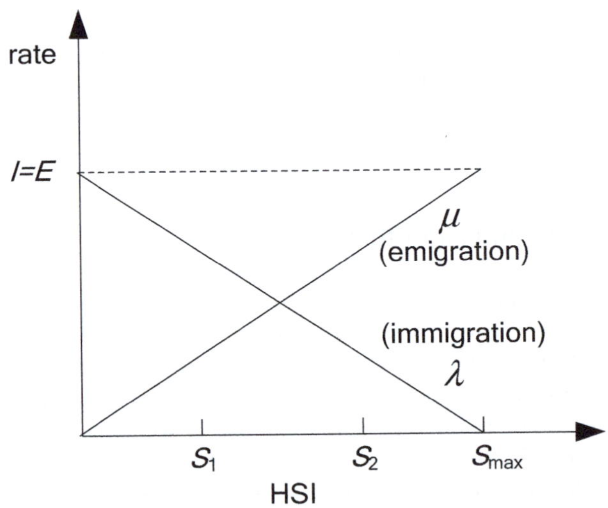

Figure 3.1. *BBO SIV-sharing relationships. S_1 represents a low HSI solution with a low probability of sharing SIVs, but a high probability of receiving SIVs from other solutions. S_2 represents a high HSI solution with a high probability of sharing SIVs, but a low probability of receiving SIVs from other solutions*

We use the emigration and immigration rates of each solution to probabilistically share information between habitats. With a certain probability, we modify each solution based on other solutions. If a given solution is selected to be modified, then we use its immigration rate λ to probabilistically decide whether or not to modify each SIV in that solution. If a given SIV in a given solution S_i is selected to be modified, then we use the emigration rates μ of the other solutions to probabilistically decide which of the solutions should migrate a randomly selected SIV to solution S_i.

3.1.2. *Mutation*

Cataclysmic events (unusually large flotsam arriving from a neighboring habitat, disease, natural catastrophes, etc.) can drastically change the HSI of a natural habitat. They can also cause the species count to differ from its equilibrium value. A habitat's HSI can therefore change suddenly due to random events, and we model such events as mutation.

Mutation tends to increase diversity among the population. Without mutation, high HSI solutions will tend to be more dominant in the population. Mutation makes low HSI solutions likely to mutate, which gives them a chance of improving. It also makes high HSI solutions likely to mutate, which gives them a chance of improving even more than they already have. We usually use elitism, which is a common EA mechanism, to save the SIVs of the habitat that has the best solution in the optimization process, so even if mutation ruins its HSI, we have saved it and can revert back to it if needed. So we use mutation on both poor solutions and good solutions. Those solutions that are average are hopefully improving already, and so we avoid mutating them, although there is still some mutation probability, except for the most probable solution.

The mutation mechanism is problem-dependent, just as it is for GAs. If a solution is selected for mutation, then we simply replace a randomly chosen SIV in the solution with a new, randomly generated SIV. We do not explore alternative mutation schemes in this chapter, but all of the mutation schemes that have been implemented for EAs could also be implemented for BBO.

3.1.3. *BBO implementation*

First we provide some definitions, and then we provide an outline of the basic BBO algorithm. We use R to refer to the set of real numbers, Z to refer to the set of integers and $\varnothing$ to refer to the empty set.

DEFINITION 3.1.– *A habitat* $x \in \mathrm{SIV}^m$ *is a vector of* m *integers that represents a feasible solution to some optimization problem.*

DEFINITION 3.2.– *A suitability index variable* $\mathrm{SIV} \in C$ *is an integer that is an allowable value in a habitat.* $C \in Z^n$ *is the set of all integers that are allowed in a habitat.*

DEFINITION 3.3.– *A habitat suitability index* HSI*:* $x \to R$ *is a measure of the goodness of the solution that is represented by the habitat.*

DEFINITION 3.4.– *An ecosystem* $\{x_1, x_2, \cdots, x_N\}$*, also denoted as* $\{x\}$*, is a group of N habitats, where N is the size of the ecosystem.*

Note that in most bio-inspired optimization algorithms, x is called a candidate solution or individual, SIV is called a decision variable or independent variable, HSI is called fitness and N is called population size.

DEFINITION 3.5.– *Immigration rate* $\lambda(\mathrm{HSI}) : R \to R$ *is a monotonically non-increasing function of* HSI. λ_i *is proportional to the likelihood that* SIVs *from neighboring habitats will migrate into habitat* x_i *.*

DEFINITION 3.6.– *Emigration rate* $\mu(\mathrm{HSI}) : R \to R$ *is a monotonically non-decreasing function of HSI.* μ_i *is proportional to the likelihood that* SIVs *from habitat* x_i *will migrate into neighboring habitats.*

In practice, we assume that λ and μ are linear with the same maximum values. However, these assumptions are made only for mathematical convenience, and better optimization performance might be attained if these assumptions are relaxed.

DEFINITION 3.7.– *Migration* $\Omega(\lambda, \mu) : \{x\} \to x$ *is a probabilistic operator that adjusts habitat* x *based on the ecosystem of candidate solutions. The probability that* x *is modified is proportional to its immigration rate* λ*, and the probability that the source of the modification comes from* x_j *is proportional to the emigration rate* μ_j.

Migration can be loosely described as follows.

For each habitat (candidate solution) x_k

  For each SIV (decision variable)

    Use λ_k to probabilistically decide whether to immigrate to x_k

    If immigrating then

      Use $\{\mu_j\}$ to probabilistically select the emigrating habitat x_j

      $x_k(\text{SIV}) \leftarrow x_j(\text{SIV})$

    End if

  Next SIV

Next habitat

Figure 3.2. *Migration in BBO. x_k is the kth candidate habitat and $x_k(\text{SIV})$ is a decision variable, or SIV, in x_k*

In Figure 3.2, the statement "Use λ_k to probabilistically decide whether to immigrate to x_k" can be implemented with the following logic, where *rand*(0, 1) is a random number uniformly distributed between 0 and 1:

If $\lambda_k < rand(0,1)$ then

    Immigration = true

Else

    Immigration = false

End if

Also in Figure 3.2, the statement "Use $\{\mu_j\}$ to probabilistically select the emigrating habitat x_j" can be implemented with any fitness-based selection method. For instance, we could use tournament selection by randomly choosing two or more habitats for a tournament, and then selecting x_k as the fittest habitat in the tournament. If we use roulette-wheel selection, then

$$\Pr(\text{emigration from } x_j) = \frac{\mu_j}{\sum_{k=1}^{N} \mu_k} \qquad [3.1]$$

Figure 3.3 illustrates BBO migration as described above.

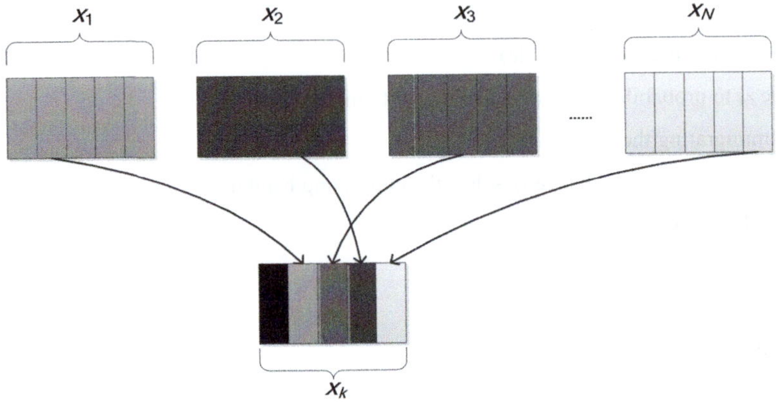

Figure 3.3. *Illustration of BBO migration for five SIVs in a habitat. SIV 1 is not selected for immigration, but SIVs 2–5 are selected for immigration. Equation [3.1] is used to select the emigrating habitats*

Figure 3.3 shows an example of habitat x_k immigrating SIVs as follows:

1) Immigration is not selected for the first SIV; that is why the first SIV in x_k remains unchanged.

2) Immigration is selected for the second SIV, and equation [3.1] is used to choose x_1 as the emigrating habitat; this is why the second SIV in x_k is replaced by the second SIV from x_1.

3) Immigration is selected for the third SIV, and equation [3.1] is used to choose x_3 as the emigrating habitat; this is why the third SIV in x_k is replaced by the third SIV from x_3.

4) Immigration is selected for the fourth SIV, and equation [3.1] is used to choose x_2 as the emigrating habitat; this is why the fourth SIV in x_k is replaced by the fourth SIV from x_2.

5) Finally, immigration is selected for the fifth SIV, and equation [3.1] is used to choose x_N as the emigrating habitat; this is why the fifth SIV in x_k is replaced by the fifth SIV from x_N.

DEFINITION 3.8.– *Mutation* $M(\lambda,\mu):x\rightarrow x$ *is a probabilistic operator that randomly modifies a habitat's SIVs based on a given mutation rate.*

Mutation can be described as follows.

For each SIV (decision variable) in habitat (candidate solution) x_i

If $rand(0, 1) < P_m$

$x_i(\text{SIV}) \leftarrow rand(L_s, U_s)$

End if

Next SIV

Figure 3.4. *Mutation in BBO. P_m is the mutation probability, and $rand(L_s, U_s)$ is a uniformly distributed random number between L_s and U_s, which are the lower and upper search bounds of the sth SIV*

In Figure 3.4, each SIV in the habitat (that is, in the population) is mutated with a probability of P_m. If mutation occurs for a given SIV, then that SIV is replaced with a random number within its search domain. This mutation is the same as is often used in other bio-inspired optimization algorithms.

DEFINITION 3.9.– *An ecosystem transition function* $\psi=(m,N,\lambda,\mu,\Omega,M):\{x\}\rightarrow\{x\}$ *is a 6-tuple that modifies the ecosystem from one optimization iteration to the next.*

An ecosystem transition function can be written as follows:

$$\psi=\{\lambda\circ\mu\circ\Omega\circ\text{HSI}\circ M\circ\text{HSI}\} \quad [3.2]$$

In others words, the ecosystem transition function begins by computing the immigration and emigration rates of each habitat. Then, migration is performed on each habitat, followed by an HSI calculation. Finally, mutation is performed, followed by an HSI recalculation for each habitat.

DEFINITION 3.10.– *A BBO algorithm* $\text{BBO}=(\vartheta,\psi,T)$ *is a 3-tuple that proposes a solution to an optimization problem.* $\vartheta:\phi\rightarrow(\{x\},\{\text{HSI}\})$ *is a function that creates an initial ecosystem of habitats and computes each corresponding HSI.* ψ *is the ecosystem transition function defined earlier, and* $T:\{x\}\rightarrow\{true, false\}$ *is a termination criterion.*

ϑ could be implemented with random number generators, heuristic solutions to the optimization problem, or some other problem-dependent procedure. T could depend on the number of ψ iterations, or the HSI of the best habitat, or some other problem-dependent termination criterion. A BBO algorithm can be broadly described as follows:

ϑ

While not T

ψ

End

The above outline of BBO can be written as follows.

Initialize a population of habitats (that is, candidate solutions) $\{x_k\}$ for $k \in [1, N]$

While not (termination criterion)

For each x_k, set emigration rate μ_k proportional to HSI (fitness) of x_k with μ_k being normalized to [0,1]

For each x_k, set immigration rate $\lambda_k = 1 - \mu_k$

$\{z_k\} \leftarrow \{x_k\}$

Perform migration for each habitat z_k as shown in Figure 3.2

Perform mutation for each habitat z_k as shown in Figure 3.4

$\{x_k\} \leftarrow \{z_k\}$

Next generation

Figure 3.5. *Outline of a basic BBO algorithm with a population size of N. $\{x_k\}$ is a population of habitats and $\{z_k\}$ is a temporary population of habitats*

Note that in Figure 3.5, migration and mutation for each habitat in the current generation occurs before any of the habitats are replaced in the population, which requires the use of the temporary population z. Borrowing from GA terminology [VAV 96], we say that Figure 3.5 depicts a *generational* BBO algorithm as opposed to a *steady-state* algorithm.

The BBO algorithm in Figure 3.5 can be informally described with the following algorithm.

1) Initialize the BBO parameters. This step includes the derivation of a method of mapping problem solutions to SIVs and habitats (see Definitions 3.1 and 3.2), which is problem dependent. We also initialize the maximum species count $S_{\max}$ and the maximum migration rates E and I (see Figure 3.1), and define the mutation rate P_m. Note that the maximum species count and the maximum migration rates are relative quantities. That is, if they all change by the same percentage, then the behavior of BBO will not change. This is because if E and I and $S_{\max}$ change, then the migration rates λ and μ and the species count S will change by the same relative amount for each solution. In practice, we often normalize E and I to 1.

2) Initialize a random set of habitats, each habitat corresponding to a potential solution to the given optimization problem. This is the implementation of the ϑ operator described above in Definition 3.10.

3) For each habitat, map the HSI to the number of species S, the immigration rate λ and the emigration rate μ (see Figure 3.1 and Definitions 3.5 and 3.6).

4) Before migration, copy the population $\{x\}$ to temporary population $\{z\}$. Probabilistically use immigration and emigration to modify each habitat, and then compute each HSI (see Definition 3.7).

5) Mutate each habitat based on the mutation rate, and recompute each HSI (see Definition 3.8). After mutation, copy temporary population $\{z\}$ to population $\{x\}$.

6) Go to step 3 for the next iteration. This loop can be terminated after a predefined number of generations, or after an acceptable problem solution has been found. This is the implementation of the T operator described above in Definition 3.10.

Note that after each habitat is modified (steps 2, 4 and 5), its feasibility as a problem solution should be verified. If it does not represent a feasible solution, then some method needs to be implemented to map it to the set of feasible solutions.

EXAMPLE 3.1.–

This simple BBO experiment is motivated by David Goldberg's "GA simulation by hand" [GOL 89]. Suppose that we want to maximize x^2, where x is encoded as a five-bit integer. We have to decide how many individuals we want in our population, and what mutation rate we want to use. We start with a randomly generated population of four individuals, and a mutation rate of 1% per bit. For each individual, we compute the fitness value x^2, and then we assign migration rates in a linear manner as shown in Figure 3.1. Migration rates should be normalized to [0, 1], but we often set the smallest value to a number slightly greater than 0, and the largest value to a number slightly less than 1. This allows some randomness (non-determinism) even for the best and worst individuals in the population. For this example, we arbitrarily decide to use $1/N$ as the minimum values for λ and μ, and $(N-1)/N$ as the maximum values, where $N = 4$ is the population size. Suppose that our random initial population is created, as shown in Table 3.1.

String number	x (binary)	x (decimal)	$f(x) = x^2$	μ	λ
1	01101	13	169	2/5	3/5
2	11000	24	576	4/5	1/5
3	01000	8	64	1/5	4/5
4	10011	19	361	3/5	2/5

Table 3.1. *Initial population for a simple BBO problem*

The first thing we do is copy the population x to temporary population z. Then we consider the possibility of immigration to each bit of the first individual in the temporary population z_1, which is equal to x_1 (01101). We order bit numbers from left to right starting with index 1. We therefore see that

$$z_1(1)=0, z_1(2)=1, z_1(3)=1, z_1(4)=0, z_1(5)=1$$

Since z_1 is the third most-fit individual, immigration rate $\lambda_1 = 3/5$, so there is a 60% chance of immigrating to each bit in z_1. We generate a random number $r \sim rand[0,1]$ for each bit in z_1 to determine whether or not we should immigrate to that bit.

1) Suppose $r = 0.7$. Since $r > \lambda_1$, we will not immigrate to $z_1(1)$, so $z_1(1)$ remains equal to 0.

2) Suppose the next random number that we generate is $r = 0.3$. Since $r < \lambda_1$, we immigrate to $z_1(2)$. We use roulette-wheel selection to choose the emigrating bit. $x_2(2)$ has the greatest probability of emigrating to $z_1(2)$, $x_4(2)$ has the second-greatest probability, $x_1(2)$ has the third-greatest probability, and $x_3(2)$ has the least probability. We could exclude $x_1(2)$ from consideration since z_1 is a copy of x_1, but this is an implementation detail that depends on the preference of the programmer. Suppose that this roulette-wheel selection process results in the choice

of $x_3(2)$ for immigration. Then, $z_1(2) \leftarrow x_3(2) = 1$. Even though we immigrated to $z_1(2)$, it did not change from its original value.

3) We continue this process for $z_1(3)$, $z_1(4)$ and $z_1(5)$. Suppose that the random numbers that are generated result in the following:

– $z_1(3) = 1$ (no immigration)

– $z_1(4) \leftarrow x_4 = 1$ (immigration)

– $z_1(5) = 1$ (no immigration)

Now we have completed migration for z_1 and have obtained $z_1 = 01111$.

4) We repeat steps 1–3 for z_2, z_3 and z_4.

5) We next consider the possibility of mutation for each bit in each temporary individual z_1, z_2, z_3 and z_4. Mutation can be implemented as shown in Figure 3.4.

6) Now that we have a modified population of $\{z_k\}$ individuals, we copy z_k to x_k for $k \in [1,4]$, and the first BBO generation is complete.

The above process continues until some convergence criterion is met. For instance, we could continue for a specified number of generations, or continue until we achieve a satisfactory fitness value, or continue until the fitness value stops changing. □

3.2. Differences between BBO and other optimization algorithms

This section first discusses the relationship between BBO and GAs to illustrate that BBO is distinctive enough to be considered a separate optimization algorithm rather than as a special type of GA. Then we show the similarity and difference between BBO and other bio-inspired optimization algorithms from the viewpoint of algorithmic features.

3.2.1. *BBO and genetic algorithms*

In GAs, one way of implementing recombination is called global uniform recombination, in which we randomly choose each child gene from one parent, where the parent population is equal to the entire GA population, and random selection is based on fitness values, for example, roulette-wheel selection. We call this approach genetic algorithm with global uniform recombination (GA/GUR), which is shown in Figure 3.6 [SIM 11b].

Initialize a population of individuals (candidate solutions) $\{x_k\}$ for $k \in [1, N]$

While not (termination criterion)

For each individual x_k

$\quad Child_k \leftarrow [0\ 0 \cdots\cdots 0] \in R^n$

For each gene (decision variable)

$\quad$ Use fitness values to probabilistically select individual x_j

$\quad Child_k(gene) \leftarrow x_j(gene)$

Next gene

$\quad$ Probabilistically mutate $Child_k$

Next individual

$\quad \{x_k\} \leftarrow \{Child_k\}$

Next generation

Figure 3.6. *Outline of genetic algorithm with global uniform recombination (GA/GUR). N is the population size, $\{x_k\}$ is the entire population of individuals, x_k is the kth individual and $x_k(gene)$ is a decision variable in x_k*

Comparing Figures 3.5 and 3.6, we see that BBO is a generalization of a specific type of GA/GUR. This is because if, rather than setting $\lambda_k = 1 - \mu_k$ in the BBO algorithm of Figure 3.5, we instead set $\lambda_k = 1$ for all k, then the BBO algorithm of Figure 3.5 would be equivalent to the GA/GUR algorithm of Figure 3.6.

We see that even though BBO and GAs have many similarities, BBO is distinctive enough to be considered a separate optimization algorithm rather than a special type of GA. There is also another more important reason to consider BBO as a separate optimization algorithm, and that is because the biogeography roots of BBO open up many avenues for extensions and modifications that would otherwise be unavailable to the researcher. We discussed some of these extensions in the next chapter.

3.2.2. *BBO and other algorithms*

BBO is a bio-inspired optimization method, which gives it certain features in common with other bio-inspired optimization algorithms, including GAs, ES, PSO, ant colony optimization (ACO) and differential evolution (DE). For example, they all adopt operators to share information between solutions. This makes BBO applicable to many problems that GAs and PSO are used for.

However, there are some distinctive characteristics of BBO compared to these other optimization algorithms. First, we note that GAs and ESs reproduce children by crossover; namely, their solutions disappear at the end of each generation, while BBO solutions are not discarded after each generation but are rather modified by migration. Second, we find that ACO generates a new set of solutions at each generation while BBO maintains its set of solutions from one generation to the next. Lastly, BBO is contrasted with PSO and DE because PSO solutions change by virtue of another variable (velocity) and DE solutions change based on differences between other solutions, while BBO solutions change directly via migration.

It is these differences between BBO and other optimization algorithms that prove to be its strength. Some open research questions are: how do these differences make the performance of BBO differ from other optimization algorithms? What do these differences say about the types of problems that are most appropriate for BBO? This chapter presents the initial explorations into BBO but leaves these questions for further study.

3.3. Simulations

In this section, we look at the performance of BBO. A representative set of 13 benchmark functions has been used for performance verification of BBO. These functions are briefly described in Table 3.2. A more detailed description of these functions can be found in Appendix A. The functions are divided into two categories: unimodal functions, including F01–F07, and multimodal functions, including F08–F13. All benchmark functions are minimization problems.

A. Performance comparisons between BBO and other algorithms

First we compare BBO with five other optimization algorithms, including ACO, DE, ES, GA and PSO, in terms of mean performance on a set of Monte Carlo simulations. For a fair comparison, we choose basic versions of the other algorithms to compare with BBO. This is because BBO is a relatively new global optimization algorithm and other algorithms have had many years to develop. Also, we use the basic version of BBO here rather than more advanced versions. For all optimization algorithms, we choose a reasonable set of tuning values and do not make any effort in finding the best parameter settings. For BBO, we use maximum migration rates $I = 1$ and $E = 1$, and mutation rate $P_m = 0.01$. For ACO, we use initial pheromone value $\tau_0 = 1\text{E}-06$, pheromone update constant $Q = 20$, exploration constant $q_0 = 1$, global pheromone decay rate $\rho_g = 0.9$, local pheromone decay rate $\rho_l = 0.5$, pheromone sensitivity $\alpha = 1$ and visibility sensitivity $\beta = 5$. For ES, we produce $\lambda = 10$ offspring in each generation, and we use standard deviation $\sigma = 0.9$ for modifying solutions. For GA, we use roulette-wheel selection, single-point crossover with a crossover probability equal to 1, and a mutation probability equal to 0.01. For DE, we use a weighting factor $F = 0.5$ and a crossover constant $CR = 0.5$. For PSO, we use inertial constant equal to 0.3, cognitive constant equal to 1, social constant for swarm interaction equal to 1, and social constant for neighborhood interaction equal to 1. For each algorithm, we use a population size $N = 50$, and an elitism parameter equal to 2, which means that we keep the two best individuals from one generation to the next. The dimension of each benchmark function is $D = 20$, and the maximum number of fitness function evaluations (NFFEs) before program termination is set to 20,000. We ran 25 Monte Carlo simulations on each benchmark to get representative performances. To evaluate the performance of the algorithms, we define the error value as $f(x)-f(x^*)$, where x^* is the global minimum of the function and x is the best value found by the algorithm. The results are given in Table 3.3, which shows the mean minimum errors found by each algorithm.

Function	Name	Search domain	Minimum
F01	Sphere Function	$-100 \le x_i \le 100$	0
F02	Schwefel's Problem 2.22	$-10 \le x_i \le 10$	0
F03	Schwefel's Problem 1.2	$-100 \le x_i \le 100$	0
F04	Schwefel's Problem 2.21	$-100 \le x_i \le 100$	0
F05	Generalized Rosenbrock's Function	$-30 \le x_i \le 30$	0
F06	Step Function	$-100 \le x_i \le 100$	0
F07	Quartic Function	$-1.28 \le x_i \le 1.28$	0
F08	Generalized Schwefel's Problem 2.26	$-500 \le x_i \le 500$	–12,569.5
F09	Generalized Rastrigin's Function	$-5.12 \le x_i \le 5.12$	0
F10	Ackley's Function	$-32 \le x_i \le 32$	0
F11	Generalized Griewank's Function	$-600 \le x_i \le 600$	0
F12	Generalized Penalized Function 1	$-50 \le x_i \le 50$	0
F13	Generalized Penalized Function 2	$-50 \le x_i \le 50$	0

Table 3.2. *Benchmark functions; for the definitions of these benchmark functions, see Appendix A*

From Table 3.3, we see that BBO performs best on 9 of the 13 benchmark functions, while PSO performs best on the other three functions (F08, F11 and F12). For function F06, both BBO and PSO perform best. Furthermore, we find that for the unimodal functions (F01–F07), BBO usually obtains the best performance. BBO performs as well as PSO for multimodal functions (F08–F13), and both algorithms perform better than the other algorithms. This indicates that BBO is a competitive optimization algorithm for solving optimization problems, including both unimodal and multimodal functions.

Function	BBO	ACO	DE	ES	GA	PSO
F01	**1.04E-04**	2.05E+01	3.16E+03	5.67E+04	1.43E-03	3.68E-01
F02	**6.29E-15**	7.12E+02	5.38E+02	4.21E+03	3.80E+02	2.25E+02
F03	**3.84E+01**	4.33E+02	2.16E+02	5.80E+02	4.43E+03	1.76E+02
F04	**6.35E-15**	7.97E+01	2.34E+01	1.22E+01	4.10E+00	6.78E-01
F05	**8.85E-01**	8.90E+02	3.41E+01	2.34E+03	6.73E+01	6.74E+00
F06	**0.00E+00**	6.54E+00	2.15E+02	6.78E+03	3.24E+00	**0.00E+00**
F07	**8.61E-09**	6.05E+02	2.38E+01	4.21E+02	5.33E+02	2.58E+00
F08	4.97E+00	6.72E+02	1.25E+01	4.32E+01	1.05E+02	**8.81E-01**
F09	**8.77E-01**	8.97E+01	1.16E+02	5.38E+02	2.76E+01	4.54E+02
F10	**4.18E-01**	9.84E-01	3.87E+00	1.45E+00	3.40E-02	7.53E-01
F11	6.53E+00	6.76E+01	2.31E+01	7.89E+01	3.76E+02	**3.30E+00**
F12	4.64E-32	6.87E-08	2.53E-05	1.12E-07	3.54E-32	**2.62E-32**
F13	**8.47E-32**	1.83E-01	2.51E-07	7.82E-07	6.06E-31	9.52E-32

Table 3.3. *Comparison of experimental results of BBO, ACO, DE, ES, GA and PSO. The best value in each row is indicated in bold*

B. Influence of population size

Next, we investigate the influence of population size on the performance of BBO. In theory, increasing the population size will increase the diversity of the solutions and promote the exploration of the search space. However, the choice of the best population size is problem-specific. In this experiment, all the parameter settings are the same as those used above except for population size. Performance results for different population sizes are shown in Table 3.4.

From Table 3.4, we see that the performance of BBO with population size $N = 50$ is better than other population sizes ($N = 20$, 100, and 200) for the majority of the functions. The results tell us that although increasing population size can increase the solution diversity and result in better exploration of the search space, a population size that is too large ($N = 100$ or 200) considerably decreases the probability of finding the best solution, and thus degrades performance. On the other hand, when the population size is too small, for example $N = 20$, the algorithm lacks diversity so that BBO is easily trapped in local minima. So we can conclude that a moderate population size is best for obtaining the global optimum.

Function	$N = 20$	$N = 50$	$N = 100$	$N = 200$
F01	6.77E-01	**1.04E-04**	8.03E+00	7.72E-02
F02	1.23E-08	**6.29E-15**	7.67E-15	2.36E-10
F03	5.31E+01	**3.84E+01**	9.53E+02	1.28E+02
F04	2.24E-07	**6.35E-15**	1.26E-10	6.44E-14
F05	6.32E+00	8.85E-01	2.93E+00	**2.46E-01**
F06	5.70E-12	**0.00E+00**	**0.00E+00**	**0.00E+00**
F07	1.38E-06	8.61E-09	8.43E-07	**5.13E-09**
F08	2.24E+00	**4.97E+00**	9.02E+00	8.09E+00
F09	6.70E-01	8.77E-01	**8.29E-02**	5.62E-01
F10	5.52E-02	4.18E-01	1.10E-01	**4.78E-03**
F11	1.86E+00	**6.53E+00**	6.72E+01	7.27E+00
F12	3.36E-14	**4.64E-32**	3.14E-20	3.43E-15
F13	2.74E-04	**8.47E-32**	5.65E-18	2.61E-12

Table 3.4. *Comparison of experimental results for population sizes N = 20, 50, 100 and 200, where boldface indicates the best value in each row*

C. Influence of dimension

In order to investigate the influence of problem dimension on the performance of BBO, in this experiment, we choose dimensions $D = 20$, 30, 50 and 100, while all other parameter settings are the same as those used above. The results are shown in Table 3.5 after $D \times 1{,}000$ NFFEs. From Table 3.5, we see that overall performance decreases when the problem dimension increases. At $D = 20$, BBO can perform best for the majority of the functions, but at $D = 100$, BBO can only locate the near-global optimum for all functions. From these results, we recognize that increasing the problem dimension makes the problem more difficult.

Function	$D = 20$	$D = 30$	$D = 50$	$D = 100$
F01	**1.04E-04**	6.70E-02	8.09E-01	6.76E-02
F02	**6.29E-15**	1.32E-04	6.73E-04	2.31E-03
F03	3.84E+01	**8.39E-01**	2.26E+02	1.45E+02
F04	**6.35E-15**	4.23E-14	5.43E-12	3.92E-10
F05	**8.85E-01**	2.16E+00	2.23E+00	1.45E+00
F06	**0.00E+00**	**0.00E+00**	**0.00E+00**	4.86E-12
F07	8.61E-09	**7.38E-15**	5.64E-10	9.02E-09
F08	**4.97E+00**	5.32E+01	7.89E+02	7.78E+02
F09	**8.77E-01**	1.89E-01	9.14E-01	2.31E+00
F10	4.18E-01	3.12E+00	5.98E-01	**8.32E-02**
F11	6.53E+00	7.78E+00	**9.61E-01**	4.43E+01
F12	4.64E-32	3.54E-30	**3.10E-32**	2.67E-20
F13	**8.47E-32**	3.56E-10	1.90E-08	5.51E-15

Table 3.5. *Comparison of experimental results for problem dimensions D = 20, 30, 50 and 100 at D × 1,000 NFFEs, where boldface indicates the best value in each row*

D. Influence of mutation rate

Here we investigate the influence of mutation rate. Selecting the best mutation rate is not easy for a specific problem, and there are no general rules. Three different mutation rates, $P_m = 0.1$, 0.01 and 0.001, are used in this experiment. All remaining parameter values are the same as those used above. Table 3.6 shows the results for different mutation rates. Based on these results, we find that mutation rate $P_m = 0.01$ gives the best performance. These results indicate that mutation helps increase diversity and increases the chances for finding a good solution, but a high mutation rate ($P_m = 0.1$) results in too much exploration and is detrimental to the search. As mutation rate decreases from 0.1 to 0.01, optimization performance increases greatly, but as the mutation rate continues to decrease to 0.001, optimization performance decreases rapidly. A small mutation rate is not able to sufficiently increase solution diversity.

Function	$P_m = 0.1$	$P_m = 0.01$	$P_m = 0.001$
F01	3.67E-03	1.04E-04	**3.48E-05**
F02	8.81E-08	**6.29E-15**	4.55E-12
F03	7.42E+01	**3.84E+01**	5.17E+01
F04	1.55E-13	**6.35E-15**	7.89E-14
F05	2.16E+00	**8.85E-01**	3.25E+00
F06	4.37E-01	**0.00E+00**	1.26E-04
F07	6.74E-05	**8.61E-09**	7.80E-07
F08	9.03E+01	**4.97E+00**	5.75E+01
F09	9.67E-01	**8.77E-01**	9.43E-01
F10	**8.47E-02**	4.18E-01	3.56E-01
F11	2.14E+00	6.53E+00	**6.53E-01**
F12	3.22E-19	**4.64E-32**	7.80E-30
F13	5.60E-15	**8.47E-32**	3.49E-31

Table 3.6. *Comparison of experimental results for mutation rates P_m = 0.1, 0.01 and 0.001, where boldface indicates the best value in each row*

E. Influence of maximum migration rate

As mentioned above, the choice of maximum migration rate may be important for the performance of BBO. In this experiment, we test some combinations of maximum immigration rate I and maximum emigration rate E to test the performance. We use three combinations: (I = 1, E = 10), and (I = 1, E = 1), and (I = 10, E = 1). All other parameter values are the same as those used above, and the results are shown in Table 3.7.

It can be seen in Table 3.7 that each combination obtains the best performance on the same number of functions, which indicates that the maximum migration rates do not have a noticeable effect on BBO performance. This is because migration rates are used in a relative sense; therefore, a simple scaling of their magnitudes does not affect BBO performance.

Function	$I = 1, E = 10$	$I = 1, E = 1$	$I = 10, E = 1$
F01	6.73E-02	1.04E-04	**3.16E-05**
F02	1.26E-10	6.29E-15	**0.00E+00**
F03	5.71E+01	3.84E+01	**6.44E+00**
F04	5.34E-08	**6.35E-15**	5.27E-10
F05	**6.38E-02**	8.85E-01	4.20E+00
F06	**0.00E+00**	**0.00E+00**	**0.00E+00**
F07	1.07E-07	**8.61E-09**	7.16E-07
F08	2.35E+01	**4.97E+00**	6.65E+00
F09	**2.27E-02**	8.77E-01	5.04E-02
F10	**5.43E-02**	4.18E-01	3.47E-01
F11	7.61E+00	**6.53E+00**	9.04E+00
F12	1.49E-15	4.64E-32	**2.12E-32**
F13	**4.27E-32**	8.47E-32	7.32E-10

Table 3.7. *Comparison of experimental results using maximum migration rates (I = 1, E = 10), and (I = 1, E = 1), and (I = 10, E = 1), where boldface indicates the best value in each row*

Discussion

The BBO algorithm is a simple, robust and novel global optimization method. BBO has good optimization performance due to its migration and mutation operators. From the experimental results we can conclude the following:

1) The BBO algorithm is an effective optimization method, and it can obtain the global optimum (or near-global optimum) for many benchmark functions.

2) Experiment A (Table 3.3) shows that BBO is competitive with other optimization algorithms on the majority of functions.

3) Experiment B (Table 3.4) shows that a moderate population size is best for obtaining the global optimum.

4) Experiment C (Table 3.5) shows that BBO has difficulty obtaining the optimum for high-dimensional functions.

5) Experiment D (Table 3.6) shows that a moderate mutation rate of about 1% gives the best performance for BBO.

6) Experiment E (Table 3.7) shows that the maximum migration rates do not greatly affect BBO performance.

3.4. Conclusion

We have seen how biogeography, the study of the geographical distribution of biological species, can be used to obtain the BBO algorithm. We have also seen that BBO has similarities with other bio-inspired optimization algorithms from the viewpoint of algorithmic features, but it is distinctive enough to be considered a separate optimization algorithm rather than a special type of some other EA. We have applied BBO to benchmark functions, and we have shown that BBO provides good performance in comparison with other optimization algorithms. The performance results show that BBO theory can be successfully applied to general optimization problems. In the next chapter, we will discuss other aspects of biogeography theory that can inspire variations in the BBO algorithm.

4

BBO Extensions

If natural biogeography is an optimization process, then it stands to reason that modeling BBO more closely after natural biogeography could result in better optimization performance. With this idea in mind, we simulate these BBO extensions on a set of benchmark functions and application problems, with results that support the hypothesis that natural biogeography is itself an optimization process.

Overview of the chapter

This chapter discusses some extensions that can be made to BBO to improve performance. We first discuss the effects of different migration curve shapes on BBO in section 4.1, and we use blended migration to improve performance in section 4.2. We then provide some alternative approaches to BBO implementation in section 4.3. Finally, we apply BBO extensions to solve some real-world optimization problems in section 4.4.

4.1. Migration curves

Up to this point, we have assumed that the BBO migration curves are linear as shown in Figure 3.1. This is a convenient assumption, and it corresponds to linear rank-based selection in GAs. But in biogeography, migration curves are nonlinear, and the exact shape of biogeography migration curves is difficult to quantify and changes from one habitat to the next. It was surmised that nonlinear migration curves might give better performance than the linear curves in the basic BBO algorithm. This led to the investigation of several different migration curves. Here we discuss three promising curves in detail, which are motivated by influence factors that are associated with biogeography, as described in Chapter 1. These curves include quadratic migration curves, sinusoidal migration curves and generalized sinusoidal migration curves, which are shown in Figure 4.1 [MA 10a, MA 11b]. It is assumed that the maximum immigration rate and maximum emigration rate are both equal to 1.

Before introducing the nonlinear migration curves, we first review the normalized linear migration rates depicted in Figure 3.1:

$$\lambda_k = 1 - r_k \\ \mu_k = r_k \qquad [4.1]$$

where r_k is the fitness rank of the kth solution in the population, and is normalized to the range [0, 1]. $r_k = 0$ denotes the least fit solution, and $r_k = 1$ denotes the most fit solution. This formulation means that the immigration rate λ and the emigration rate μ are linear functions of solution fitness rank. Although linear migration curves do not exist in natural biogeography, they exhibit migration features and properties that are much simpler than those exhibited by general, nonlinear migration curves.

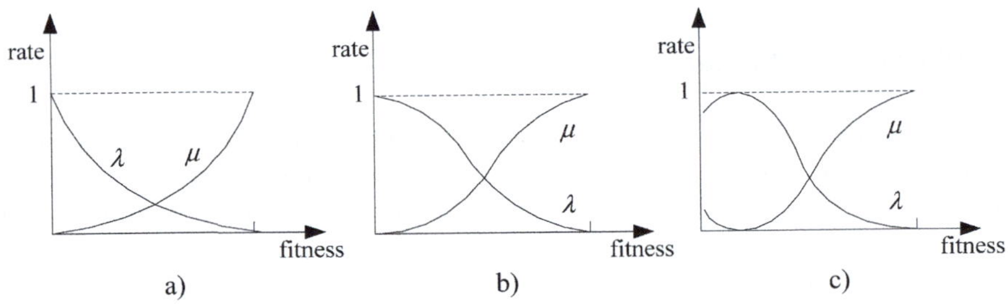

Figure 4.1. *Three nonlinear migration curves, where a), b) and c), respectively, show a quadratic migration curve, sinusoidal migration curve and generalized sinusoidal migration curve. λ is the immigration rate and μ is the emigration rate, and it is assumed that the maximum immigration rate and the maximum emigration rate are both equal to 1*

The first nonlinear curve that we discuss is the quadratic migration curve, which assigns the migration rates as:

$$\lambda_k = (1 - r_k)^2 \\ \mu_k = (r_k)^2 \qquad [4.2]$$

Migration rates λ and μ are concave quadratic functions of solution fitness rank, as shown in Figure 4.1(a). This curve is inspired by island biogeography, which was developed to explain the species distribution of biological habitats. Based on an experimentally tested theory of island biogeography [WHI 98], we know that migration in a single habitat follows a quadratic function of the size of the habitat and its geographical proximity to other habitats. According to equation [4.2], when the solution fitness rank is small, the immigration rate rapidly decreases from its maximum while the emigration rate slowly increases from zero. When the solution fitness rank is large, the immigration rate gradually decreases to zero and the emigration rate rapidly increases to its maximum.

The second nonlinear curve is the sinusoidal migration curve, which assigns the migration rates as:

$$\begin{aligned} \lambda_k &= \frac{1}{2}\left(\cos\left(r_k \cdot \pi\right)+1\right) \\ \mu_k &= \frac{1}{2}\left(-\cos\left(r_k \cdot \pi\right)+1\right) \end{aligned} \qquad [4.3]$$

Migration rates λ and μ are sinusoidal functions of solution fitness rank, as shown in Figure 4.1(b). This curve takes into account species mobility, the evolution of particular species and population size. These factors make the migration curves look like sinusoids. Based on equation [4.3], when the solution fitness rank is small or large, the immigration rate and the emigration rate both change slowly from their extremes, and when the solution fitness rank is medium, the migration rates change rapidly.

Classical island biogeography theory indicates that the immigration rate decreases and the emigration rate increases as the number of species increases in a habitat. In BBO, this corresponds to a monotonic decrease in immigration rate and a monotonic increase in emigration rate as solution fitness rank increases, as shown in the previous two migration curves, although their shapes are different. This means that as a solution becomes more fit, the probability of incorporating features from other solutions decreases. However, recent advances in biogeography indicate that for some pioneer species, at least for plants, an initial increase in species count results in an initial increase in immigration rate and an initial decrease in emigration rate [WU 95]. This is because the original unfavorable environmental conditions of the island are ameliorated by the first colonists, which make it more hospitable to additional species. That is, the positive effect of increased diversity due to initial immigration overcomes the negative effect of increased species count. In BBO, this corresponds to an initial increase in immigration rate as a very poor candidate solution initially improves its fitness. This can be viewed as a temporary positive feedback mechanism in BBO. A very poor candidate solution accepts features from other solutions, increasing its fitness, which subsequently increases its likelihood of accepting even more features from other solutions. This idea can also be incorporated into other bio-inspired optimization algorithms [MÜH 93], but its motivation comes from biogeography. These migration curves are depicted in Figure 4.1(c), and are expressed as:

$$\lambda_k = \frac{1}{2}\left(\cos\left(r_k \cdot \pi + \beta\right) + 1\right)$$
$$\mu_k = \frac{1}{2}\left(-\cos\left(r_k \cdot \pi + \beta\right) + 1\right) \quad [4.4]$$

where β is a negative trigonometric offset angle (typically between $-\pi/2$ and 0) that denotes the degree of temporary positive immigration rate feedback. With this curve, fitness rank is normalized to $[0, 1-\beta/\pi]$. This is called the generalized sinusoidal migration curve. This curve shows that immigration initially increases with solution fitness rank. It gives improving solutions the momentum that they need to continue improving. As a solution continues to become fitter after the initial increase in immigration, immigration begins to decrease to give less fit solutions relatively greater opportunities to immigrate good solution features.

To test the performance of the proposed nonlinear migration curves, they are compared with the linear migration curve using 13 benchmark functions, which are described in Table 3.2, where F01–F07 are unimodal functions and F08–F13 are multimodal functions. For a fair comparison, we use the nonlinear migration curves instead of the linear migration curve in the basic BBO algorithm. The other parameters of BBO are the same as described in section 3.3 of Chapter 3. In addition, the dimension of each benchmark function is set to 30, and the maximum number of fitness function evaluations (NFFEs) is set to 20,000. The parameter β of the generalized sinusoidal migration curve is set to $-\pi/2$.

Table 4.1 summarizes the performance on 13 benchmark functions for the four migration curves. It is seen from this table that the three nonlinear migration curves generally perform better than the linear migration curve, and the generalized sinusoidal migration curve performs the best for most of the functions. The results indicate that migration curves that are closely associated with biogeography improve the optimization ability of BBO.

Function	Linear	Quadratic	Sinusoidal	Generalized
F01	2.17E–02	8.24E–02	6.38E–02	**8.24E–03**
F02	1.84E–03	2.39E–04	8.03E–04	**1.47E–04**
F03	6.33E–02	7.93E+03	1.31E–02	**3.08E–03**
F04	5.68E–14	4.31E–14	4.34E–14	**4.64E–14**
F05	9.24E–01	**5.14E–01**	3.47E+00	8.36E–01
F06	**0.00E+00**	1.19E–02	**0.00E+00**	8.21E–15
F07	1.37E–15	1.84E–04	1.16E–15	**0.00E+00**
F08	2.63E–06	2.72E–07	5.08E–06	**9.38E–09**
F09	1.55E–13	1.56E–01	**1.21E–14**	4.38E–04
F10	**0.00E+00**	5.38E–01	9.71E–01	**0.00E+00**
F11	7.49E–01	**2.98E–01**	1.97E–01	3.65E–01
F12	2.26E–30	2.87E–22	**4.11E–30**	7.81E–25
F13	1.28E–10	**5.54E–32**	7.36E–11	9.05E–11

Table 4.1. *Comparison of experimental results of linear migration curve, quadratic migration curve, sinusoidal migration curve and generalized sinusoidal migration curve. Bold face indicates the mean minimum error value for each function*

4.2. Blended migration

In biogeography, migration is the movement of species between habitats. In BBO, migration is a probabilistic operator that adjusts each solution x_k by sharing features between solutions. In the basic BBO algorithm, the probability that the solution x_k is selected as the immigrating solution is proportional to its immigration rate λ_k, and the probability that the solution x_j is selected as the emigrating solution is proportional to the emigration rate μ_j. Migration can be expressed as:

$$x_k(\text{SIV}) \leftarrow x_j(\text{SIV}) \quad [4.5]$$

where an SIV is a solution feature, equivalent to a gene in GAs. In other words, an SIV is a search variable and the set of all possible SIVs is the search space from which an optimal solution will be selected. Equation [4.5] means that a solution feature of solution x_k is replaced by a feature from solution x_j.

Here we propose a migration operator called blended migration [MA 10b, MA 11a], which is a generalization of the basic BBO migration operator, and which is motivated by blended crossover in GAs. Blended crossover is frequently used in GAs [MÜH 93]. In blended crossover, instead of copying a parent's gene to a child chromosome, the offspring are obtained by combining parents' genes. In blended migration in BBO, a solution feature of solution x_k is not simply replaced by a feature from solution x_j. Instead, a new solution feature in a BBO solution is comprised of two components: the migration of a feature from another solution, and the migration of a feature from itself. Blended migration is defined as:

$$x_k(\text{SIV}) \leftarrow \alpha x_k(\text{SIV}) + (1-\alpha)x_j(\text{SIV}) \qquad [4.6]$$

where α is a real number between 0 and 1. It could be random or deterministic, or it could be proportional to the relative fitness of the solutions x_k and x_j. Equation [4.6] means that the new solution feature (SIV) of x_k comes from a combination of its own SIV and the emigrating solution's SIV.

The core idea of the proposed blended migration operator is based on two considerations. First, the operator is easily used with continuous-domain optimization problems. Second, blended combination operators have been widely and successfully used in other bio-inspired optimization algorithms. Blended migration is an attractive BBO modification from a couple of different viewpoints. On the one hand, good solutions will be less likely to be degraded due to migration. On the other hand, poor solutions can still accept a lot of new features from good solutions. That is, if the solution x_k is much more fit than the solution x_j, it would make sense to have α close to 1; but if the solution x_k is much less fit than the solution x_j, it would make sense to have α close to 0. Blended migration is similar to the blended crossover approach of the breeder GA [MÜH 93] and ES [MCT 08], in which both of the parents can contribute characteristics to a single feature of an offspring.

To explore the effect of blended migration on BBO performance, we test the basic BBO (α=0) and blended BBO (with α=0.5 and 0.8) on a set of benchmark functions. We use the same BBO parameters as described in the previous experiment, and the results are shown in Table 4.2.

Function	Basic BBO (α=0)	Blended BBO (α=0.5)	Blended BBO (α=0.8)
F01	2.17E–02	**3.11E–03**	3.25E–02
F02	1.84E–03	**5.26E–06**	7.74E–05
F03	6.33E–02	**1.38E–03**	2.36E–02
F04	5.68E–14	2.87E–14	**1.84E–14**
F05	9.24E–01	**3.74E–01**	6.25E+00
F06	**0.00E+00**	**0.00E+00**	**0.00E+00**
F07	1.37E–15	1.96E–15	**8.65E–16**
F08	2.63E–06	**4.35E–07**	3.32E–06
F09	1.55E–13	**1.06E–13**	8.74E–13
F10	**0.00E+00**	6.32E–01	5.62E–01
F11	7.49E–01	**4.14E–01**	2.97E+00
F12	2.26E–30	**4.43E–32**	1.65E–30
F13	1.28E–10	**2.54E–11**	3.11E–10

Table 4.2. *Comparison of experimental results of basic BBO, and blended BBO with α = 0.5 and 0.8, where boldface indicates the mean minimum error value for each function*

Table 4.2 shows that blended BBO with $\alpha = 0.5$ performs the best on nine functions, blended BBO with $\alpha = 0.8$ performs the best on two functions, and basic BBO ($\alpha = 0$) performs the best on one function. For function F06, all the three algorithms attain the global optimum. This result indicates that the value of α is influential on BBO performance. Blended BBO with $\alpha > 0$ generally outperforms basic BBO ($\alpha = 0$), which means that blended migration can significantly improve the optimization ability of BBO. Also, blended BBO with $\alpha = 0.5$ is better than blended BBO with $\alpha = 0.8$, which indicates that the best migration option is for a new solution feature to be contributed equally from itself and the selected emigrating solution.

4.3. Other approaches to BBO

The basic BBO algorithm presented in Figure 3.5 in Chapter 3 is called partial immigration-based BBO. The word "partial" means that only one solution feature is considered at a time for immigration. That is, for temporary solution z_k, λ_k is tested against a random number once for every feature to decide whether or not to replace that feature in z_k. The term "immigration-based" means that λ_k is first used to decide whether or not to immigrate to z_k; then the μ_j variables are used to choose the emigrating solution, but only if immigration was selected as described in the standard BBO algorithm.

However, there are also other ways that could be used to implement migration. Instead of testing λ_k against a random number once for each solution feature, we could test λ_k against a random number only once for each solution, and then if immigration were selected, we could replace all of the solution features in z_k. We call this total immigration-based BBO, an outline of which is presented in Figure 4.2.

```
Initialize a population of candidate solutions { x_k } for k ∈ [1, N]
While not (termination criterion)
    For each x_k, set emigration rate μ_k proportional to the fitness of x_k, where μ_k
    is normalized to [0, 1]
    For each x_k, set immigration rate λ_k = 1 − μ_k
    { z_k } ← { x_k }
    For each solution z_k
        Use λ_k to probabilistically decide whether to immigrate to z_k
        If immigrating then
            For each solution feature SIV
                Use {μ_i} to probabilistically select the emigrating solution x_j
                z_k(SIV) ← x_j(SIV)
            Next solution feature
        End if
        Probabilistically decide whether to mutate z_k
    Next solution
    { x_k } ← { z_k }
Next generation
```

Figure 4.2. *Outline of total immigration-based BBO with a population size of N. $\{x_k\}$ is the population of solutions and $\{z_k\}$ is a temporary population of solutions. x_k is the kth candidate solution, and $x_k(SIV)$ is the solution feature SIV of x_k*

As a third option, we could first use μ_k to decide whether or not to emigrate a solution feature from a given solution. Then, if emigration were selected, the λ_j values could be used to select the immigrating solution. This idea results in partial emigration-based BBO, an outline of which is shown in Figure 4.3.

```
Initialize a population of candidate solutions { x_k } for k ∈ [1, N]
While not (termination criterion)
    For each x_k , set emigration rate μ_k proportional to the fitness of x_k , where μ_k is
    normalized to [0,1]
    For each x_k , set immigration rate λ_k = 1 − μ_k
    { z_k } ← { x_k }
    For each solution x_r (r = 1 to N )
      For each solution feature SIV
        Use μ_r to probabilistically decide whether to emigrate from x_r
        If emigrating then
          Use {λ_j} to probabilistically select the immigrating solution z_k
          z_k(SIV) ← x_r(SIV)
        End if
      Next solution feature
    Next solution
    For each z_k in the population, probabilistically decide whether to mutate z_k
    { x_k } ← { z_k }
Next generation
```

Figure 4.3. *Outline of partial emigration-based BBO with a population size of N. $\{x_k\}$ is the population of solutions and $\{z_k\}$ is a temporary population of solutions. x_k is the kth candidate solution, and $x_k(SIV)$ is the solution feature SIV of x_k*

Finally, instead of testing μ_k against a random number once for each solution feature, we could test μ_k against a random number only once for each solution, and then if emigration were selected, all solution features could be emigrated from x_k. This is called total emigration-based BBO, an outline of which is shown in Figure 4.4.

```
Initialize a population of candidate solutions { x_k } for k ∈ [1, N]
While not (termination criterion)
    For each x_k, set emigration rate μ_k proportional to the fitness of x_k, where μ_k is
    normalized to [0,1]
    For each x_k, set immigration rate λ_k = 1 − μ_k
    { z_k } ← { x_k }
    For each solution x_r (r = 1 to N)
        Use μ_r to probabilistically decide whether to emigrate from x_r
        If emigrating then
            For each solution feature SIV
                Use {λ_j} to probabilistically select the immigrating solution z_k
                z_k(SIV) ← x_r(SIV)
            Next solution feature
        End if
    Next solution
    For each z_k in the population, probabilistically decide whether to mutate z_k
    { x_k } ← { z_k }
Next generation
```

Figure 4.4. *Outline of total emigration-based BBO with a population size of N. $\{x_k\}$ is the population of solutions and $\{z_k\}$ is a temporary population of solutions. x_k is the kth candidate solution, and $x_k(SIV)$ is the solution feature SIV of x_k*

These four combinations of partial/total and immigration/emigration are inspired by the original philosophy of BBO migration [SIM 11a, MA 13a, MA 13b]. In addition, each of these approaches could be combined with the nonlinear migration curves as discussed in section 4.1, and/or blended migration as discussed in section 4.2. As with any other EA, we can also implement elitism or other strategies, although these procedures are not shown in the outlines of the algorithms.

To explore the effect of all the four BBO migration options, we use the same BBO parameters as described in the previous experiment. In addition, we use linear migration curves as described in Figure 3.1. Table 4.3 summarizes the performance of all the four BBO migration options on the 13 benchmark functions.

Function	Partial immigration BBO	Total immigration BBO	Partial emigration BBO	Total emigration BBO
F01	2.17E–02	7.26E–02	3.57E–05	**1.41E–05**
F02	1.84E–03	1.04E–03	7.86E–05	**5.93E–05**
F03	6.33E–02	7.82E–01	9.16E–04	**1.02E–04**
F04	5.68E–14	9.65E–15	6.45E–19	**7.44E–20**
F05	9.24E–01	3.78E–01	**1.46E–02**	1.85E–02
F06	**0.00E+00**	**0.00E+00**	**0.00E+00**	**0.00E+00**
F07	1.37E–15	5.37E–14	9.23E–18	**4.79E–19**
F08	**2.63E–06**	2.90E–06	6.14E–04	3.26E–04
F09	**1.55E–13**	7.13E–12	8.91E–11	9.08E–11
F10	**0.00E+00**	8.45E–11	2.93E–11	2.96E–11
F11	7.49E–01	**0.00E+00**	8.24E–11	4.33E-10
F12	2.26E–30	**1.98E–30**	8.75E–15	2.97E–14
F13	**1.28E–10**	3.26E–10	3.81E–06	1.04E–05

Table 4.3. *Benchmark results for four BBO migration options, where bold face indicates the mean minimum error value for each function*

For unimodal functions F01–F07, total emigration-based BBO performs the best, except for function F06, for which all the four algorithms attain the global optimum, and function F05, for which partial emigration-based BBO performs the best. For multimodal functions F08–F13, partial immigration-based BBO performs the best on four functions (F08, F09, F10 and F13), and total immigration-based BBO performs the best on the other two functions (F11 and F12). The results indicate that emigration-based BBO algorithms are better than immigration-based BBO algorithms for unimodal functions, and immigration-based BBO algorithms are better than emigration-based BBO algorithms for multimodal functions. These results show that migration approaches can affect BBO performance, and the four migration options of partial/total and immigration/emigration can be adapted for different classes of benchmark functions.

4.4. Applications

In this section, we investigate the performance of BBO extensions on some real-world optimization problems from the 2011 IEEE Congress on Evolutionary Computation. These problems are briefly summarized in Table 4.4, and a more detailed description can be found in Das and Suganthan [DAS 10]. We compare total emigration-based BBO, combined with the generalized sinusoidal migration curve with $\beta = -\pi/2$ and blended migration with $\alpha = 0.5$, which generally provided better performance than the other BBO extensions in the previous experiments, with some advanced versions of bio-inspired optimization algorithms. These other algorithms include stud GA (which we call SGA) [KHA 98], standard PSO 2007 (which we call SPSO 07) [BRA 07, QU 12] and adaptive DE (which we call ADE) [DAS 11a, DAS 11b, GHO 11, GON 11]. We compare with SGA because SGA is an improvement of the classic GA and uses the best individual at each generation for crossover. We compare with PSO because it often offers good performance and is itself a relatively new evolutionary algorithm. We use the current standard PSO 2007, obtained from Particle Swarm Central (http://www.particleswarm.info/). We compare with DE because it is one of the most powerful evolutionary algorithms and has demonstrated excellent performance on many problems. We use the adaptive DE proposed by Asafuddoula *et al.* [ASA 11], where control parameter settings are gradually adapted according to the learning progress, and which uses center-based differential-exponential crossover and incorporates local search to improve its efficiency.

The other parameters used in BBO in this experiment are the same as those used in the previous experiment. For the SGA, we use real coding, roulette-wheel selection, single-point crossover with a crossover probability of 1 and a mutation probability of 0.001. For SPSO 07, we use an inertia weight of 0.8, a cognitive constant of 0.5, a social constant for swarm interaction of 1.0 and a social constant for neighborhood interaction of 1.0. For ADE, we use an adaptive scaling factor (F) with the range [0.1–0.5], and an adaptive crossover rate (CR) with the range [0.8–0.98]. Each algorithm has a population size of 50 and a maximum of 100,000 fitness function evaluations. The results of solving these real-world optimization problems are given in Table 4.5. All results are computed from 25 Monte Carlo simulations.

Problem	Dimension	Comments
P01	6	Parameter estimation for frequency-modulated (FM) sound waves
P02	30	Lennard-Jones potential problem
P03	1	Bifunctional catalyst blend optimal control problem
P04	1	Optimal control of a nonlinear stirred tank reactor
P05	30	Tersoff potential function minimization problem (instance 1)
P06	30	Tersoff potential function minimization problem (instance 2)
P07	20	Spread spectrum radar polyphase code design
P08	7	Transmission network expansion planning problem
P09	126	Large-scale transmission pricing problem
P10	12	Circular antenna array design problem
P11.1	120	Dynamic economic dispatch problem (instance 1)
P11.2	216	Dynamic economic dispatch problem (instance 2)
P11.3	6	Static economic load dispatch problem (instance 1)
P11.4	13	Static economic load dispatch problem (instance 2)
P11.5	15	Static economic load dispatch problem (instance 3)
P11.6	40	Static economic load dispatch problem (instance 4)
P11.7	140	Static economic load dispatch problem (instance 5)
P11.8	96	Hydrothermal scheduling problem (instance 1)
P11.9	96	Hydrothermal scheduling problem (instance 2)
P11.10	96	Hydrothermal scheduling problem (instance 3)
P12	26	Spacecraft trajectory optimization problem (Messenger)
P13	22	Spacecraft trajectory optimization problem (Cassini 2)

Table 4.4. *Problem set descriptions. More details about these problems can be found in Das and Suganthan [DAS 10]*

According to Table 4.5, BBO performs best on eight problems (P02, P05, P06, P10, P11.2, P11.8, P11.9 and P11.10), ADE performs best on seven problems (P01, P09, P11.4, P11.6, P11.7, P12 and P13), SGA performs best on three problems (P07, P11.1 and P11.5) and SPSO 07 performs best on problem P04. In addition, we see that for problems P03 and P08, all the four algorithms attain the same optimum, and for problem P11.3, SGA, ADE and BBO, all attain the same optimum. These results indicate that BBO performs similarly to ADE, and is significantly better than SGA and SPSO 07.

Problem	SGA	SPSO 07	ADE	BBO
P01	7.44E–18	2.60E–02	**0.00E+00**	7.35E–17
P02	–2.62E+01	–2.81E+01	–2.60E+01	**–2.83E+01**
P03	**1.15E–05**	**1.15E–05**	**1.15E–05**	**1.15E–05**
P04	2.03E+01	**1.37E+01**	2.54E+01	1.43E+01
P05	–3.68E+01	–3.27E+01	–3.66E+01	**–3.69E+01**
P06	–2.91E+01	–2.68E+01	–2.91E+01	**–2.92E+01**
P07	**5.00E–01**	5.08E–01	5.08E–01	9.97E–01
P08	**2.20E+02**	**2.20E+02**	**2.20E+02**	**2.20E+02**
P09	3.05E+02	1.60E+03	**4.62E+01**	1.04E+03
P10	–2.01E+01	–2.09E+01	–2.17E+01	**–2.18E+01**
P11.1	**4.79E+04**	1.50E+05	5.73E+04	5.25E+04
P11.2	1.81E+07	6.10E+06	1.06E+06	**1.05E+06**
P11.3	**1.54E+04**	1.58E+04	**1.54E+04**	**1.54E+04**
P11.4	1.80E+04	1.82E+04	**1.79E+04**	1.89E+04
P11.5	**3.26E+04**	3.27E+04	3.27E+04	3.29E+04
P11.6	1.21E+05	1.37E+05	**1.20E+05**	1.32E+05
P11.7	1.95E+06	2.13E+06	**1.70E+06**	1.91E+06
P11.8	9.54E+05	1.16E+06	9.31E+05	**9.23E+05**
P11.9	9.39E+05	1.60E+06	1.23E+06	**9.30E+05**
P11.10	9.51E+05	1.21E+06	9.29E+05	**9.24E+05**
P12	7.89E+00	1.38E+01	**7.07E+00**	1.64E+01
P13	8.65E+00	8.78E+00	**8.61E+00**	1.43E+01

Table 4.5. *Comparison of real-world optimization results for SGA, SPSO 07, ADE and BBO. The best result in each row is shown in bold*

If we use more advanced versions of GA, PSO and DE, it might be possible to obtain better results than those here. However, the same could be said for other improvements of BBO. The purpose of these comparisons is not to tune our algorithms to obtain the best possible performance for specific problems, but rather to show that BBO is a competitive algorithm for real-world optimization problems.

4.5. Conclusion

We have seen that BBO is actually a family of algorithms, and so it could be called a metaheuristic. It includes the options shown in Table 4.6 based on migration curves, migration blending and migration approaches. We have also seen that BBO is a competitive algorithm for benchmark functions and real-world optimization problems. A more intensive study of the combinations of the options in Table 4.6 and other BBO extensions inspired by other aspects of biogeography remains as a task for future research. In the next chapter, we will discuss some of the theoretical aspects of BBO.

Migration curves	Migration blending	Migration approaches
Linear	None ($\alpha = 0$)	Partial immigration-based
Quadratic	$\alpha = 0.5$	Total immigration-based
Sinusoidal	α = some other constant	Partial emigration-based
Generalized sinusoidal	$\alpha \propto$ fitness	Total emigration-based

Table 4.6. *BBO implementation options. BBO can be implemented with the combination of any choice from column 1, any choice from column 2 and any choice from column 3*

5

BBO as a Markov Process

The study of bio-inspired optimization algorithms has often been *ad hoc*, simulation-based and non-analytic. Historically, engineers have been more concerned with the applications of algorithms than with their mathematical analyses. Gradually, engineers have begun to focus more on the questions of how and why. It is important for engineers who want to become well informed and well rounded in the area of evolutionary computation research to understand and answer these questions. Markov theory has become a fundamental area of mathematics with applications in physics, chemistry, computer science, social science, engineering, biology and other areas. Markov theory is a good way to answer theoretical questions about bio-inspired optimization algorithms, and it might also lead to unexpected and new avenues of research.

Overview of the chapter

In this chapter, we will see that Markov theory provides insight into BBO behavior. Section 5.1 gives an overview of Markov theory definitions and notations, and is the foundation of the BBO analysis in this chapter. Section 5.2 develops a Markov model for the basic BBO algorithm, and section 5.3 analyzes the convergence properties of BBO for binary problems. Section 5.4 develops and discusses Markov models of BBO extensions.

5.1. Markov definitions and notations

Markov models have been a valuable theoretical tool to analyze bio-inspired optimization algorithms, including simple genetic algorithms [DAV 93, MA 15a, MA 16b, NIX 92, REE 03, SIM 13a, SIM 13b, SUZ 95, SUZ 98] and simulated annealing [LUN 86]. A Markov chain is a random process which has a discrete set of possible state values $S = \{s_i\}$ for $i = 1, 2, \ldots, T$. For instance, the weather might be described by the set of states S = {rainy, nice, snowy}. We use the notation $S(t)$ to denote the state at time step t. The initial state is $S(0)$, the state at the next time step is $S(1)$, and so on. The system state might change from one time step to the next, or it might remain the same from one time step to the next. The transition from one state to another is entirely probabilistic. In a first-order Markov process, also called a first-order Markov chain, the probability that the system transitions to any given state at the next time step depends only on the current state; that is, the probability is independent of all previous states. The probability that the system transitions from state s_i at time step t to s_j is given by the probability $p_{ij}(t)$, which is called a transition probability. If the transition probability is independent of t, that is, $p_{ij}(t_1) = p_{ij}(t_2)$ for all $i, j \in [1, T]$ and for all t_1 and t_2, then the Markov chain is said to be homogeneous. The $T \times T$ matrix $P = [p_{ij}]$ is called the transition matrix. Therefore,

$$\sum_{j=1}^{T} p_{ij} = 1 \quad \text{for } i = 1, \cdots, T \qquad [5.1]$$

where p_{ij} is the element in the ith row and jth column of matrix P, which is also called the probability matrix, or the stochastic matrix of the Markov process. Equation [5.1] indicates that the sum of the elements of each row of P is 1.

EXAMPLE 5.1.–

According to [KEN 74], the Land of Oz never has two nice days in a row. If it is nice one day, then the next day has a 50% chance of rain and a 50% chance of snow. If it rains, then the next day has a 50% chance of rain again, a 25% chance of snow and a 25% chance of nice weather. If it snows, then the next day has a 50% chance of snow again, a 25% chance of rain and a 25% chance of nice weather. We see that the weather forecast for a given day depends solely on the weather of the previous day. If we assign states R, N and S, to rain, nice weather and snow, respectively, then we can form a Markov matrix that represents the probability of various weather transitions:

$$\begin{array}{c} \\ P = \begin{array}{c} R \\ N \\ S \end{array} \end{array} \begin{array}{c} \begin{array}{ccc} R & N & S \end{array} \\ \begin{bmatrix} 0.5 & 0.25 & 0.25 \\ 0.5 & 0 & 0.5 \\ 0.25 & 0.25 & 0.5 \end{bmatrix} \end{array} \qquad [5.2]$$

Suppose a Markov process begins in state s_i at time 0. We know from the previous discussion that the probability that the process is in state s_k at time 1 is given by $\Pr(S(1)=s_k \,|S(0)=s_i)=p_{ik}$. Next, we consider the following time step. We can use the total probability theorem [MIT 05] to find the probability that the process is in state s_j at time 2 as:

$$\begin{aligned}
&\Pr(S(2)=s_j \,|S(0)=s_i) \\
&\quad = \Pr(S(1)=s_1 \,|S(0)=s_i)\Pr(S(2)=s_j \,|S(1)=s_1)+ \\
&\qquad \Pr(S(1)=s_2 \,|S(0)=s_i)\Pr(S(2)=s_j \,|S(1)=s_2)+\cdots+ \\
&\qquad \Pr(S(1)=s_T \,|S(0)=s_i)\Pr(S(2)=s_j \,|S(1)=s_T) \\
&\quad = p_{i1}p_{1j}+p_{i2}p_{2j}+\cdots+p_{iT}p_{Tj} \\
&\quad = \sum_{k=1}^{T} p_{ik}\; p_{kj}
\end{aligned} \qquad [5.3]$$

But this is equal to the element in the ith row and jth column of the square of P; that is,

$$\Pr(S(2)=s_j \,|S(0)=s_i)=\left[P^2\right]_{ij} \qquad [5.4]$$

Continuing this line of reasoning in an inductive manner, we find that:

$$\Pr(S(t)=s_j \,|S(0)=s_i)=\left[P^t\right]_{ij} \qquad [5.5]$$

That is, the probability that the Markov process transitions from state s_i to state s_j after t time steps is equal to the element in the ith row and jth column of P^t.

In Example 5.1, we can compute P^t for various values of t to obtain:

$$P^2=\begin{bmatrix} 0.4375 & 0.1875 & 0.3750 \\ 0.3750 & 0.2500 & 0.3750 \\ 0.3750 & 0.1875 & 0.4375 \end{bmatrix}$$

$$P^4=\begin{bmatrix} 0.4023 & 0.1992 & 0.3984 \\ 0.3984 & 0.2031 & 0.3984 \\ 0.3984 & 0.1992 & 0.4023 \end{bmatrix} \qquad [5.6]$$

$$P^8=\begin{bmatrix} 0.4000 & 0.2000 & 0.4000 \\ 0.4000 & 0.2000 & 0.4000 \\ 0.4000 & 0.2000 & 0.4000 \end{bmatrix}$$

Now suppose that we do not know the initial state of the Markov process, but we do know the probabilities for each state; the probability that the initial state $S(0)$ is equal to s_k is given by $p_k(0), k \in [1, T]$. Then, we can use the total probability theorem [MIT 05] to obtain:

$$\begin{aligned} &\Pr(S(1) = s_i) \\ &= \Pr(S(0) = s_1)\, p_{1i} + \Pr(S(0) = s_2)\, p_{2i} + \cdots + \Pr(S(0) = s_T)\, p_{Ti} \\ &= \sum_{k=1}^{T} \Pr(S(0) = s_k) p_{ki} \\ &= \sum_{k=1}^{T} p_{ki}\, p_k(0) \end{aligned} \quad [5.7]$$

Generalizing the above equation, we obtain:

$$\begin{bmatrix} \Pr(S(1) = s_1) \\ \cdots \\ \Pr(S(1) = s_T) \end{bmatrix}^T = p^T(0)P \quad [5.8]$$

where $p(0)$ is the column vector comprised of $p_k(0), k \in [1, T]$. Note that in equation [5.8], the subscript T denotes the dimension number and the superscript T denotes the matrix transpose. Generalizing this development for multiple time steps, we obtain:

$$p^T(t) \begin{bmatrix} \Pr(S(t) = s_1) \\ \cdots \\ \Pr(S(t) = s_T) \end{bmatrix}^T = p^T(0)P^t \quad [5.9]$$

Equation [5.9] indicates that a Markov chain is completely specified by $p(0)$ and P. For Markov chains, we have the following theorems.

THEOREM 5.1.– *A regular $T \times T$ transition matrix P, also called a primitive transition matrix, is one for which all elements of P^t are non-zero for some t. If P is a regular transition matrix, then*:

1) $\lim_{t \to \infty} P^t = \pi$;

2) *All rows of π are identical and are denoted as π_i;*

3) *Each element of π_i is positive;*

4) *The probability that the Markov process is in the jth state after an infinite number of transitions is equal to the jth element of* π_i; *that is,* $\lim_{t\to\infty}\Pr(S(t)=s_j)=\lim_{t\to\infty}p_{ij}^t=\pi_{ij}$;

5) *Each row* π_i *is the unique probability vector which satisfies* $\pi_i P=\pi_i$. *That is,* π_i^T *is the eigenvector of* P^T *corresponding to the eigenvalue 1, normalized so that its elements sum to 1, namely,* $\sum_{i=1}^{T}\pi_i=1$;

6) *If we form the matrices* P_j, $j\in[1,T]$, *by replacing the jth column of P with zeros, then the jth element of* π_i *is given as*:

$$\pi_{ij}=\frac{|P_j-I|}{\Sigma_{i=1}^{T}|P_i-I|} \qquad [5.10]$$

where I is the $T\times T$ *identity matrix and* $|\cdot|$ *is the determinant operator.*

The first five properties above are also called the Perron–Frobenius theorem, and the last property is called the Davis–Principe theorem in books on Markov chains. They comprise the fundamental limit theorem for regular Markov chains, and are proven in [GRI 97, Chapter 1] and [DAV 93].

The first four properties of Theorem 5.1 can be combined as follows [IOS 80, p. 123].

THEOREM 5.2.– *Let P be a primitive transition matrix of order T; that is, all of the elements of* P^t *are positive for some integer t. Then,* P^t *converges as* $t\to\infty$ *to a stochastic matrix which has positive entries. That is, for all i, j* ∈ *[1, T],*

$$P^{\infty}=\lim_{t\to\infty}\left(p_{ij}\right)^t=\begin{pmatrix}\pi_i\\ \vdots\\ \pi_i\end{pmatrix}_{T\times T} \qquad [5.11]$$

where $\pi_i=(\pi_{i1},\ldots,\pi_{iT})$ *and* $\pi_{ij}>0$ *for* $1\le j\le T$.

We also have the following theorem [IOS 80, p. 126].

THEOREM 5.3.– *Let P be a transition matrix of order T with the structure:*

$$P = \begin{bmatrix} C & 0 \\ R & Q \end{bmatrix} \quad [5.12]$$

where C is a primitive stochastic matrix of order m, and R, Q ≠ 0. Then, P^t converges as t→ ∞ to a stochastic matrix. That is,

$$P^{\infty} = \lim_{t \to \infty} \left(p_{ij} \right)^t = \begin{pmatrix} \pi_i \\ \vdots \\ \pi_i \end{pmatrix}_{T \times T} \quad [5.13]$$

where $\pi_i = (\pi_{i1}, \ldots, \pi_{im}, 0, \ldots, 0)$ and $\pi_{ij} > 0$ for $1 \leq j \leq m < T$.

Later in this chapter, we will use Theorem 5.1 to derive important properties of the BBO Markov transition matrix, and we will use Theorems 5.2 and 5.3 to derive BBO convergence properties.

EXAMPLE 5.2.–

Today's weather forecast in Oz is 80% sun and 20% snow. What is the weather forecast for two days from now?

From equation [5.9], $p^T(2) = p^T(0)P^2$, where P is given in Example 5.1 and $p(0) = [0.0 \quad 0.8 \quad 0.2]^T$. This gives $p(2) = [0.3750 \quad 0.2375 \quad 0.3875]^T$. That is, two days from now there is a 37.5% chance of rain, a 23.75% chance of sun and a 38.75% chance of snow.

EXAMPLE 5.3.–

Using equation [5.6] and applying Theorem 5.1 to Example 5.1, we see that any given day in the distant future has a 40% probability of rain, a 20% probability of sun and a 40% probability of snow. Therefore, 40% of the days in Oz are rainy, 20% are sunny and 40% are snowy. Furthermore, we can find the eigenvalues of P^T as 1, –0.25 and 0.25. The eigenvector corresponding to the eigenvalue 1 is $[0.4 \quad 0.2 \quad 0.4]^T$.

Markov models can be valuable tools for analyzing EAs because they give us exact results. We can run simulations to investigate the performance of various EAs, but simulations can be misleading. For instance, a set of Monte Carlo simulations might happen to give misleading results due to the particular sequence of random numbers generated during the simulation. In addition, the random number generator using in the simulation may be incorrect, which occurs more often than we would like to think, and which would give misleading results [SAV 08]. Finally, the number of Monte Carlo simulations required to estimate highly improbable outcomes might be so high as to not be attainable in a reasonable amount of computational time. Markov model results avoid all of these pitfalls and give exact results.

Next, we define more notation that we will use later to derive a Markov model for BBO, which could also be suitable for Markov models for other EAs.

We will focus on EAs with a population size N operating in a discrete search space. The set of candidate solutions is the set of all bit strings x_i consisting of q bits each. Therefore, the cardinality of the search space is $n = 2^q$. We use v to denote the population vector, where the component v_i is the number of candidate solutions x_i in the population. We see that:

$$\sum_{i=1}^{n} v_i = N \qquad [5.14]$$

This equation simply means that the total number of solutions in the population is equal to N. We use y_k to denote the kth solution in the population:

$$Y = \{y_1, \cdots y_N\} = \{\underbrace{x_1, x_1, \cdots, x_1}_{v_1 \text{ copies}}, \underbrace{x_2, x_2, \cdots, x_2}_{v_2 \text{ copies}}, \cdots \underbrace{x_n, x_n, \cdots, x_n}_{v_n \text{ copies}}\} \qquad [5.15]$$

where the y_k solutions have been ordered to group identical solutions. We use T to denote the total number of possible populations Y. That is, T is the number of $n \times 1$ integer vectors v such that $\Sigma_{i=1}^{n} v_i = N$ and $v_i \in [0, N]$.

We order y_k in the same order as x_i. That is:

$$y_k = \begin{cases} x_1, \text{ for } k = 1, \cdots, v_1 \\ x_2, \text{ for } k = v_1 + 1, \cdots, v_1 + v_2 \\ x_3, \text{ for } k = v_1 + v_2 + 1, \cdots, v_1 + v_2 + v_3 \\ \vdots \\ x_n, \text{ for } k = \sum_{i=1}^{n-1} v_i + 1, \cdots, N \end{cases} \qquad [5.16]$$

This is also shown in equation [5.15] and can be written more compactly as:

$$y_k = x_{m(k)} \qquad \text{for } k = 1, 2, \cdots, N \tag{5.17}$$

where $m(k)$ is defined as:

$$m(k) = \min r \text{ such that } \sum_{i=1}^{r} v_i \geq k \tag{5.18}$$

EXAMPLE 5.4.–

Suppose we have a two-bit optimization problem ($q = 2$, $n = 4$) with a population size $N = 3$. The search space consists of the bit strings $x = \{x_1, x_2, x_3, x_4\} = \{00, 01, 10, 11\}$. Suppose that the candidate solutions in the current population are $y = \{x_2, x_4, x_4\} = \{01, 11, 11\}$. Then, we have $v_1 = 0$, $v_2 = 1$, $v_3 = 0$ and $v_4 = 2$.

How many possible EA populations exist for a population size N in a search space of cardinality n? That is, what is the value of T? This number can be calculated in several different ways. In [NIX 92], it is shown that:

$$T = \binom{n+N-1}{N} \tag{5.19}$$

We can also use the multinomial theorem [CHU 92, SIM 11d] to find T. The multinomial theorem can be stated in several ways, including the following. Given K classes of objects, the number of different ways that N objects can be selected (independent of order) while choosing from each class no more than M times is the coefficient q_N in the polynomial:

$$\begin{aligned} q(x) &= (1 + x + x^2 + \cdots + x^M)^K \\ &= 1 + q_1 x + q_2 x^2 + \cdots + q_N x^N + \cdots + x^{MK} \end{aligned} \tag{5.20}$$

Recall that the population vector v is an n-element vector such that each element is an integer between 0 and N (inclusive), and the sum of its elements is N. T is the number of unique population vectors v. Thus, T is the number of ways that N objects can be selected (independent of order) from n classes of objects while choosing from each class no more than N times. Applying the multinomial theorem to this problem gives:

$$\begin{aligned} T &= q_N \\ q(x) &= (1 + x + x^2 + \cdots + x^N)^n \\ &= 1 + q_1 x + q_2 x^2 + \cdots + x^{Nn} \end{aligned} \tag{5.21}$$

A different form of the multinomial theorem can also be used to find T. The multinomial theorem can be stated as:

$$(x_1 + x_2 + \cdots + x_N)^n = \sum_{S(k)} \frac{n!}{\prod_{j=0}^{N} k_j!} \prod_{j=0}^{N} x_j^{k_j}$$
$$= \sum_{S(k)} \prod_{i=0}^{N} \binom{\sum_{j=0}^{i} k_j}{k_i} \prod_{j=0}^{N} x_j^{k_j} \qquad [5.22]$$
$$S(k) = \left\{ k \in R^N : k_j \in \{0,1,\ldots,n\}, \sum_{j=0}^{N} k_j = n \right\}$$

Now, consider the polynomial $(x^0 + x^1 + x^2 + \cdots + x^N)^n$. From equation [5.22] of the multinomial theorem, we see that the coefficient of $[(x^0)^{k_0}(x^1)^{k_1}(x^2)^{k_2}\cdots(x^N)^{k_N}]$ is given by:

$$\prod_{i=0}^{N} \binom{\sum_{j=0}^{i} k_j}{k_i} \qquad [5.23]$$

If we sum up these terms for all k_j such that:

$$\sum_{j=0}^{N} jk_j = N \qquad [5.24]$$

then we obtain the coefficient of x^N. However, equation [5.21] shows that T is equal to the coefficient of x^N. Therefore,

$$T = \sum_{S'(k)} \prod_{i=0}^{N} \binom{\sum_{j=0}^{i} k_j}{k_i}$$
$$S'(k) = \left\{ k \in R^{N+1} : k_j \in \{0,1,\ldots,n\}, \sum_{j=0}^{N} k_j = n, \sum_{j=0}^{N} jk_j = n \right\} \qquad [5.25]$$

Equations [5.19], [5.21] and [5.25] give three equivalent expressions for T.

EXAMPLE 5.5.–

Suppose that our population consists of 2-bit candidate solutions (q = 2, n = 4) and a population size N = 4. Equation [5.19] gives:

$$T = \binom{7}{4} = 35 \qquad [5.26]$$

Equation [5.21] gives:

$$\begin{aligned} q(x) &= (1+x+x^2+x^3+x^4)^4 \\ &= 1+\cdots+35x^4+\cdots+x^{16} \\ T &= q_4 = 35 \end{aligned} \qquad [5.27]$$

Equation [5.25] gives:

$$\begin{aligned} T &= \sum_{S'(k)} \prod_{i=0}^{4} \binom{\sum_{j=0}^{i} k_j}{k_i} \\ S'(k) &= \left\{ k \in R^5 : k_j \in \{0,1,\ldots,4\}, \sum_{j=0}^{4} k_j = 4, \sum_{j=0}^{4} jk_j = 4 \right\} \\ &= \left\{ \begin{matrix} (3 \quad 0 \quad 0 \quad 0 \quad 1),(2 \quad 1 \quad 0 \quad 1 \quad 0),(2 \quad 0 \quad 2 \quad 0 \quad 0), \\ (1 \quad 2 \quad 1 \quad 0 \quad 0),(0 \quad 4 \quad 0 \quad 0 \quad 0) \end{matrix} \right\} \\ T &= 4+12+6+12+1 = 35 \end{aligned} \qquad [5.28]$$

As expected, all three methods for the calculation of T give the same result.

Finally, we define another notation that we will later use to derive a Markov chain model for BBO; this idea is called generalized multinomial probability [BEA 99]. Suppose that an experiment has n possible outcomes $\{x_1, x_2, \cdots, x_n\}$ and that the experiment is repeated N times. Suppose that the probability of obtaining outcome x_i on the kth trial is equal to P_{ki}. Let $C = [C_1, C_2, \cdots, C_n]$ be a vector of random variables, where C_i denotes the total number of times that x_i occurs in N trials, and let $\gamma = [\gamma_1, \gamma_2, \cdots, \gamma_n]$ be a realization of C. Define:

$$Y(\gamma) = \left\{ J \in R^{N \times n} : J_{ki} \in \{0,1\}, \sum_{i=1}^{n} J_{ki} = 1 \text{ for all } k, \sum_{k=1}^{N} J_{ki} = \gamma_i \text{ for all } i \right\} \qquad [5.29]$$

Note that the cardinality of $Y(\gamma)$ is:

$$|Y(\gamma)| = \frac{N!}{\gamma_1!\cdots\gamma_n!} \qquad [5.30]$$

Then, the generalized multinomial theorem gives the following probability that the repeated experiment results in the outcome vector γ:

$$\Pr(C=\gamma) = \sum_{J\in Y(\gamma)} \prod_{k=1}^{N}\prod_{i=1}^{n} P_{ki}^{J_{ki}} \qquad [5.31]$$

EXAMPLE 5.6.–

Here we use a simple example to clarify generalized multinomial probability. Consider a simple EA experiment in which a trial can result in one of four possible solutions x_1, x_2, x_3 and x_4 with probabilities Q_{i1}, Q_{i2}, Q_{i3} and Q_{i4}, respectively. Suppose that the total number of trials is equal to 2. Suppose that the probabilities are given as:

$$\begin{aligned}&\text{Trial 1: } Q_{11}=0.1, Q_{12}=0.3, Q_{13}=0.5, Q_{14}=0.1\\ &\text{Trial 2: } Q_{21}=0.1, Q_{22}=0.1, Q_{23}=0.1, Q_{24}=0.7\end{aligned} \qquad [5.32]$$

In this example, we calculate the probability that x_1 and x_4 occur after two trials. In order to calculate this probability, let $C=[C_1, C_2, \cdots, C_n]$ denote a vector of random variables, where C_i is the total number of times that x_i occurs after two trials. Based on equation [5.31], we set $C_1 = 1$, $C_2 = 0$, $C_3 = 0$ and $C_4 = 1$ to obtain:

$$\Pr(C_1=1, C_2=0, C_3=0, C_4=1) = \sum_{J\in Y}\prod_{k=1}^{2}\prod_{i=1}^{4}[Q_{ki}]^{J_{ki}} \qquad [5.33]$$

where:

$$Y = \left\{ J\in R^{2\times 4} : J_{ki}\in\{0,1\}, \sum_{i=1}^{4} J_{ki} = 1 \text{ for all } k, \sum_{k=1}^{2} J_{ki} = C_i \text{ for all } i \right\} \qquad [5.34]$$

According to equation [5.34], J belongs to Y if it satisfies the following conditions:

1) J is a 2×4 matrix;

2) Each element of J is either 0 or 1;

3) The elements in each row of J add up to 1;

4) The elements in the ith column of J add up to C_i.

Note from equation [5.30] that the cardinality of Y is:

$$|Y| = N!/(C_1!\cdots C_n!) \quad [5.35]$$

So there are a total of $N!/(C_1!\cdots C_n!) = 2!/(1!1!0!1!) = 2$ matrices $J^{(t)}$ that satisfy these conditions, and they are found as:

$$J^{(1)} = \begin{bmatrix} 1 & 0 & 0 & 0 \\ 0 & 0 & 0 & 1 \end{bmatrix} \quad J^{(2)} = \begin{bmatrix} 0 & 0 & 0 & 1 \\ 1 & 0 & 0 & 0 \end{bmatrix} \quad [5.36]$$

Substituting these matrices in equation [5.33] gives:

$$\begin{aligned} &\Pr(C_1 = 1, C_2 = 0, C_3 = 0, C_4 = 1) \\ &= \sum_{t=1}^{2} \left(Q_{11}^{J_{11}^{(t)}} Q_{12}^{J_{12}^{(t)}} Q_{13}^{J_{13}^{(t)}} Q_{14}^{J_{14}^{(t)}} \right) \times \left(Q_{21}^{J_{21}^{(t)}} Q_{22}^{J_{22}^{(t)}} Q_{23}^{J_{23}^{(t)}} Q_{24}^{J_{24}^{(t)}} \right) \\ &= Q_{11}Q_{24} + Q_{14}Q_{21} = 0.08 \end{aligned} \quad [5.37]$$

5.2. Markov model of BBO

This section defines a Markov model for a basic BBO algorithm [SIM 11b, SIM 11d, SIM 11c]. A Markov model of BBO provides us with the transition probability p_{ij} from the ith population distribution to the jth population distribution, where the states of the system include all possible population distributions (that is, all possible combinations of candidate solutions). In BBO, two main steps are significant, migration and mutation, so the transition probability includes the migration probability and the mutation probability for one generation.

Before the formal derivation, we make some assumptions. First, all of the new BBO solutions are created before any solutions are replaced in the population; that is, we use a generational BBO algorithm rather than a steady-state BBO algorithm, which is clearly shown in Figure 3.5 in Chapter 3.

Second, a solution can emigrate a bit to itself. This means that, in the statement "Use $\{\mu_j\}$ to probabilistically select the emigrating solution x_j" in Figure 3.2,

j might be chosen to be equal to k. that is, when a bit is replaced via migration in a given solution z_k, the new bit might be chosen to come from z_k itself. In this case, the bit is not actually replaced in z_k. However, the probabilistic choice of the emigrating solution allows this self-migration to happen on occasion.

A. Migration

In the basic BBO algorithm, we use λ_i to denote the immigration probability of x_i and μ_i to denote the emigration probability of x_i. Note that μ_i is proportional to the fitness of x_i, and λ_i decreases with the fitness of x_i. We use the notation $x_i(s)$ to denote the sth bit of solution x_i, and use the notation $\varsigma_i(s)$ to denote the set of population indices j such that the sth bit of x_j is equal to the sth bit of x_i. That is:

$$\varsigma_i(s) = \{j : x_j(s) = x_i(s)\} \qquad [5.38]$$

EXAMPLE 5.7.–

To clarify equation [5.38], an example is presented based on Example 5.3. First, we explain how to calculate $\varsigma_1(1)$. We arbitrarily number bits from left to right; that is, in any given bit string, bit 1 is the left-most bit and bit 2 is the right-most bit. From the definition of $\varsigma_i(s)$, we see that:

$$\varsigma_1(1) = \{j : x_j(1) = x_1(1)\} \qquad [5.39]$$

Since $x_1 = 00$, we see that $x_1(1) = 0$ (that is, the left-most bit is 0). Then, equation [5.39] can be written as:

$$\varsigma_1(1) = \{j : x_j(1) = 0\} \qquad [5.40]$$

But $x_j(1) = 0$ for $x_j \in \{00, 01\}$, which in turn indicates that $x_j(1) = x_1(1)$ for $j \in [1, 2]$; therefore, $\varsigma_1(1) = \{1, 2\}$. Continuing this process, we see that:

$$\begin{aligned} &\varsigma_1(1) = \{1, 2\},\ \varsigma_1(2) = \{1, 3\},\ \varsigma_2(1) = \{1, 2\},\ \varsigma_2(2) = \{2, 4\} \\ &\varsigma_3(1) = \{3, 4\},\ \varsigma_3(2) = \{1, 3\},\ \varsigma_4(1) = \{3, 4\},\ \varsigma_4(2) = \{2, 4\} \end{aligned} \qquad [5.41]$$

During migration, if the sth feature of y_k is not selected for immigration during generation t, then:

$$y_k(s)_{t+1} = x_{m(k)}(s), \qquad (\text{if no immigration to } y_{k,t}) \qquad [5.42]$$

That is, $y_k(s)$ does not change from generation t to generation $t + 1$. However, if the sth feature of y_k is selected for immigration during generation t, the probability that $y_k(s)_{t+1}$ is equal to $x_i(s)$ is proportional to the sum of the emigration rates of all solutions whose sth feature is equal to $x_i(s)$. This probability can be written as:

$$\Pr\left(y_k(s)_{t+1} = x_i(s) \middle| \text{immigration}\right) = \frac{\sum_{j \in \varsigma_i(s)} v_j \mu_j}{\sum_{j=1}^{n} v_j \mu_j} \qquad [5.43]$$

We can combine equations [5.42] and [5.43], along with the fact that the probability of immigration to $y_k(s)$ is equal to $\lambda_{m(k)}$ to obtain the total probability as follows:

$$\begin{aligned}
&\Pr\left(y_{k,t+1}(s) = x_i(s)\right) \\
&\quad = \Pr\left(\text{no immigration to } y_{k,t}\right)\Pr\left(y_{k,t+1}(s) = x_i(s) \middle| \text{no immigration}\right) + \\
&\quad\quad \Pr\left(\text{immigration to } y_{k,t}\right)\Pr\left(y_{k,t+1}(s) = x_i(s) \middle| \text{immigration}\right) \\
&\quad = \left(1 - \lambda_{m(k)}\right) 1_0\left(x_{m(k)}(s) - x_i(s)\right) + \lambda_{m(k)} \frac{\sum_{j \in \varsigma_i(s)} v_j \mu_j}{\sum_{j=1}^{n} v_j \mu_j}
\end{aligned} \qquad [5.44]$$

where 1_0 is the indicator function on the set $\{0\}$.

Now, recall that there are q bits in each solution. Use $M_{ki}(v)$ to denote the probability that immigration results in $y_{k,t+1} = x_i$, given that the population distribution is described by the population count vector v. $M_{ki}(v)$ can be written as:

$$\begin{aligned}
M_{ki}(v) &= \Pr\left(y_{k,t+1} = x_i\right) \\
&= \prod_{s=1}^{q} \left[\left(1 - \lambda_{m(k)}\right) 1_0\left(x_{m(k)}(s) - x_i(s)\right) + \lambda_{m(k)} \frac{\sum_{j \in \varsigma_i(s)} v_j \mu_j}{\sum_{j=1}^{n} v_j \mu_j} \right]
\end{aligned} \qquad [5.45]$$

Note that $M_{ki}(v)$ can be computed for each $k \in [1, N]$ and each $i \in [1, n]$ in order to form the $N \times n$ matrix $M(v)$. The kth row of $M(v)$ corresponds to the kth iteration of the outer loop in algorithm description in Figure 3.5 (there are N iterations of the outer loop in Figure 3.5). The ith column of $M(v)$ corresponds to the probability of obtaining solution x_i during each outer loop iteration; that is, $M_{ki}(v)$ gives the probability of obtaining the ith outcome on the kth immigration trial.

B. Mutation

Only migration is considered in equation [5.45]. Next, we consider the possibility of mutation. U is used to denote the $n \times n$ mutation matrix, where U_{ji} is the probability that x_j mutates to x_i. As in typical EAs, mutation operates independently on each candidate solution by probabilistically reversing each bit in each candidate solution. Suppose that the event that each bit of a candidate solution is flipped is stochastically independent and occurs with probability $p_m \in (0, 1)$. Then, the probability U_{ji} can be written as:

$$U_{ji} = \Pr\left(x_j \to x_i\right) = p_m^{H_{ij}} \left(1 - p_m\right)^{q - H_{ij}} \qquad [5.46]$$

where H_{ij} represents the Hamming distance between bit strings x_i and x_j.

So the probability that the kth immigration trial, followed by mutation, results in x_i is denoted as $P_{ki}(v)$. This can be written as:

$$\begin{aligned} P_{ki}(v) &= \sum_{j=1}^{n} M_{kj}(v) U_{ji} \\ P(v) &= M(v) U \end{aligned} \qquad [5.47]$$

where the elements of $M(v)$ are given in equation [5.45]. $P(v)$ contains the probabilities when both migration and mutation are considered. u is defined as the population count vector after migration and mutation are completed for a given generation, where the component u_i is the number of solutions x_i in the population. The transition probability $\Pr(u|v)$ that we obtain a population count vector u after one generation, given that we started with a population count vector v, can be obtained from the generalized multinomial theorem described in section 5.1 as:

$$\begin{aligned} &\Pr(u|v) = \sum_{J \in Y} \prod_{k=1}^{N} \prod_{i=1}^{n} \left[P_{ki}(v)\right]^{J_{ki}}, \\ &\text{where } Y = \left\{ J \in R^{N \times n} : J_{ki} \in \{0,1\}, \sum_{i=1}^{n} J_{ki} = 1 \text{ for all } k, \sum_{k=1}^{N} J_{ki} = u_i \text{ for all } i \right\} \end{aligned} \qquad [5.48]$$

Equation [5.48] can be used to find the transition matrix for the basic BBO algorithm with migration and mutation as described in Figure 3.5.

The Markov transition matrix, denoted as Q, is obtained by computing equation [5.48] for each possible v and each possible u. Each element of Q will give the transition probability from one population count vector v to another population count vector u after one generation. Note that Q is a $T \times T$ matrix, where T is the total number of possible populations. T can be calculated by several different methods as described in section 5.1. Once we calculate the transition matrix Q, a wealth of Markov tools [GRI 97] can be applied to explore the statistical properties of BBO, including the limiting probability of each possible BBO population as the generation count approaches infinity.

EXAMPLE 5.8.–

Here we confirm the BBO Markov model with simulation. We use the 3-bit one-max problem with a search space cardinality of eight and a population size of four. The one-max problem has a fitness function that is proportional to the number of ones in the population member, and is a popular test function in EA research. The fitness values of the 3-bit one-max problem are given as:

$$\begin{aligned} &f(000)=1,\ f(001)=2,\ f(010)=2,\ f(011)=3, \\ &f(100)=2,\ f(101)=3,\ f(110)=3,\ f(111)=4. \end{aligned} \qquad [5.49]$$

From equation [5.19] in section 5.1, we calculate the total number of possible populations as:

$$T=\binom{n+N-1}{N}=\binom{8+4-1}{4}=330 \qquad [5.50]$$

We can use equation [5.48] to calculate the probability of transitioning between each of the 330 population distributions, which gives us a 330 × 330 transition matrix *P*. We can then calculate the limiting probability of each possible population distribution by Theorem 5.1. This is the probability, in the limit as the generation count approaches infinity, that the BBO population consists of any particular set of individuals.

Table 5.1 shows the probabilities of obtaining an all-optimal population, along with the probabilities of obtaining a no-optimal population. The population count vector "All optimal" in Table 5.1 denotes the population that contains all optimal solutions, which is the population count vector (0 0 0 0 0 0 0 4). The population count vector "No optimal" in Table 5.1 denotes the population that does not contain any optimal solutions, which is the vector (* * * * * * * 0), where * denotes "don't care bit". Note that the notation $(v_1, v_2, \cdots, v_8)$ indicates the numbers of solutions that are equal to $(000, 001, \cdots, 111)$. The Markov model and simulation results match well, which confirms the model. Table 5.1 shows that a high mutation rate of 10% per bit results in too much exploration, so the uniform optimal population is not one of the most probable populations. With this high 10% mutation rate, the probability that the population does not have any optimal solutions is 30%, as shown in the table. However, as the mutation rate decreases to the more typical values of 1% and 0.1%, the probabilities that the population is composed entirely of optimal solutions increase to 53% and 86%, respectively, and the probabilities that the population has no optimal solutions decrease to 11% and 9%, respectively.

Mutation rate	Population count vector	Probability	
		Markov	Simulation
0.1	All optimal	0.0290	0.0285
	No optimal	0.2999	0.3026
0.01	All optimal	0.5344	0.5322
	No optimal	0.1134	0.1138
0.001	All optimal	0.8605	0.8437
	No optimal	0.0923	0.1092

Table 5.1. *BBO Markov model and simulation results for the 3-bit one-max problem. The table shows the probabilities of obtaining an all-optimal population (0 0 0 0 0 0 0 4) and the probabilities of obtaining a no-optimal population (* * * * * * * 0), where * denotes "don't care bit". Simulation results are the averages of 100 Monte Carlo simulations*

Figure 5.1 shows typical simulation results of 20,000 generations of BBO for the 3-bit one-max problem with a mutation rate of 1% per bit. It is seen that the simulation results closely match the Markov results in Table 5.1. The simulation results are approximate, will vary from one run to the next and will equal the Markov results only as the number of generations approaches infinity.

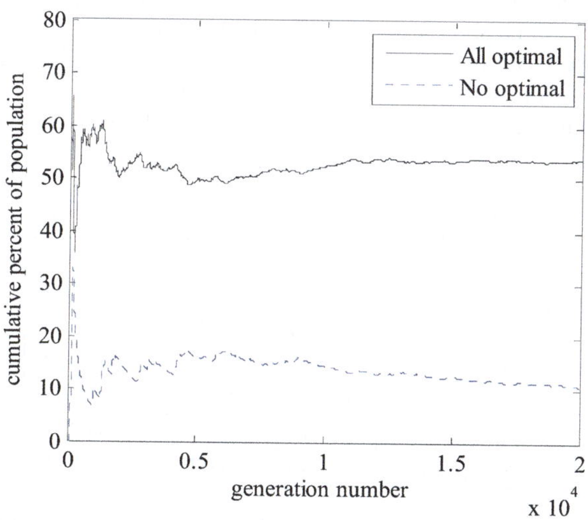

Figure 5.1. *Typical BBO simulation results for a 3-bit one-max optimization problem with a mutation rate of 1% per bit. The plot shows the probability of obtaining an all-optimal population and the probability of obtaining a no-optimal population*

EXAMPLE 5.9.–

In this example, we compare Markov model results for the basic BBO algorithm with a GA with global uniform recombination (GA/GUR), as shown in Figure 3.6 in Chapter 3. In Figure 3.6, BBO reduces to GA/GUR when BBO has a constant immigration rate of $\lambda = 1$. We use the following fitness values:

$$f(000) = 5,\ f(001) = 2,\ f(010) = 2,\ f(011) = 3,$$
$$f(100) = 2,\ f(101) = 3,\ f(110) = 3,\ f(111) = 4. \qquad [5.51]$$

These fitness values are the same as those in equation [5.49], except that we made the 000 bit string the most fit solution. This is called a deceptive problem because usually when we add a 1 bit to one of the above solutions, its fitness increases. The exception is that 111 is not the most fit solution, but rather 000 is the most fit solution.

Table 5.2 shows comparisons between Markov model results for GA/GUR and BBO. The table shows the probability of obtaining a population in which all solutions are optimal, and the probability of obtaining a population in which no solutions are optimal. It is seen from this table that the Markov model and simulation results match well for GA/GUR and BBO, which confirms the model. Furthermore, in every Markov performance comparison in the table, BBO performs significantly better than GA/GUR. This is especially true when the mutation rate is low (0.1% per bit), in which case BBO performs better than GA/GUR in its higher probability of obtaining a population with all optimal solutions (90% vs. 63%), and in its lower probability of obtaining a population with no optimal solutions (7% vs. 34%).

Mutation rate	Population count vector	Probability			
		BBO		GA/GUR	
		Markov	Simulation	Markov	Simulation
0.1	All optimal	**0.0120**	0.0118	0.0109	0.0107
	No optimal	**0.7954**	0.7913	0.8325	0.8314
0.01	All optimal	**0.6506**	0.6487	0.4760	0.4715
	No optimal	**0.1915**	0.1891	0.4103	0.4004
0.001	All optimal	**0.9074**	0.9022	0.6383	0.6279
	No optimal	**0.0730**	0.0721	0.3482	0.3472

Table 5.2. *BBO and GA/GUR Markov model and simulation results for the 3-bit deceptive problem. The table shows the probabilities of obtaining an all-optimal population and the probabilities of obtaining a no-optimal population. The best Markov performance is shown in bold in each row*

5.3. BBO convergence

In the previous section, we obtained the migration probability and mutation probability, which can be calculated by equations [5.45] and [5.46], respectively: $M_{kj} = \Pr(y_{k,t+1} = x_j) \geq 0$ and $U_{ji} = \Pr(x_j \rightarrow x_i) > 0$. Therefore, M is a non-negative stochastic matrix; although it is not a transition matrix since it is not square, each row still sums to 1. We also see that U is a positive left stochastic matrix; that is, each of its columns sum to 1. We now present two theorems that show that there is a positive probability of obtaining any solution in the search space from any solution in a BBO population after migration and mutation. This means that there is a positive probability of transitioning from any population vector u to any other population vector v in one generation, which means that the BBO transition matrix is primitive.

THEOREM 5.4.– *If M is a positive stochastic matrix and U is a positive left stochastic matrix, then the product MU is positive.*

PROOF.– If M is positive and stochastic, then every entry of M is positive; that is, $M_{kj} > 0$ for $k \in [1, N]$ and $j \in [1, n]$, and $\sum_{j=1}^{n} M_{kj} = 1$ for all k. Similarly, if U is positive, then every entry of U is positive; that is, $U_{ji} > 0$ for $i, j \in [1, n]$. Therefore, by matrix multiplication, $(MU)_{ki} = \sum_{j=1}^{n} M_{kj} U_{ji} > 0$ for $k \in [1, N]$ and $i \in [1, n]$.

THEOREM 5.5.– *The transition matrix of BBO with migration and mutation is primitive.*

PROOF.– From equation [5.47], we know that if $P_{ki}(v) = [MU]_{ki} > 0$ for all $k \in [1, N]$ and $i \in [1, n]$, then $p_{ij} = \Pr(u|v) = \sum_{J \in Y} \prod_{k=1}^{N} \prod_{i=1}^{n} [P_{ki}(v)]^{J_{ki}} > 0$ for $i, j \in [1, T]$, where Y is given in equation [5.47]. So the transition matrix $P = (p_{ij})$ of BBO is positive. Therefore, since every positive transition matrix is primitive, P is primitive.

COROLLARY 5.1.– *There exists a unique limiting distribution for the states of the BBO Markov chain. Also, the probability that the Markov chain is in the ith state at any time is positive for all $i \in [1, T]$.*

PROOF.– Corollary 5.1 is an immediate consequence of Theorems 5.2 and 5.5.

Before we obtain the convergence properties of BBO, some precise definitions of the term *convergence* are required [GUO 01, MA 14a, MA 14b]. Assume that the search space of a global optimization problem is I with cardinality $|I| = n$. Further assume that the BBO algorithm with population size N consists of both migration and mutation, as shown in Figure 3.5 in Chapter 3.

DEFINITION 5.1.– *Let* $A(t)=\{a_i(t) | i\in[1, N], a_i(t)\in I\}$ *be the population at generation t, where N is the population size, $a_i(t)$ denotes a candidate solution in the search space I and A(t) may contain duplicate elements;* $f: I \to R$ *denotes a fitness function assigning real values to solutions;* $I^* = \{a^* \mid a^* = \arg\max\{f(a) | a\in I\}\}$ *is a subset in the search space, each member of which has the globally maximum fitness; and the best solutions in the population at generation t are* $I^*(t)=\{a_j^*(t)\}\subset A(t)$, *where* $f(a_j^*(t))\ge f(a_i(t))$ *for all* $a_j^*(t)\in I^*(t)$ *and for all* $i\in[1, N]$.

We use the notation $a^*(t)$ to denote an arbitrary element of $I^*(t)$ (that is, one of the best solutions in the population at generation t). Because of migration and mutation, $a^*(t)$ and its fitness will change randomly over time. As $t\to\infty$, the convergence, or lack of convergence, of $a^*(t)$ to the subset I^* indicates whether or not the BBO algorithm is globally convergent. That is, BBO is said to converge if:

$$\Pr\left(\lim_{t\to\infty} a^*(t)\in I^*\right)=1 \Leftrightarrow \Pr\left(a^*\in \lim_{t\to\infty} A(t)\right)=1 \quad [5.52]$$

Note that $a^*(t)$ is not necessarily unique. However, Definition 5.1 states that the BBO algorithm is globally convergent if and only if [5.52] holds for every $a^*(t)$. Clearly, the evolution of $a^*(t)$ is a homogeneous finite Markov chain, which we call an $a^*(t)$-chain.

Now we sort all the states of I in order of descending fitness; that is, $I=\{I_1,\cdots,I_n\}$ and $f(I_1)\ge f(I_2)\ge\cdots\ge f(I_n)$. We define S as the set of indices of I; that is, $S=\{1, 2,\cdots,n\}$. Furthermore, we define S^* as the elements $\{j\}$ of S such that $I_j\in I^*$; that is, $I_j\in I^*$ for all $j\in S^*$. This leads to the following definition.

DEFINITION 5.2.– Let $\hat{P} = (\hat{p}_{ij})$ be the transition matrix of an $a^*(t)$-chain, where $\hat{p}_{ij}$ for $i, j \in [1, n]$ is the probability that $a^*(t) = I_i$ transitions to $a^*(t+1) = I_j$. The BBO algorithm converges to a global optimum if and only if $a^*(t)$ transitions from any state $i \in I$ to I^* as $t \to \infty$ with probability one, that is, if:

$$\lim_{t \to \infty} \sum_{j \in S^*} \left(\hat{P}^t\right)_{ij} = 1 \qquad \text{for all} \quad i \in S \tag{5.53}$$

As noted earlier, there may be more than one $a^*(t)$-chain since more than one element of the search space may have a globally maximum fitness. Definition 5.2 states that the BBO algorithm converges to a global optimum if and only if equation [5.53] holds for every $a^*(t)$-chain. Also note that $\hat{P}$ depends on the other solutions in the population at generation *t*. Definition 5.2 states that the BBO algorithm converges to a global optimum if and only if [5.53] holds for every possible $\hat{P}$ transition matrix for every $a^*(t)$-chain.

THEOREM 5.6.– *If the transition matrix $\hat{P} = (\hat{p}_{ij})$ of an $a^*(t)$-chain is a positive stochastic matrix, then BBO with migration and mutation does not converge to any of the global optima.*

PROOF.– Since every positive matrix is also a primitive one, it follows by Theorem 5.2 that the limiting distribution of $\hat{P}$ is unique with all positive entries. Therefore, for any $i \in S$,

$$\lim_{t \to \infty} \sum_{j \in S^*} \left(\hat{P}^t\right)_{ij} = 1 - \lim_{t \to \infty} \sum_{j \in S - S^*} \left(\hat{P}^t\right)_{ij} = 1 - \sum_{j \in S - S^*} \lim_{t \to \infty} \left(\hat{P}^t\right)_{ij} = 1 - \sum_{j=|S^*|+1}^{|S|} \pi_j < 1 \tag{5.54}$$

where we use the notation $S - S^*$ to denote all elements of *S* that do not belong to S^*. We see that equation [5.53] is not satisfied, which completes the proof.

THEOREM 5.7.– *If the transition matrix $\hat{P} = (\hat{p}_{ij})$ of an $a^*(t)$-chain is a stochastic matrix with the structure:*

$$\hat{P} = \begin{bmatrix} C & 0 \\ R & Q \end{bmatrix} \tag{5.55}$$

where C is a positive stochastic matrix of order $\left|S^*\right|$*, and* $R, Q \neq 0$*, then the BBO algorithm converges to one or more of the global optima.*

PROOF.– From Theorem 5.3, we see that for all $i, j \in S$,

$$\hat{P}^{\infty} = \lim_{t \to \infty} \left(\hat{p}_{ij} \right)^t = \begin{pmatrix} \pi_i \\ \vdots \\ \pi_i \end{pmatrix}_{|S| \times |S|} \quad [5.56]$$

where $\pi_i = \left(\pi_{i1}, \cdots, \pi_{i|S^*|}, 0, \cdots, 0 \right)$, $\pi_{ij} > 0$ for $1 \leq j \leq \left|S^*\right|$, and $\sum_{j=1}^{|S^*|} \pi_{ij} = 1$. It follows directly that for any $i \in S$,

$$\lim_{t \to \infty} \sum_{j \in S^*} \left(\hat{P}^t \right)_{ij} = \sum_{j \in S^*} \lim_{t \to \infty} \left(\hat{P}^t \right)_{ij} = \sum_{j=1}^{|S^*|} \pi_{ij} = 1 \quad [5.57]$$

We see that equation [5.53] is satisfied, which completes the proof.

Theorems 5.6 and 5.7 can be applied directly to determine the global convergence of BBO if the structure of the transition matrix of the Markov chain can be determined, as we will show in the remainder of this section. In particular, we will formalize the observation that the transition matrix of BBO without elitism satisfies the conditions of Theorem 5.6 (as stated in Theorem 5.5). We will further show that the transition matrix of the $a^*(t)$-chain of BBO with elitism satisfies the conditions of Theorem 5.7.

A. Elitism

We now discuss a modified BBO which uses *elitism*, an idea which is also implemented in many other EAs. There are many ways to implement elitism, but here we define elitism as the preservation of the best solution at each generation in a separate partition of the population space. This enlarges the population size by one solution; the elite solution increases the population size from N to $N + 1$. However, note that the population size is still constant (that is, equal to $N + 1$) from one generation to the next. The elite solution does not take part in recombination or mutation, but is maintained separately from the other N members of the population.

At each generation, if a solution in the N-member main population is better than the elite solution, then the elite solution is replaced with a copy of the better solution.

Relative to a standard N-member BBO population, elite BBO increases the number of possible population distributions by a factor of n, which is the search space size. That is, each possible population distribution of the N-member main population could also include one of n elite solutions. The number of possible population distributions increases by a factor of n, from T to nT. We order these new states so that each group of n states has the same elite solution. Also, the elite solution in the mth group of n states is the mth best solution in the search space for $m = 1, \ldots, n$.

The elitist-preserving process can be represented by an upgrade transition matrix O, which contains the probabilities that each population distribution of the $(N+1)$-member population transitions to some other population distribution after the elitist-preserving step. That is, the element in the ith row and jth column of O, denoted as $O(i, j)$, is the probability that the ith population distribution transitions to the jth population distribution after the step in which the elite solution is replaced with the best solution from the N-member main population. The upgrade matrix is similar to the one in [RUD 94]. It does not include the effects of migration or mutation, but only includes the elitism-preserving step. The upgrade matrix only includes the probability of changing the elite solution; it does not include the probability of changing the N-member main population, since it does not include migration or mutation. If there are no solutions in the N-member main population that are better than the elite solution, then the elite solution does not change. The structure of the upgrade matrix O can be written as:

$$O = \begin{pmatrix} O_{11} & 0 & \cdots & 0 \\ O_{21} & O_{22} & \ddots & \vdots \\ \vdots & \ddots & \ddots & 0 \\ O_{n1} & \cdots & O_{n,n-1} & O_{nn} \end{pmatrix} \quad [5.58]$$

where each O_{ij} matrix is $T \times T$, where T is the number of population distributions in BBO with a population size of N and search space cardinality of n. O_{11} is the identity matrix since the first n population distributions have the global optimum as their elite solution, and the elite solution can never be improved from the global optimum. Matrices O_{aa} with $a \geq 2$ are diagonal matrices composed of all zeros and ones. Since the population distributions are ordered by grouping common elite solutions, and since elite solutions in the population distribution ordering are in order of decreasing fitness, the super block diagonals in O are zero matrices as shown in equation [5.58]; that is, there is zero probability that the ith population distribution transitions to the jth population distribution if $i < j$. So the Markov chain of elite BBO can be described as:

$$
\begin{aligned}
P^{+} &= \operatorname{diag}(P)O \\
&= \begin{pmatrix} P & 0 & \cdots & 0 \\ 0 & P & \ddots & \vdots \\ \vdots & \ddots & \ddots & 0 \\ 0 & \cdots & 0 & P \end{pmatrix} \begin{pmatrix} O_{11} & 0 & \cdots & 0 \\ O_{21} & O_{22} & \ddots & \vdots \\ \vdots & \ddots & \ddots & 0 \\ O_{n1} & \cdots & O_{n,n-1} & O_{nn} \end{pmatrix} \\
&= \begin{pmatrix} PO_{11} & 0 & \cdots & 0 \\ PO_{21} & PO_{22} & \ddots & \vdots \\ \vdots & \ddots & \ddots & 0 \\ PO_{n1} & \cdots & PO_{n,n-1} & PO_{nn} \end{pmatrix}
\end{aligned} \qquad [5.59]
$$

where P is the $T \times T$ transition matrix described in equation [5.48] in section 5.2.

EXAMPLE 5.10.–

To explain the update matrix O described in equation [5.58], a simple example is presented. Suppose there exists a search space consisting of $n = 3$ candidate solutions which are given as $x = \{x_1, x_2, x_3\}$ where the fitness of x_1 is lowest and the fitness of x_3 is highest. Suppose the main population size is $N = 1$, so the elitist population size is $N + 1 = 2$. Thus, there are nine possible populations before the elitist-preserving step, which are $\{C_1, C_2, C_3, C_4, C_5, C_6, C_7, C_8, C_9\} = \{\{x_3, x_3\}, \{x_3, x_2\}, \{x_3, x_1\}, \{x_2, x_3\}, \{x_2, x_2\}, \{x_2, x_1\}, \{x_1, x_3\}, \{x_1, x_2\}, \{x_1, x_1\}\}$. Note that the first element in each population is the elite solution, and the last N elements ($N = 1$ in this example) is the main population. Also note that the populations are ordered in such a way that the first three have the most fit solution as their elite solution, the next three have the second most fit solution as their elite solution, and the last three have the least fit solution as their elite solution. The update matrix O is a 9×9 matrix.

The population $C_1 = \{x_3, x_3\}$ transitions to the population $C_1 = \{x_3, x_3\}$ with probability 1; that is, $O(1, 1) = 1$. Population C_1 cannot transition to any other population $C_i (i \neq 1)$; that is, $O(1, i) = 0$ for $i \neq 1$. Similarly, population $C_2 = \{x_3, x_2\}$ transitions to C_2 with probability 1 since the elite x_3 is better than the main-member population x_2; therefore, $O(2, 2) = 1$, and $O(2, i) = 0$ for $i \neq 2$. Continuing with this reasoning, we obtain the O matrix as follows:

$$O = \begin{bmatrix} O_{11} & O_{12} & O_{13} \\ O_{21} & O_{22} & O_{23} \\ O_{31} & O_{32} & O_{33} \end{bmatrix} = \begin{bmatrix} 1 & & & & & & & & \\ 0 & 1 & & & & & & & \\ 0 & 0 & 1 & & & & & & \\ 1 & 0 & 0 & 0 & & & & & \\ 0 & 0 & 0 & 0 & 1 & & & & \\ 0 & 0 & 0 & 0 & 0 & 1 & & & \\ 1 & 0 & 0 & 0 & 0 & 0 & 0 & & \\ 0 & 0 & 0 & 0 & 1 & 0 & 0 & 0 & \\ 0 & 0 & 0 & 0 & 0 & 0 & 0 & 0 & 1 \end{bmatrix} \qquad [5.60]$$

where each O_{ij} matrix is 3×3, and the blank elements above the diagonal are each 0.

Now we consider the convergence of the $a^*(t)$-chain, which is the sequence of elite solutions in the elite BBO algorithm. If the elite solution is equal to the global optimum, we call this an absorbing state of the $a^*(t)$-chain. Recall that the elite solution in elite BBO can only be replaced by one with better fitness. Therefore, the $a^*(t)$-chain of elite BBO contains three classes of states: 1) at least one absorbing state, 2) non-absorbing states which transition to absorbing states in one step and 3) non-absorbing states which transition to non-absorbing states in one step. So the transition matrix $\hat{P}$ of the $a^*(t)$-chain, which we introduced in equations [5.53]–[5.57], can be written as:

$$\hat{P} = \begin{pmatrix} I_k & 0 \\ R & Q \end{pmatrix} \qquad [5.61]$$

where I_k is a $k \times k$ unit matrix corresponding to optimal solutions (k is the number of optima), R is a matrix of order $(|S| - k) \times k$ corresponding to non-absorbing states that transition to absorbing states ($|S|$ is the cardinality of the state space S, so $|S| - k$ is the number of non-absorbing states), and Q is a matrix of order $(|S| - k) \times (|S| - k)$ corresponding to non-absorbing states that transition to non-absorbing states. The matrix $\hat{P}$ of equation [5.61] has the same structure as $\hat{P}$ in Theorem 5.7. It follows from Theorem 5.7 that the $a^*(t)$-chain of elite BBO is globally convergent.

These results are similar to the canonical GA [RUD 94], which is proven to never converge to the global optimum, but elitist variants of which are proven to converge to the global optimum. We sum up these results in the following corollary.

COROLLARY 5.2.– *BBO with migration and mutation does not converge to any of the global optima, but elite BBO, which preserves the best solution at each generation, converges to the global optimum.*

PROOF.– This is an immediate consequence of Theorems 5.5 and 5.6 (the non-convergence of BBO without elitism), Theorem 5.7 (the convergence of BBO with elitism) and the discussion above.

B. Convergence rate

The previous subsection analyzed the convergence properties of elite BBO, and this subsection discusses its convergence rate. The transition matrix of elite BBO after t steps can be found from equation [5.61] as:

$$\hat{P}^t = \begin{pmatrix} I_k & 0 \\ N_t R & Q^t \end{pmatrix} \quad [5.62]$$

where $N_t = I + Q + Q^2 + \cdots + Q^{t-1} = (I - Q)^{-1}(I - Q^t)$. If $\|Q\| < 1$, then the limiting distribution of the Markov chain of BBO can be found from $\hat{P}^\infty$, which can be written as:

$$\hat{P}^\infty = \lim_{t \to \infty} \hat{P}^t = \begin{bmatrix} I_k & 0 \\ (I - Q)^{-1} R & 0 \end{bmatrix} \quad [5.63]$$

Modified BBO with elitism has been proven to converge to a global optimum in the previous subsection, and there exists a limiting distribution $\pi^* = \pi(0)\hat{P}^\infty$, where $\pi(0) = [\pi_1(0), \ldots, \pi_k(0), \pi_{k+1}(0), \ldots, \pi_{|S|}(0)] = [\bar{\pi}_1(0), \bar{\pi}_2(0)]$. (Recall that k is the number of global optima.) The convergence rate estimate of elite BBO can be obtained as follows.

THEOREM 5.8.– If $\|Q\| = \rho < 1$, *the convergence rate of elite BBO satisfies* $\|\pi(t) - \pi^*\| \le O(\rho^t)$.

PROOF.–

$$
\begin{aligned}
\left\|\pi(t)-\pi^*\right\| &= \left\|\pi(0)\hat{P}^t - \pi(0)\hat{P}^\infty\right\| \\
&= \left\|\pi(0)\left[\begin{bmatrix} I_k & 0 \\ N_t R & Q^t \end{bmatrix} - \begin{bmatrix} I_k & 0 \\ (I-Q)^{-1} R & 0 \end{bmatrix}\right]\right\| \\
&= \left\|\begin{bmatrix} \bar{\pi}_1(0) & \bar{\pi}_2(0) \end{bmatrix}\begin{bmatrix} 0 & 0 \\ \left(N_t-(I-Q)^{-1}\right)R & Q^t \end{bmatrix}\right\| \\
&= \left\|\bar{\pi}_2(0)\left[N_t\left(N_t-(I-Q)^{-1}\right)R \quad Q^t\right]\right\| \\
&\leq \left\|\bar{\pi}_2(0)\right\|\left(\left\|N_t-(I-Q)^{-1}\right\|\cdot\|R\|+\left\|Q^t\right\|\right) \\
&= \left\|\bar{\pi}_2(0)\right\|\left(\left\|(I-Q)^{-1}\left(I-Q^t\right)-(I-Q)^{-1}\right\|\cdot\|R\|+\left\|Q^t\right\|\right) \\
&= \left\|\bar{\pi}_2(0)\right\|\left(\left\|(I-Q)^{-1}-(I-Q)^{-1}Q^t-(I-Q)^{-1}\right\|\cdot\|R\|+\left\|Q^t\right\|\right) \\
&\leq \left\|\bar{\pi}_2(0)\right\|\left(\left\|-(I-Q)^{-1}\right\|\cdot\|R\|+1\right)\left\|Q^t\right\| \leq \left\|\bar{\pi}_2(0)\right\|\left(\frac{\|R\|}{1-\|Q\|}+1\right)\|Q\|^t \\
&= O\left(\rho^t\right)
\end{aligned}
\qquad [5.64]
$$

Note that elite BBO is guaranteed to converge to a global optimum regardless of the initial state. In addition, note that we can improve the convergence rate bound by decreasing the parameter ρ. That is, reducing the number of non-absorbing states which transition to other non-absorbing states can accelerate the convergence of elite BBO. In spite of differences between GAs and BBO [SIM 11d], we see from Theorem 5.8 that the convergence rate of BBO with elitism is very similar to that of GAs [HE 99, Theorem 5].

EXAMPLE 5.11.–

Theorem 5.8 gives the upper bound of the convergence rate estimate of elite BBO. In this example, we use simulations to confirm this theorem. Note that in equation [5.64] the parameter ρ is a norm: $\rho=\|Q\|$. Here we define $\|.\|$ as the infinity norm $\|.\|_\infty$; that is,

$$
\|Q\|_\infty = \max_i\left(\sum_{j=1}^{n} q_{ij}\right) \qquad [5.65]
$$

where q_{ij} is the element in the ith row and the jth column of matrix Q. Now note that the transition matrix $\hat{P}$ in equation [5.61] can be obtained from equations [5.48] and [5.59] using elementary matrix transformations. We can thus use Theorem 5.7 to check for BBO convergence, and we can use Theorem 5.8 to estimate the convergence rate of BBO. That is, we define $\pi(t)-\pi^*$ as the error between a BBO population distribution and a distribution that includes at least one optimal solution. We then define the convergence criterion as an arbitrarily small error (for example, $\|\pi(t)-\pi^*\|=10^{-6}$). We can then estimate the time t to convergence from equation [5.64] as follows:

$$t \approx \log_{\rho} 10^{-6} \qquad [5.66]$$

Test functions in this section are limited to 3-bit problems with a search space cardinality of eight and a population size of four. The fitness functions include the one-max problem described by equation [5.49] and the deceptive problem described by equation [5.51]. In addition, we add a multimodal problem, which is given as:

$$f = \begin{pmatrix} 4 & 2 & 2 & 3 & 2 & 3 & 3 & 4 \end{pmatrix} \qquad [5.67]$$

In equation [5.67], fitness values are listed in binary order, so the first element of the fitness function corresponds to the bit string 000, the second element corresponds to the bit string 001, and so on. For the BBO parameters, we use a maximum immigration rate and maximum emigration rate of 1, and we use linear migration curves. We test elite BBO with three different mutation rates which are applied to each bit in each solution at each generation: 0.1, 0.01 and 0.001. Note that we do not test with a zero mutation rate because the theory in this section requires that the mutation rate be positive (see Theorem 5.4). Convergence is not guaranteed unless the mutation rate is positive.

Numerical calculations show that the transition matrices for these three problems satisfy the convergence conditions of Theorem 5.7, which indicates that the BBO algorithm converges to one or more of the global optima. As a heuristic test of Theorem 5.8, we use simulations to record the first generation number of obtaining a population in which all solutions are optimal, and all results are computed from 25 Monte Carlo simulations. Tables 5.3–5.5 show comparisons of the theoretical convergence time t, the corresponding parameter ρ and the first generation number of finding an all-optimal population, averaged over 25 Monte Carlo simulations. For the one-max problem, the all-optimal population is (0 0 0 0 0 0 0 4), for the deceptive problem, the all-optimal population is (4 0 0 0 0 0 0 0), and for the multimodal problem, the all-optimal population is (* 0 0 0 0 0 0 *), where the sum of the * terms in the vector is 4.

Tables 5.3–5.5 show time to convergence, and time to finding an optimum for BBO. The tables confirm the statement following Theorem 5.8 that the convergence behavior of BBO is similar to that of GA. The tables show that GA converges slightly faster than BBO for high mutation rates, but BBO converges slightly faster for low mutation rates. The latter behavior is more important in practice because low mutation rates provide faster convergence.

Mutation rate	Theoretical analysis				Average generation number using simulation	
	Parameter ρ		Convergence time t			
	BBO	GA	BBO	GA	BBO	GA
0.1	0.87	0.85	99.20	90.45	87.58	82.39
0.01	0.68	0.73	35.82	44.16	42.12	50.18
0.001	0.30	0.44	11.47	16.81	11.63	17.37

Table 5.3. *Convergence rate comparison for the 3-bit one-max problem. The table shows the convergence time t in seconds, the corresponding ρ and the first generation number of finding an all-optimal population using BBO and GA, averaged over 25 Monte Carlo simulations*

Mutation rate	Theoretical analysis				Average generation number using simulation	
	Parameter ρ		Convergence time t			
	BBO	GA	BBO	GA	BBO	GA
0.1	0.96	0.95	338.43	270.13	289.55	256.37
0.01	0.48	0.52	18.82	21.49	20.65	22.44
0.001	0.25	0.29	9.97	11.14	9.92	11.08

Table 5.4. *Convergence rate comparison for the 3-bit deceptive problem. The table shows the convergence time t in seconds, the corresponding ρ and the first generation number of finding an all-optimal population using BBO and GA, averaged over 25 Monte Carlo simulations*

Mutation rate	Theoretical analysis				Average generation number using simulation	
	Parameter ρ		Convergence time t			
	BBO	GA	BBO	GA	BBO	GA
0.1	0.61	0.59	27.95	25.16	26.09	24.87
0.01	0.11	0.17	6.26	7.95	6.17	7.72
0.001	0.04	0.05	4.29	4.76	4.18	4.63

Table 5.5. *Convergence rate comparison for the 3-bit multimodal problem. The table shows the convergence time t in seconds, the corresponding ρ and the first generation number of finding an all-optimal population using BBO and GA, averaged over 25 Monte Carlo simulations*

Several things are notable about the results in Tables 5.3–5.5. First, the mutation rate affects the convergence rate of BBO. For all test problems, the convergence rate improves when the mutation rate decreases. We can accelerate the convergence of BBO by decreasing the mutation rate. This may provide practical guidance for BBO tuning for real-world problems. Second, by analyzing the relationship of the parameter ρ and the convergence time t in Tables 5.3–5.5, we see that the convergence time t is exponentially related to the parameter ρ as predicted by Theorem 5.8. Third, the theoretical results and simulation results match well for most of the test problems, which confirms the convergence rate estimate provided by Theorem 5.8.

5.4. Markov models of BBO extensions

Section 5.2 set up the Markov model of the basic BBO algorithm, which is also called partial immigration-based BBO, and which considers the immigration of each solution feature as separate probabilistic trials. This section derives Markov models for the three variations of BBO as described in Figures 4.2, 4.3 and 4.4 in Chapter 4, including total immigration-based BBO, partial emigration-based BBO and total emigration-based BBO [MA 13a, MA 13b].

A. Total immigration-based BBO

Total immigration-based BBO (Figure 4.2) bases migration on the immigration rate for each solution, and probabilistically decides whether or not to immigrate all solution features (that is, all independent variables) to a given solution. This is different from the basic BBO algorithm, which considers immigration of one solution feature at a time. For total immigration-based BBO, given that the population distribution at generation t is equal to v, the probability $P_{ki}(v)$ that immigration results in $y_{k,t+1} = x_i$ at generation t+1 can be obtained as follows:

$$\begin{aligned} M_{ki}(v) &= \Pr\left(y_{k,t+1} = x_i\right) \\ &= \Pr\left(\text{no immigration to } y_{k,t}\right)\Pr\left(y_{k,t+1} = x_i \,|\text{no immigration}\right) + \\ &\quad \Pr\left(\text{immigration to } y_{k,t}\right)\Pr\left(y_{k,t+1} = x_i \,|\text{immigration}\right) \\ &= \left(1-\lambda_{m(k)}\right)1_0\left(x_{m(k)} - x_i\right) + \lambda_{m(k)}\prod_{s=1}^{q}\frac{\sum_{j\in\varsigma_i(s)} v_j\mu_j}{\sum_{j=1}^{n} v_j\mu_j} \end{aligned} \quad [5.68]$$

Note that the first term of the right side of equation [5.68] denotes the probability when immigration does not occur; that is, when y_k is not selected for immigration.

The second term on the right side of equation [5.68] denotes the probability when immigration occurs, and it is proportional to the product of the summed emigration rates of all solutions whose bits are equal to those of x_i.

Now suppose that the mutation probability is the same as that in the basic BBO algorithm. That is, $P_{ki}(v)$ is defined as in equation [5.47], except that we use $M_{ki}(v)$ from equation [5.68] instead of $M_{ki}(v)$ from equation [5.47]. Then, the transition matrix for total immigration-based BBO is calculated as shown in equation [5.48].

B. Partial emigration-based BBO

Partial emigration-based BBO (Figure 4.3) bases migration on emigration rates for each solution, and probabilistically decides whether or not to emigrate each solution feature. If emigration is selected, the immigrating solution is probabilistically selected based on immigration rates. For all BBO variations, the probability that y_k is equal to some specific value includes the probability that immigration does not occur and the probability that immigration occurs. Suppose r denotes the current emigration trial number (there are N emigration trials of the outer loop in Figure 4.3). If the sth feature of y_k is not selected for immigration during generation t, then:

$$y_k(s)_{t+1} = x_{m(k)}(s) \quad \left(\text{if no immigration to } y_{k,t}(s) \text{ on the } r\text{th emigration trial}\right) \qquad [5.69]$$

To calculate the probability that immigration does not occur, first consider the emigration probability of $x_{m(r)}$ on the rth emigration trial, which can be written as:

$$\Pr(\text{emigration on the } r\text{th emigration trial}) = \mu_{m(r)} \qquad [5.70]$$

where the meaning of $m(r)$ is similar to $m(k)$ in equation [5.18]. The immigration probability of y_k is proportional to its immigration rate. So the probability that $x_{m(r)}(s)$ immigrates to $y_k(s)$ on the rth emigration trial can be written as:

$$\begin{aligned}
&\Pr(\text{immigration to } y_k \text{ on the } r\text{th emigration trial}) \\
&= \Pr(\text{emigration on the } r\text{th emigration trial}) \\
&\quad \times \Pr(\text{immigration to } y_k \,|\, \text{emigration on the } r\text{th emigration trial}) \\
&= \mu_{m(r)} \frac{\lambda_{m(k)}}{\sum_{j=1}^{n} v_j \lambda_j}
\end{aligned} \qquad [5.71]$$

The probability that immigration does not occur for $y_k(s)$ on the rth emigration trial can be written as:

$$\begin{aligned}&\Pr(\text{no immigration to } y_k \text{ on the } r\text{th emigration trial})\\&\quad = 1 - \Pr(\text{immigration to } y_k \text{ on the } r\text{th emigration trial})\\&\quad = 1 - \mu_{m(r)} \frac{\lambda_{m(k)}}{\sum_{j=1}^{n} v_j \lambda_j}\end{aligned} \qquad [5.72]$$

However, if the sth feature of y_k is selected for immigration on the rth emigration trial during generation t, then in order to have $y_k(s) = x_i(s)$, the sth feature of the emigrating solution $x_{m(r)}$ must be equal to the sth feature of x_i; that is,

$$\begin{aligned}&y_k(s)_{t+1} = x_{m(r)}(s) = x_i(s)\\&\quad (\text{if immigration to } y_{k,t}(s) \text{ on the } r\text{th emigration trial})\end{aligned} \qquad [5.73]$$

Equations [5.69], [5.70], [5.71] and [5.72] are combined to obtain the total migration probability of one bit of y_k after R emigration trials. Here use R to denote the total number of emigration trials; this differs from r, which indicates a specific emigration trial. So the total migration probability can be written as:

$$\begin{aligned}&\Pr(y_{k,t+1}(s) = x_i(s) \text{ after } R \text{ emigration trials})\\&= \Pr(\text{no immigration to } y_{k,t} \text{ on } R\text{th emigration trial})\\&\quad \times \Pr(y_{k,t+1}(s) = x_i(s) \text{ after } R\text{-1 emigration trials})\\&+ \Pr(\text{immigration to } y_{k,t} \text{ on } R\text{th emigration trial})\\&\quad \times \Pr(y_{k,t+1}(s) = x_i(s) \mid \text{immigration on } R\text{th emigration trial})\\&= \left(1 - \mu_{m(R)} \frac{\lambda_{m(k)}}{\sum_{j=1}^{n} v_j \lambda_j}\right) \Pr\left(x_{m(k)}(s) = x_i(s) \text{ after } (R-1) \text{ emigration trials}\right)\\&+ \left(\mu_{m(R)} \frac{\lambda_{m(k)}}{\sum_{j=1}^{n} v_j \lambda_j}\right) 1_0\left(x_{m(R)}(s) - x_i(s)\right)\end{aligned} \qquad [5.74]$$

Note that the above equation applies to one bit in one solution after R emigration trials. For example, consider the simple case of one emigration trial (R = 1). In this case, the probability that $y_k(s) = x_i(s)$ at generation t+1 can be written as:

$$\begin{aligned}
&\Pr(y_{k,t+1}(s) = x_i(s) \text{ after 1 emigration trial}) \\
&= \Pr(\text{no immigration to } y_{k,t} \text{ on 1st emigration trial}) \\
&\quad \times \Pr(y_{k,t+1}(s) = x_i(s) \mid \text{no immigration}) \\
&+ \Pr(\text{immigration to } y_{k,t} \text{ on 1st emigration trial}) \\
&\quad \times \Pr(y_{k,t+1}(s) = x_i(s) \mid \text{immigration on 1st emigration trial}) \\
&= \left(1 - \mu_{m(1)} \frac{\lambda_{m(k)}}{\sum_{j=1}^{n} v_j \lambda_j}\right) 1_0\left(x_{m(k)}(s) - x_i(s)\right) + \left(\mu_{m(1)} \frac{\lambda_{m(k)}}{\sum_{j=1}^{n} v_j \lambda_j}\right) 1_0\left(x_{m(1)}(s) - x_i(s)\right)
\end{aligned} \quad [5.75]$$

To express this in a more compact form, we introduce the notation $\eta_{m(k)}$ for $\lambda_{m(k)} \Big/ \sum_{j=1}^{n} v_j \lambda_j$ in the following equations. So after two emigration trials, we obtain:

$$\begin{aligned}
&\Pr(y_{k,t+1}(s) = x_i(s) \text{ after 2 emigration trials}) \\
&= \Pr(\text{no immigration to } y_{k,t} \text{ on 2nd emigration trial}) \\
&\quad \times \Pr(y_{k,t+1}(s) = x_i(s) \text{ after 1 emigration trial}) \\
&+ \Pr(\text{immigration to } y_{k,t} \text{ on 2nd emigration trial}) \\
&\quad \times \Pr(y_{k,t+1}(s) = x_i(s) \mid \text{immigration on 2nd emigration trial}) \\
&= \left(1 - \mu_{m(2)} \eta_{m(k)}\right) \Pr\left(x_{m(k)}(s) = x_i(s) \text{ after 1 emigration trial}\right) \\
&\quad + \left(\mu_{m(2)} \eta_{m(k)}\right) 1_0\left(x_{m(2)}(s) - x_i(s)\right) \\
&= \left(1 - \mu_{m(2)} \eta_{m(k)}\right) \begin{pmatrix} \left(1 - \mu_{m(1)} \eta_{m(k)}\right) 1_0\left(x_{m(k)}(s) - x_i(s)\right) \\ + \left(\mu_{m(1)} \eta_{m(k)}\right) 1_0\left(x_{m(1)}(s) - x_i(s)\right) \end{pmatrix} \\
&\quad + \left(\mu_{m(2)} \eta_{m(k)}\right) 1_0\left(x_{m(2)}(s) - x_i(s)\right) \\
&= \left(1 - \mu_{m(2)} \eta_{m(k)}\right)\left(1 - \mu_{m(1)} \eta_{m(k)}\right) 1_0\left(x_{m(k)}(s) - x_i(s)\right) \\
&\quad + \left(1 - \mu_{m(2)} \eta_{m(k)}\right)\left(\mu_{m(1)} \eta_{m(k)}\right) 1_0\left(x_{m(1)}(s) - x_i(s)\right) + \left(\mu_{m(2)} \eta_{m(k)}\right) 1_0\left(x_{m(2)}(s) - x_i(s)\right)
\end{aligned} \quad [5.76]$$

Note that the first term of the right side of the above equation denotes the probability when immigration does not occur on either of the two emigration trials, the second term denotes the probability when immigration occurs on the first emigration trial but not on the second emigration trial, and the third term denotes the probability when immigration occurs on the second emigration trial. After N emigration trials (recall that the population size is N), we can use induction to see that the probability can be written as:

$$\begin{aligned}
&\Pr(y_{k,t+1}(s)=x_i(s) \text{ after } N \text{ emigration trials})\\
&=\Pr(\text{no immigration to } y_{k,t} \text{ on } N\text{th emigration trial})\\
&\quad\times\Pr(y_{k,t+1}(s)=x_i(s) \text{ after } N\text{-1 emigration trials})\\
&+\Pr(\text{immigration to } y_{k,t} \text{ on } N\text{th emigration trial})\\
&\quad\times\Pr(y_{k,t+1}(s)=x_i(s) \mid \text{immigration on } N\text{th emigration trial})\\
&=\left(1-\mu_{m(N)}\eta_{m(k)}\right)\Pr\left(x_{m(k)}(s)=x_i(s) \text{ after } (N-1) \text{ emigration trials}\right)\\
&+\left(\mu_{m(N)}\eta_{m(k)}\right)1_0\left(x_{m(N)}(s)-x_i(s)\right)\\
&=\left(1-\mu_{m(N)}\eta_{m(k)}\right)\left(1-\mu_{m(N-1)}\eta_{m(k)}\right)\cdots\\
&\quad\left(1-\mu_{m(3)}\eta_{m(k)}\right)\left(1-\mu_{m(2)}\eta_{m(k)}\right)\left(1-\mu_{m(1)}\eta_{m(k)}\right)1_0\left(x_{m(k)}(s)-x_i(s)\right)\\
&+\left(1-\mu_{m(N)}\eta_{m(k)}\right)\left(1-\mu_{m(N-1)}\eta_{m(k)}\right)\cdots\\
&\quad\left(1-\mu_{m(3)}\eta_{m(k)}\right)\left(1-\mu_{m(2)}\eta_{m(k)}\right)\left(\mu_{m(1)}\eta_{m(k)}\right)1_0\left(x_{m(1)}(s)-x_i(s)\right)\\
&+\left(1-\mu_{m(N)}\eta_{m(k)}\right)\left(1-\mu_{m(N-1)}\eta_{m(k)}\right)\cdots\left(1-\mu_{m(3)}\eta_{m(k)}\right)\left(\mu_{m(2)}\eta_{m(k)}\right)1_0\left(x_{m(2)}(s)-x_i(s)\right)\\
&\quad\cdots\\
&+\left(\mu_{m(N)}\eta_{m(k)}\right)1_0\left(x_{m(N)}(s)-x_i(s)\right)
\end{aligned} \qquad [5.77]$$

Note that the first term of the right side of the above equation denotes the probability when immigration does not occur on any of the N emigration trials, the second term denotes the probability when immigration occurs on the first emigration trial but none of the later emigration trials, the third term denotes the probability when immigration occurs on the second emigration trial but none of the later trials, and so on. A more compact form of equation [5.77] is:

$$\begin{aligned}
&\Pr(y_{k,t+1}(s)=x_i(s) \text{ after } N \text{ emigration trials})\\
&=\prod_{l=1}^{N}\left(1-\mu_{m(l)}\eta_{m(k)}\right)1_0\left(x_{m(k)}(s)-x_i(s)\right)\\
&+\sum_{L=1}^{N-1}\left[\prod_{l=L}^{N-1}\left(1-\mu_{m(l+1)}\eta_{m(k)}\right)\left(\mu_{m(L)}\eta_{m(k)}\right)1_0\left(x_{m(L)}(s)-x_i(s)\right)\right]\\
&+\left(\mu_{m(N)}\eta_{m(k)}\right)1_0\left(x_{m(N)}(s)-x_i(s)\right)
\end{aligned} \qquad [5.78]$$

We again use $P_{ki}(v)$ to denote the probability that immigration results in $y_{k,t+1} = x_i$ given that the population distribution at generation t is equal to v. Recalling that there are q bits in each solution, this can be written as:

$$\begin{aligned} M_{ki}(v) &= \Pr\left(y_{k,t+1} = x_i\right) \\ &= \prod_{s=1}^{q} \left(\Pr(y_{k,t+1}(s) = x_i(s) \text{ after } N \text{ emigration trials})\right) \end{aligned} \quad [5.79]$$

The incorporation of mutation probability is the same as for basic BBO, so $P_{ki}(v)$ is defined the same as equation [5.47], except we use $M_{ki}(v)$ from equation [5.79]. Then, the transition matrix for total immigration-based BBO is calculated as shown in equation [5.48].

C. Total emigration-based BBO

Total emigration-based BBO (Figure 4.4) bases migration on the emigration rate for each solution, and probabilistically decides whether or not to emigrate all solution features from each solution. This differs from partial emigration-based BBO, which considers emigration of one solution feature at a time. But in total emigration-based BBO, the immigration probability of one bit of y_k after R emigration trials is the same as in partial emigration-based BBO. This can be seen by carefully comparing Figures 4.3 and 4.4; the probability of immigrating to $z_k(s)$ is the same for both algorithms. Therefore, the Markov transition matrix is the same for total emigration-based BBO as it is for partial emigration-based BBO.

EXAMPLE 5.12.–

In this example, we use simulation to confirm the Markov models of three BBO variations: total immigration-based BBO, partial emigration-based BBO and total emigration-based BBO. To provide some comparisons, we also use the Markov model of the basic BBO algorithm in this example. For convenience, partial immigration-based BBO (the basic BBO) and total immigration-based BBO are called immigration-based BBO algorithms, and partial emigration-based BBO and total emigration-based BBO are called emigration-based BBO algorithms. The Markov transition matrices derived in sections 5.1 and 5.4 are used to obtain the probability, in the limit as the generation count approaches infinity, that the BBO population consists of a particular set of solutions. Test problems and population size are the same as those in Example 5.9.

Tables 5.6, 5.7 and 5.8 show Markov and simulated results for the four BBO algorithms with mutation rates of 0.1, 0.01 and 0.001 per bit per generation. The tables show the probability of obtaining a population in which all solutions are optimal, and the probability of obtaining a population in which no solutions are optimal.

Mutation rate	Population count vector	Probability							
		Partial immigration BBO (basic BBO)		Total immigration BBO		Partial emigration BBO		Total emigration BBO	
		Markov	Simulation	Markov	Simulation	Markov	Simulation	Markov	Simulation
0.1	All Optimal	0.0245	0.0251	0.0258	0.0265	**0.0630**	0.0616	**0.0630**	0.0631
	No Optimal	0.2998	0.2970	0.3108	0.3232	**0.2361**	0.2358	**0.2361**	0.2312
0.01	All Optimal	0.5343	0.5314	0.5330	0.5244	**0.7551**	0.7596	**0.7551**	0.7510
	No Optimal	0.1134	0.1309	0.1175	0.1160	**0.0328**	0.0342	**0.0328**	0.0362
0.001	All Optimal	0.8605	0.8590	0.8601	0.8587	**0.9542**	0.9547	**0.9542**	0.9527
	No Optimal	0.0923	0.0975	0.0927	0.0900	**0.0221**	0.0230	**0.0221**	0.0267

Table 5.6. *Optimization results for the 3-bit one-max problem. The table shows the probabilities of obtaining an all-optimal population and the probabilities of obtaining a no-optimal population using BBO Markov models and simulations. The best Markov performance is in bold in each row*

Mutation rate	Population count vector	Probability							
		Partial immigration BBO (basic BBO)		Total immigration BBO		Partial emigration BBO		Total emigration BBO	
		Markov	Simulation	Markov	Simulation	Markov	Simulation	Markov	Simulation
0.1	All Optimal	0.0315	0.0318	**0.0328**	0.0328	0.0313	0.0340	0.0313	0.0323
	No Optimal	**0.3751**	0.3765	0.3837	0.3889	0.4116	0.4261	0.4116	0.4201
0.01	All Optimal	**0.6206**	0.6209	0.6185	0.6142	0.5695	0.5510	0.5695	0.5704
	No Optimal	**0.1362**	0.1333	0.1399	0.1383	0.2250	0.2249	0.2250	0.2196
0.001	All Optimal	**0.8771**	0.8770	0.8766	0.8753	0.7817	0.7792	0.7817	0.7842
	No Optimal	**0.0937**	0.0927	0.0942	0.0952	0.1941	0.1917	0.1941	0.1896

Table 5.7. *Optimization results for the 3-bit deceptive problem. The table shows the probabilities of obtaining an all-optimal population and the probabilities of obtaining a no-optimal population using BBO Markov models and simulations. The best Markov performance is in bold in each row*

Mutation rate	Population count vector	Probability							
		Partial immigration BBO (basic BBO)		Total immigration BBO		Partial emigration BBO		Total emigration BBO	
		Markov	Simulation	Markov	Simulation	Markov	Simulation	Markov	Simulation
0.1	All Optimal	0.0485	0.0472	**0.0506**	0.0546	0.0487	0.0473	0.0487	0.0431
	No Optimal	0.1819	0.1825	**0.1780**	0.1716	0.2090	0.2060	0.2090	0.2038
0.01	All Optimal	0.6872	0.6937	0.6863	0.6813	**0.7057**	0.7195	**0.7057**	0.7172
	No Optimal	**0.0484**	0.0418	0.0499	0.0474	0.0846	0.0859	0.0846	0.0825
0.001	All Optimal	**0.9352**	0.9309	0.9350	0.9362	0.9070	0.8983	0.9070	0.9046
	No Optimal	**0.0337**	0.0323	0.0339	0.0318	0.0701	0.0857	0.0701	0.0693

Table 5.8. *Optimization results for the 3-bit multimodal problem. The table shows the probabilities of obtaining an all-optimal population and the probabilities of obtaining a no-optimal population using BBO Markov models and simulations. The best Markov performance is in bold in each row*

Several things are notable about the results in Tables 5.6, 5.7 and 5.8. First, the mutation rate affects the performance for all the four BBO algorithms. For all the three problems, the performance of the four algorithms improves as the mutation rate decreases; that is, the probability of obtaining an all-optimal population increases, and the probability of obtaining a no-optimal population decreases. The tables show that a high mutation rate of 0.1 per bit results in too much exploration so the probability of obtaining an all-optimal population is low, and the probability of obtaining a no-optimal population is relatively high. However, as the mutation rate decreases to 0.01 and 0.001, the probability of obtaining an all-optimal population significantly increases, and the probability of obtaining a no-optimal population significantly decreases. The higher the mutation rate, the lower the probability that the optimum is found and kept for the next generation, which gives worse performance, as shown in the tables.

Second, for the one-max problem in Table 5.6, the emigration-based BBO algorithms outperform the immigration-based BBO algorithms for all mutation rates; that is, emigration-based BBO algorithms have a higher probability of obtaining an all-optimal population and a lower probability of obtaining a no-optimal population. For example, for a mutation rate of 0.001 per bit, the best performance is obtained by emigration-based BBO algorithms in their high probability of obtaining an all-optimal population (95.43%), and in their low probability of obtaining a no-optimal population (2.22%). Partial immigration-based BBO and total immigration-based BBO probabilities are 86.05% and 86.01%, respectively, for obtaining an all-optimal population, and 9.23% and 9.28%, respectively, for obtaining a no-optimal population.

Third, for the deceptive problem in Table 5.7, the immigration-based BBO algorithms outperform the emigration-based BBO algorithms for all mutation rates, with partial immigration-based BBO slightly better than total immigration-based BBO. For example, for a mutation rate of 0.001 per bit, the best performance is obtained by partial immigration-based BBO in its high probability of obtaining an all-optimal population (87.71%), and in its low probability of obtaining a no-optimal population (9.38%). Total immigration-based BBO and emigration-based BBO algorithms are 87.67% and 78.18%, respectively, for obtaining an all-optimal population, and 9.43% and 19.41%, respectively, for obtaining a no-optimal population.

Fourth, for the multimodal problem in Table 5.8, the probability of obtaining an all-optimal population and the probability of obtaining a no-optimal population are very similar for all the four BBO algorithms. Specifically, total immigration-based BBO outperforms the other three algorithms when the mutation rate is 0.1 per bit, and partial immigration-based BBO outperforms the other three algorithms when the mutation rate is 0.001 per bit, but the emigration-based BBO algorithms outperform the immigration-based BBO algorithms when the mutation rate is 0.01 per bit.

All of these results show that different variations of BBO provide different optimization performance for different types of problems. For the unimodal problem, the emigration-based BBO algorithms are better than the immigration-based BBO algorithms. For the deceptive problem, the immigration-based BBO algorithms are better than the emigration-based BBO algorithms. For the multimodal problem, the emigration-based BBO algorithms perform similarly to the immigration-based BBO algorithms. Tables 5.6–5.8 further show that the performance of partial immigration-based BBO and total immigration-based BBO are similar for all test problems. These results are summarized in Table 5.9.

Finally, Tables 5.6–5.8 show that the Markov model results and the simulation results match well for all of the test problems, which confirms the Markov models of the proposed BBO variations.

Problem	Best algorithm	Other notes
Unimodal problem	Emigration-based BBO	The two emigration-based BBO algorithms perform identically
Deceptive problem	Immigration-based BBO	The two immigration-based BBO algorithms perform similarly
Multimodal problem	No significant difference between BBO algorithms	All BBO algorithms perform better with low mutation rates

Table 5.9. *Conclusions from the Markov study of the four BBO algorithms on 3-bit problems. The "best algorithm" for each problem is determined from an inspection of Tables 5.6–5.8*

5.5. Conclusions

In this chapter, we derived Markov models for BBO and its variations, and we discussed convergence based on the Markov models. These models give theoretically exact results, whereas simulations change from one run to the next due to initialization and the random number generator that is used for migration and mutation. Unfortunately, the dimension of the Markov model grows factorially with the population size and the search space size. This limits its application to very small problems. However, Markov models are still useful for giving exact results without the need to rely on the random nature of stochastic simulations.

In this concluding section, we discuss the curse of dimensionality of the Markov model. The curse of dimensionality is a phrase that was originally used in the context of dynamic programming [BEL 61; SIM 13a, p. 81]. However, it applies even more appropriately to Markov models of population-based optimization algorithms, including BBO. The size of the transition matrix of a Markov model of BBO is $T \times T$, where T is calculated by equation [5.19]. The transition matrix dimensions for a few combinations of population size N, and search space cardinality n, which is equal to 2^q for q-bit search spaces, are shown in Table 5.10. We see that the transition matrix dimension grows ridiculously large for problems of even modest dimension. This seems to indicate that Markov modeling is interesting only from a theoretical viewpoint and does not have any practical applications. However, there are a couple of reasons that such a response may be premature.

Number of bits q	$n = 2^q$	N	T
10	2^{10}	10	10^{23}
10	2^{10}	20	10^{42}
20	2^{20}	20	10^{102}
50	2^{50}	50	10^{688}

Table 5.10. *Markov transition matrix dimensions for various search space cardinalities n and population sizes N. Adapted from [REE 03, p. 131]*

First, although we cannot apply Markov models to realistically sized problems, Markov models still give us exact probabilities for small problems. This allows us to look at the advantages and disadvantages of different EAs for small problems, assuming that we have Markov models for EAs other than BBO. A lot of research in EAs today is focused on simulations. The problem with simulations is that their outcomes depend on implementation details and on the specific random number generator that is used. Also, if some event has a very small probability of occurring, then it would take many simulations to discover that probability. Simulation results are useful and necessary, but they must always be taken with a dash of skepticism and a pinch of salt.

Second, the dimension of the Markov transition matrices can be reduced. Our Markov models include T states, but many of these states are very similar to each other. For example, consider BBO with a search space cardinality of 10 and a population size of 10. Table 5.10 shows us that the Markov model has 10^{23} states, but these include the states:

$$\begin{aligned} v(1) &= \{5,5,0,0,0,0,0,0,0,0\} \\ v(2) &= \{4,6,0,0,0,0,0,0,0,0\} \\ v(3) &= \{6,4,0,0,0,0,0,0,0,0\} \end{aligned} \qquad [5.80]$$

These three states are so similar that it makes sense to group them together and consider them as a single state. We can do this with many other states to get a new Markov model with a reduced state space. Each state in the reduced-order model consists of a group of the original states. The transition matrix would then specify the probability of transitioning from one group of original states to another group of original states. This idea was proposed in [SPE 97] and is further discussed in [REE 03]. It is hard to imagine how to group states to reduce a $10^{23} \times 10^{23}$ matrix to a manageable size, but this idea may allow us to handle larger problems than we would be able to otherwise.

6

Dynamic System Models of BBO

In the last chapter, we saw that the transition matrix for a BBO Markov model is found by calculating the cumulative probabilities of migration and mutation. We now develop this construction in more detail to examine the dynamics of a population as it transitions from generation to generation. This will give us a more complete model, which is called a dynamic system model. The dynamic system model is based on the Markov model, but the application is quite different. The Markov model gives the steady-state probability of each possible population as the generation count approaches infinity. The dynamic system model gives the time-varying proportion of each possible solution in the search space as the population size approaches infinity.

Overview of the chapter

This chapter develops a dynamic system model for the basic BBO algorithm. Section 6.1 presents the basic notation that we will use in later sections. Section 6.2 derives the BBO dynamic system model and some of its properties, based on the Markov model. Section 6.3 uses our dynamic system model to solve some benchmark problems.

6.1. Basic notation

This section introduces the notation used in the BBO dynamic system model. Some of this notation may be more general or more specific in other contexts. The definitions indicated here are not universal, but nevertheless are commonly used, and more importantly for our purposes, are specifically used in this chapter.

First, we note a difference between the notations of Markov processes and dynamic systems. A Markov process is a process whose state at time step $(t + 1)$ depends only on the state at time t. The transition of the state from one time step to the next is probabilistic. A dynamic system is a process whose state at time step $(t + 1)$ depends only on the state at time t, but the transition of the state from one time step to the next is deterministic.

Second, the Markov model gives us the probability of occurrence of each possible population distribution as the number of generations approaches infinity. The dynamic system model is quite different; it gives us the proportion of each possible solution in the population as a function of time as the population size approaches infinity.

Third, in the Markov model, each state represents a possible population distribution, that is, a possible distribution of solutions in the search space. The probability that the system transitions from state i to state j is given by the probability p_{ij}, which is the probability that the population transitions from the ith possible population distribution to the jth possible population distribution in one generation. The states in the dynamic system model are not the same as the states in the Markov model. Dynamic system model states represent the proportion of each possible solution in the population.

Fourth, the number of states is different in the Markov model and in the dynamic system model. Recall from equation [5.19] in Chapter 5 that for a Markov model with a population size of N and a search space of cardinality n, the number of states T is given by:

$$T = \binom{n + N - 1}{N} \qquad [6.1]$$

For the dynamic system model, the number of states is equal to the search space cardinality n of the optimization problem. The number of states of the dynamic system model is thus only a small fraction of the number of states of the Markov model. This makes the dynamic system model applicable to larger problems than the Markov model.

Finally, both the Markov model and the dynamic system model allow us to obtain exact results without the need to rely on the random nature of stochastic simulations. The results of the two models are more precise than simulation. The dynamic system model for EAs in general is explained in detail in [REE 03], [NIX 92], [VOS 90] and [VOS 91].

EXAMPLE 6.1.–

We use an example to illustrate the population distribution in the Markov model and the candidate solution proportions in the dynamic system model. Suppose there exists a search space consisting of $n = 3$ possible solutions which are $x = \{x_1, x_2, x_3\} = \{00, 01, 10\}$ with a population size $N = 2$. So there are six possible population distributions from equation [6.1], which are $\{\{00, 00\}, \{00, 01\}, \{00, 10\}, \{01, 01\}, \{01, 10\}, \{10, 10\}\}$, and the corresponding population count vectors are $\{\{2, 0, 0\}, \{1, 1, 0\}, \{1, 0, 1\}, \{0, 2, 0\}, \{0, 1, 1\}, \{0, 0, 2\}\}$. Suppose that the solutions in the current population are $y = \{x_2, x_3\} = \{01, 10\}$. Then, we have the population count vector $v = \{v_1, v_2, v_3\} = \{0, 1, 1\}$, and the solution proportion vector $p = v/N = \{0, 0.5, 0.5\}$.

6.2. Dynamic system models of BBO

In this section, we use the Markov model of the previous chapter to derive a dynamic system model of the basic BBO algorithm. Therefore, some parameters in this chapter are the same as those in the previous chapter. The view of a BBO as a dynamic system was originally published in [SIM 11a], which forms the basis of this chapter.

First, recall the probability $P_{ki}(v)$ that the kth migration results in $y_k = x_i$ at generation $(t + 1)$, and is described as:

$$
\begin{aligned}
P_{ki}(v) &= \Pr\left(y_{k,t+1} = x_i\right) \\
&= \prod_{s=1}^{q}\left[\left(1-\lambda_{m(k)}\right)1_0\left(x_{m(k)}(s)-x_i(s)\right)+\lambda_{m(k)}\frac{\sum_{j\in\varsigma_i(s)} v_j\mu_j}{\sum_{j=1}^{n} v_j\mu_j}\right]
\end{aligned}
\quad [6.2]
$$

In equation [6.2], y_k is the kth solution in the population:

$$
\{y_1,\cdots y_N\} = \{\underbrace{x_1,x_1,\cdots,x_1}_{v_1 \text{ copies}},\underbrace{x_2,x_2,\cdots,x_2}_{v_2 \text{ copies}},\cdots\underbrace{x_n,x_n,\cdots,x_n}_{v_n \text{ copies}}\} \quad [6.3]
$$

where the y_k solutions have been ordered to group identical solutions. That is,

$$y_k = \begin{cases} x_1, \text{ for } k = 1, \cdots, v_1 \\ x_2, \text{ for } k = v_1 + 1, \cdots, v_1 + v_2 \\ x_3, \text{ for } k = v_1 + v_2 + 1, \cdots, v_1 + v_2 + v_3 \\ \vdots \\ x_n, \text{ for } k = \sum_{i=1}^{n-1} v_i + 1, \cdots, N \end{cases} \qquad [6.4]$$

where N is the population size. So equation [6.3] can be rewritten more compactly as:

$$y_k = x_{m(k)} \qquad \text{for } k = 1, 2, \cdots, N \qquad [6.5]$$

where $m(k)$ is defined as:

$$m(k) = \min r \text{ such that } \sum_{i=1}^{r} v_i \geq k \qquad [6.6]$$

In equation [6.2], v is the population count vector at the tth generation, which is:

$$v = \{v_1, v_2, \cdots, v_n\} \qquad [6.7]$$

where v_i is the number of copies of solution x_i in the population, so that:

$$\sum_{i=1}^{n} v_i = N \qquad [6.8]$$

In order to derive a dynamic system model for BBO, we make a slight change in the basic BBO algorithm of Figure 3.5 in Chapter 3. We still cycle through the immigration loop N times; however, instead of deterministically cycling through each population member y_k for immigration, we randomly select a population member for immigration each of the N times through the loop. Therefore, each time we cycle through the immigration loop, each y_k has a $1/N$ chance of being selected for immigration. Note that as $N \to \infty$, this is equivalent to the algorithm of Figure 3.5. With this change, the basic BBO algorithm is modified to become the algorithm shown in Figure 6.1.

Initialize a population of candidate solutions $\{x_k\}$ for $k \in [1, N]$

While not (termination criterion)

For each x_k, set emigration rate μ_k proportional to the fitness of x_k, where μ_k is normalized to [0, 1]

For each x_k, set immigration rate $\lambda_k = 1 - \mu_k$

$\{z_k\} \leftarrow \{x_k\}$

For $h = 1,\ldots, N$

Randomly select one of the z_k solutions

For each solution feature SIV

Use λ_k to probabilistically decide whether to immigrate to z_k

If immigrating then

Use $\{\mu_i\}$ to probabilistically select the emigrating solution x_j

z_k(SIV) $\leftarrow x_j$(SIV)

End if

Next solution feature

Probabilistically decide whether to mutate z_k

Next h

$\{x_k\} \leftarrow \{z_k\}$

Next generation

Figure 6.1. *Outline of the BBO algorithm with random selection of the immigrating solution. $\{x_k\}$ is the population of solutions and $\{z_k\}$ is a temporary population of solutions. x_k is the kth candidate solution and $x_k(SIV)$ is the solution feature SIV of x_k. For simplicity, we use s to denote SIV in the text*

For the algorithm described in Figure 6.1, the probability that the hth migration trial M_h results in x_i is:

$$\Pr(M_h = x_i) = \frac{1}{N}\sum_{k=1}^{N} \Pr(y_k = x_i) \qquad [6.9]$$

Now note that:

$$\Pr(y_{k_1} = x_i) = \Pr(y_{k_2} = x_i) \quad \text{if } y_{k_1} = y_{k_2} \qquad [6.10]$$

Combining equations [6.4], [6.5], [6.6] and [6.10], we get:

$$\Pr(y_{k_1} = x_i) = \Pr(y_{k_2} = x_i) \quad \text{if } (k_1 = k_2) \in \left[\sum_{i=1}^{m(k-1)} v_i + 1, \sum_{i=1}^{m(k)} v_i \right] \tag{6.11}$$

Therefore, equation [6.9] can be written as:

$$\Pr(M_h = x_i) = \frac{1}{N} \sum_{k=1}^{n} v_k \left\{ \prod_{s=1}^{q} \left[(1-\lambda_k) 1_0 \left(x_k(s) - x_i(s) \right) + \lambda_k \frac{\sum_{j \in \varsigma_i(s)} v_j \mu_j}{\sum_{j=1}^{n} v_j \mu_j} \right] \right\} \tag{6.12}$$

Now we define the population-count proportionality vector p as:

$$p = \frac{v}{N} \tag{6.13}$$

That is, p_i is the proportion of x_i solutions in the population for $i \in [1, n]$, and the elements of p add up to 1. Equation [6.12] can then be written as:

$$\begin{aligned} \Pr(M_h = x_i) &= \sum_{k=1}^{n} p_k \left\{ \prod_{s=1}^{q} \left[(1-\lambda_k) 1_0 \left(x_k(s) - x_i(s) \right) + \lambda_k \frac{\sum_{j \in \varsigma_i(s)} v_j \mu_j}{\sum_{j=1}^{n} v_j \mu_j} \right] \right\} \\ &= \sum_{k=1}^{n} \frac{p_k}{(p^T \mu)^q} \left\{ \prod_{s=1}^{q} \left[p^T \mu (1-\lambda_k) 1_0 \left(x_k(s) - x_i(s) \right) + \lambda_k \sum_{j \in \varsigma_i(s)} v_j \mu_j \right] \right\} \end{aligned} \tag{6.14}$$

The quantities on the right side of equation [6.14] are defined at the tth generation. The left side of equation [6.14] then gives probability of obtaining x_i at generation $(t + 1)$.

Next, we introduce a theorem to represent the basic population dynamics relationship [REE 03, p. 144, VOS 99, Theorem 2].

THEOREM 6.1.–*If the current population is given by the population vector* p*, then the next population expected is* $f(p)$*, where* $f(\cdot)$ *is a generational operator.*

According to Theorem 6.1, we see that the left side of equation [6.14] is equal to the proportion of x_i solutions in the population at generation $(t + 1)$:

$$p_i(t+1) = \sum_{k=1}^{n} \frac{p_k(t)}{(p^T(t)\mu)^q} \left\{ \prod_{s=1}^{q} \left[p^T(t) \mu (1-\lambda_k) 1_0 \left(x_k(s) - x_i(s) \right) + \lambda_k \sum_{j \in \varsigma_i(s)} v_j \mu_j \right] \right\} \tag{6.15}$$

where p is explicitly shown as a function of t. Based on equation [6.15] for $i \in [1, n]$, we can see the nonlinear dynamic system for the evolution of the population-count proportionality vector:

$$p(t+1) = f(p(t)) \quad [6.16]$$

Next, we incorporate mutation into our model and denote the $n \times n$ mutation matrix as U, where U_{ji} is the probability that x_j mutates to x_i. Suppose that the event that each bit of a candidate solution is flipped is stochastically independent and occurs with probability $u \in (0, 1)$. Then, the probability U_{ji} can be written as:

$$U_{ji} = u^{H_{ij}} (1-u)^{q-H_{ij}} \quad [6.17]$$

where H_{ij} represents the Hamming distance between bit strings x_i and x_j.

This gives the dynamic system model equation:

$$p(t+1) = f(p(t))U \quad [6.18]$$

If mutation is not used in the BBO algorithm, then U is the identity matrix and equation [6.18] reduces to equation [6.16].

EXAMPLE 6.2.–

To verify the dynamic system model of BBO, we consider a simple 3-bit problem ($n = 8$) with a per-bit mutation rate $u = 0.2$. The fitness values are given as follows:

$$\begin{aligned} &f(000)=8, f(001)=1, f(010)=1, f(011)=1, \\ &f(100)=1, f(101)=1, f(110)=1, f(111)=9. \end{aligned} \quad [6.19]$$

This is a relatively difficult optimization problem because $x_1 = 000$ has a high fitness, and every time we add a 1 bit to it the fitness decreases dramatically, but the solution with all 1's has the highest fitness. We begin with an initial population with proportionality vector:

$$p(0) = [0.8 \quad 0.1 \quad 0.1 \quad 0 \quad 0 \quad 0 \quad 0 \quad 0]^T \quad [6.20]$$

Figure 6.2 shows the dynamic system model result and simulation result for BBO with a population size of 1,000. The plots provide confirmation for the dynamic system model of BBO. The simulation result oscillates around its mean value, which is expected because of the mutation operator. The simulation result will vary from one simulation to the next, and will never exactly match the theory due to the stochastic nature of the simulations. That is why the dynamic system model can be more useful than simulation; the model is exact while simulation result is only approximate.

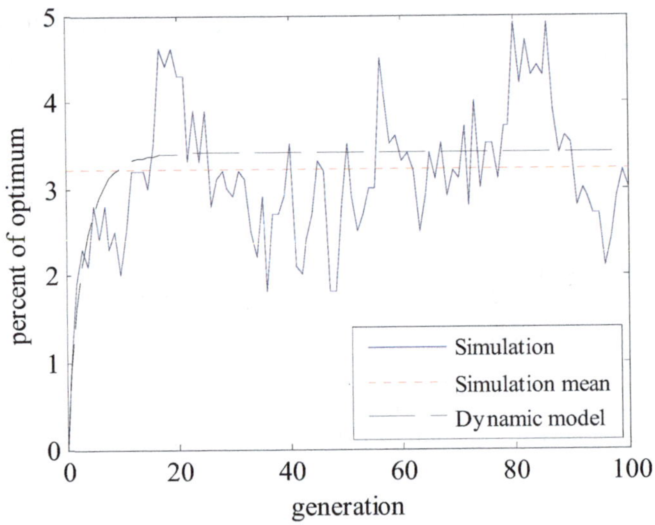

Figure 6.2. *BBO dynamic system model result giving confirmation that simulation matches theory. The traces show the proportion of the optimal solutions for a typical simulation, the mean of that proportion over all generations and the proportion according to the dynamic system model. Reprinted from [SIM 11a] with permission from Elsevier. For a color version of this figure, see www.iste.co.uk/ma-simon/evolutionary.zip*

A. Special case: $\lambda = 0$

We now consider the dynamic system model of BBO when $\lambda_k = 0$ for all k. In this case, there is no possibility of immigration and equation [6.15] reduces to:

$$p_i(t+1) = \sum_{k=1}^{n} p_k(t) \left\{ \prod_{s=1}^{q} 1_0 \left(x_k(s) - x_i(s) \right) \right\} \qquad [6.21]$$

Since each x_i is distinct, we see that:

$$\prod_{s=1}^{q} 1_0\left(x_k(s)-x_i(s)\right)=1_0\left(k-i\right) \quad [6.22]$$

which gives:

$$p_i(t+1)=\sum_{k=1}^{n} p_k(t) 1_0\left(k-i\right)=p_i(t) \quad [6.23]$$

That is, with no immigration and no mutation, the proportionality vector does not change from one generation to the next, which agrees with intuition.

B. Special case: $\lambda=1$ and random feature selection

Next, we consider the dynamic system model of BBO when $\lambda_k=1$ for all *k*. The BBO algorithm of Figure 6.1 becomes a special type of a genetic algorithm with global uniform recombination (GA/GUR), which has been described in section 3.2 in Chapter 3. GA/GUR can be implemented in many different ways, but if it is implemented with the entire population as potential contributors to the next generation, and with fitness-based selection for each solution feature in each offspring, then it is equivalent to BBO with $\lambda_k=1$ for all *k*. In this case, immigration takes place for all solutions in the population, and the new solution that results from each immigration can be thought of as an offspring of the previous generation. Suppose also that in addition to $\lambda_k=1$ for all *k*, each immigration trial migrates one randomly selected bit. Then, the BBO algorithm of Figure 6.1 becomes the GA/GUR algorithm of Figure 6.3.

The probability that y_k at generation $(t+1)$ is equal to x_i, given that solution feature *s* was selected for migration, can be written as:

$$\Pr\left(y_{k,t+1}=x_i \mid s\right)=\Pr\left[y_{k,t}(r: r\neq s)=x_i(r: r\neq s)\right]\Pr\left(y_{k,t+1}(s)=x_i(s)\right) \quad [6.24]$$

The first term on the right side of equation [6.24] is the proportion of the population which has all bits *r*, such that $r\neq s$, equal to the corresponding bits in x_i. We denote the indices of these solutions as $\tau_i(s)$:

$$\tau_i(s)=\left\{j: x_j(r: r\neq s)=x_i(r: r\neq s)\right\}, \quad i\in[1,n] \quad [6.25]$$

Note that $|\tau_i(s)| = 2$ for all (*i*, *s*). Now we can write equation [6.24] as:

$$\Pr(y_{k,t+1} = x_i | s) = \left(\frac{\sum_{j\in\tau_i(s)} v_j}{\sum_{j=1}^{n} v_j} \right) \left(\frac{\sum_{j\in\varsigma_i(s)} v_j \mu_j}{\sum_{j=1}^{n} v_j \mu_j} \right) \quad [6.26]$$

Initialize a population of candidate solutions $\{ x_k \}$ for $k \in [1, N]$

While not (termination criterion)

  For each x_k, set emigration rate μ_k proportional to the fitness of x_k, where μ_k is normalized to [0, 1]

  $\{ z_k \} \leftarrow \{ x_k \}$

  For $h = 1,\ldots, N$

   Randomly select one of the z_k solutions

    Randomly select a solution feature SIV

  Use $\{\mu_i\}$ to probabilistically select the emigrating solution x_j

  $z_k(\text{SIV}) \leftarrow x_j(\text{SIV})$

    Probabilistically decide whether to mutate z_k

  Next *h*

  $\{ x_k \} \leftarrow \{ z_k \}$

Next generation

Figure 6.3. *Outline of GA/GUR with random selection of the immigrating solution and random selection of the migrating solution feature. $\{ x_k \}$ is the population of solutions and $\{ z_k \}$ is a temporary population of solutions. x_k is the kth candidate solution and $x_k(SIV)$ is the solution feature SIV of x_k. For simplicity, we use s to denote SIV in the text*

Combining equations [6.8] and [6.13], equation [6.26] can be written as:

$$\Pr(y_{k,t+1} = x_i | s) = \sum_{j\in\tau_i(s)} p_j \frac{\sum_{j\in\varsigma_i(s)} p_j \mu_j}{\sum_{j=1}^{n} p_j \mu_j} \quad [6.27]$$

Figure 6.3 shows that each bit $s \in [1, q]$ has a $1/q$ probability of being selected as the migrating feature. Therefore,

$$\Pr(y_{k,t+1} = x_i) = \frac{1}{qp^T \mu} \sum_{s=1}^{q} \left(\sum_{j \in \tau_i(s)} p_j \right) \left(\sum_{j \in \varsigma_i(s)} p_j \mu_j \right) \quad [6.28]$$

This is a quadratic function of the p_i terms and can thus be written as:

$$\Pr(y_{k,t+1} = x_i) = \sum_{a=1}^{n} \sum_{b=1}^{n} Y_{i,ab} p_a p_b \quad [6.29]$$

Equation [6.28] shows that the p_m^2 coefficient on the right side of equation [6.29] for $m \in [1, n]$ can be written as:

$$Y_{i,mm} = \sum_{s=1}^{q} \mu_m 1_0 \left(m \in \varsigma_i(s) \right) 1_0 \left(m \in \tau_i(s) \right) = \sum_{s=1}^{q} \mu_m 1_0 \left(m \in \left(\varsigma_i(s) \cap \tau_i(s) \right) \right) \quad [6.30]$$

From the definition of $\tau_i(s)$ in equation [6.25], we know that $i \in \tau_i(s)$. We also know that there is only one other element in $\tau_i(s)$. The other element in $\tau_i(s)$, say α, has a bit string such that $x_\alpha(r) = x_i(r)$ for all $r \neq s$. But since $\alpha \neq i$, we know that $x_\alpha(s) \neq x_i(s)$, which means that $\alpha \notin \varsigma_i(s)$. Therefore,

$$\varsigma_i(s) \cap \tau_i(s) = \{i\} \quad \text{for all } s \quad [6.31]$$

Equation [6.30] can therefore be written as:

$$Y_{i,mm} = \sum_{s=1}^{q} \mu_m 1_0 (m - i) = q \mu_m 1_0 (m - i) \quad [6.32]$$

We can use equation [6.28] to show that the $p_m p_k$ coefficient $(k \neq m)$ on the right side of equation [6.29] can be written as:

$$\begin{aligned} Y_{i,mk} + Y_{i,km} &= \mu_m \sum_{s=1}^{q} 1_0 \left(m \in \varsigma_i(s) \right) 1_0 \left(k \in \tau_i(s) \right) \\ &\quad + \mu_k \sum_{s=1}^{q} 1_0 \left(k \in \varsigma_i(s) \right) 1_0 \left(m \in \tau_i(s) \right) \quad \text{for } m \neq k \end{aligned} \quad [6.33]$$

So the GA/GUR dynamic system model can be written as the following set of n coupled quadratic equations:

$$p_i(t+1) = p^T(t)Y_i p(t), \quad i \in [1, n] \quad [6.34]$$

where each element of Y_i is $Y_{i,mk}$, which denotes the mth row and kth column of Y_i. If mutation is included in the GA/GUR algorithm, then:

$$p(t+1) = \mathrm{diag}\left(p^T(t)Y_i p(t)\right)U \quad [6.35]$$

where $\mathrm{diag}\left(p^T(t)Y_i p(t)\right)$ is the $n \times n$ diagonal matrix consisting of:

$$p^T(t)Y_1 p(t), \ldots, p^T(t)Y_n p(t) \quad [6.36]$$

EXAMPLE 6.3.–

To verify the dynamic system model of BBO for the special case of $\lambda = 1$ and random feature selection, which is equivalent to GA/GUR, we use the same example described in Example 6.2 and the same initial population vector proportionality. Figure 6.4 shows the dynamic system model result and simulation result for GA/GUR. The plots provide confirmation for the dynamic system model of GA/GUR.

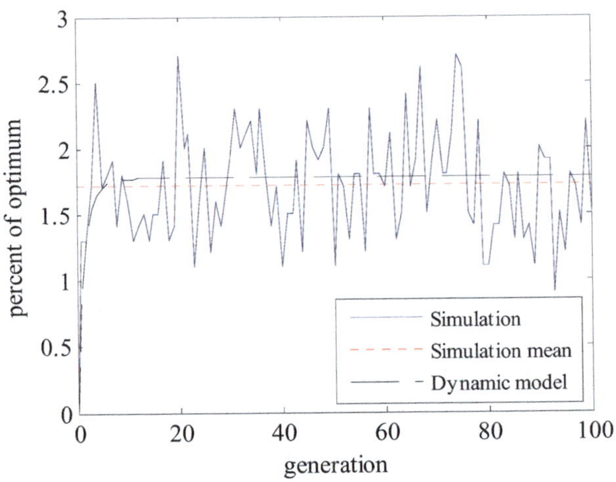

Figure 6.4. *GA/GUR dynamic system model result giving confirmation that simulation matches theory. The traces show the proportion of the optimal solutions for a typical simulation, the mean of that proportion over all generations and the proportion according to the dynamic system model. Reprinted from [SIM 11a] with permission from Elsevier. For a color version of this figure, see www.iste.co.uk/ma-simon/evolutionary.zip*

EXAMPLE 6.4.–

In this example, we compare dynamic system model results between BBO, GA/GUR and GA with single-point crossover (GA/SP) and roulette-wheel selection, which was originally developed in [VOS 99]. GA/SP is a classic EA for a dynamic system model and is summarized in [REE 03] as:

$$p_i(t+1) = \frac{p^T(t)\text{diag}(f)UC(i)U^T\text{diag}(f)\,p(t)}{\left(f^T p(t)\right)^2} \qquad [6.37]$$

where $\text{diag}(f)$ is the $n \times n$ diagonal matrix consisting of the fitness values of each candidate solution, and U is the mutation matrix given in equation [6.17]. $C(i)$ is an $n \times n$ matrix such that the element in its mth row and kth column is the probability that x_m and x_k crossover to produce x_i.

We consider a problem whose fitness values are given as:

$$f_i = \begin{cases} 8 & \text{for } x_i = (0...0) \\ 9 & \text{for } x_i = (1...1) \\ 1 & \text{for all other } x_i \end{cases} \qquad [6.38]$$

This is the same as equation [6.19] except that it is generalized for an arbitrary number of bits. The proportionality vector of the initial population is given as:

$$p(0) = \frac{1}{n-1}[1 \quad \dots \quad 1 \quad 0]^T \qquad [6.39]$$

That is, the initial population is evenly distributed among the sub-optimal solutions, and there are no optimal solutions. Figures 6.5–6.7 show steady-state dynamic system model results for three different search space cardinalities, plotted as functions of mutation rate. Figure 6.5 shows that BBO is much better at achieving a high percentage of optimal solutions than GA/GUR and GA/SP for small problems. Figures 6.6 and 6.7 show that as the problem dimension gets larger, BBO performance gets worse relative to the GAs for large mutation rates. However, BBO remains many orders of magnitude better than the GAs for small mutation rates, which are more typical for real-world problems.

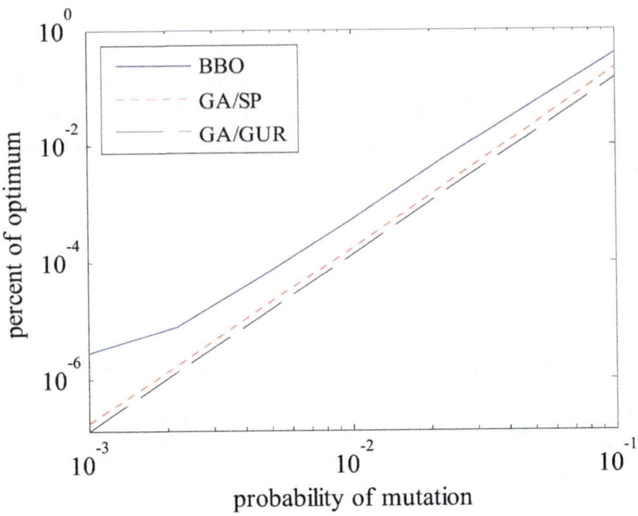

Figure 6.5. *Dynamic system model results for a 3-bit problem (search space cardinality n = 8) showing the steady-state proportion of optimal solutions. Reprinted from [SIM 11a] with permission from Elsevier. For a color version of this figure, see www.iste.co.uk/ma-simon/evolutionary.zip*

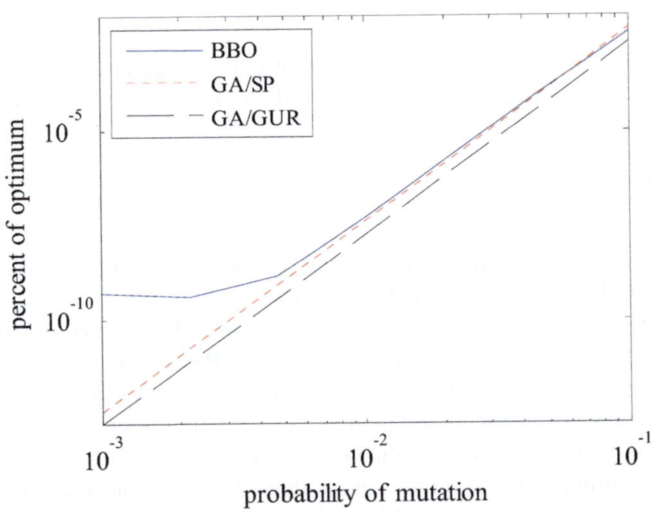

Figure 6.6. *Dynamic system model results for a 5-bit problem (search space cardinality n = 32) showing the steady-state proportion of optimal solutions. Reprinted from [SIM 11a] with permission from Elsevier. For a color version of this figure, see www.iste.co.uk/ma-simon/evolutionary.zip*

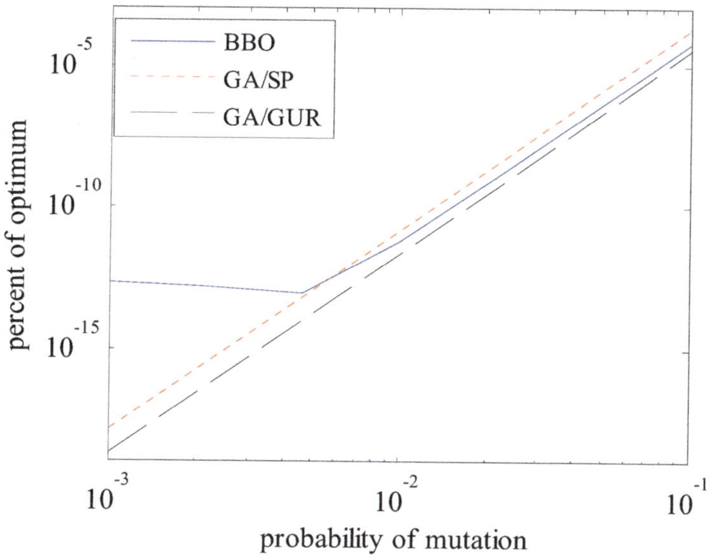

Figure 6.7. *Dynamic system model results for a 7-bit problem (search space cardinality n = 128) showing the steady-state proportion of optimal solutions. Reprinted from [SIM 11a] with permission from Elsevier. For a color version of this figure, see www.iste.co.uk/ma-simon/evolutionary.zip*

Figures 6.8–6.10 depict the same information as that shown in Figures 6.5–6.7, but presented in a different way. Figures 6.8–6.10 show dynamic system model results for three different mutation rates, plotted as functions of the problem dimension. Figure 6.8 shows that BBO is much better than the GAs for all problem dimensions if the mutation rate is low, as is typical of real-world problems. Figures 6.9 and 6.10 show that as the mutation rate increases, BBO remains better than the GAs for small problem dimensions, but becomes worse than GA/SP as the problem dimension increases.

Figure 6.8 shows that with realistic mutation rates, BBO is much better than the GAs for all problem dimensions. Furthermore, the relative advantage of BBO increases as the problem dimension increases, which is consistent with the conclusions presented in [SIM 11b] and [SIM 11d].

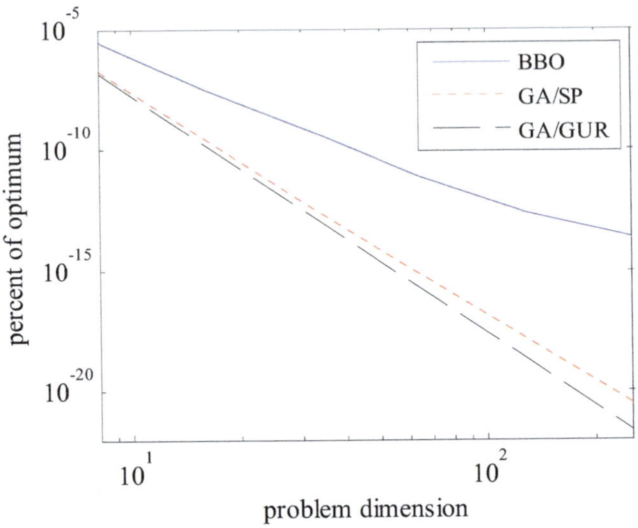

Figure 6.8. *Dynamic system model results for mutation rate = 0.1% per bit showing the steady-state proportion of optimal solutions. Reprinted from [SIM 11a] with permission from Elsevier. For a color version of this figure, see www.iste.co.uk/ma-simon/evolutionary.zip*

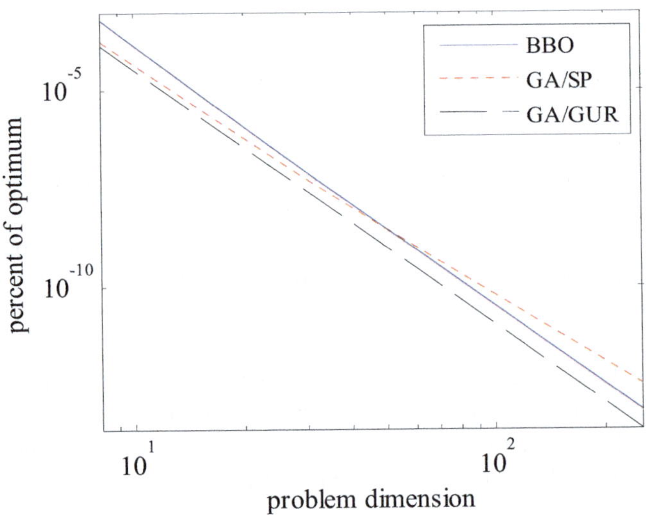

Figure 6.9. *Dynamic system model results for mutation rate = 1% per bit showing the steady-state proportion of optimal solutions. Reprinted from [SIM 11a] with permission from Elsevier. For a color version of this figure, see www.iste.co.uk/ma-simon/evolutionary.zip*

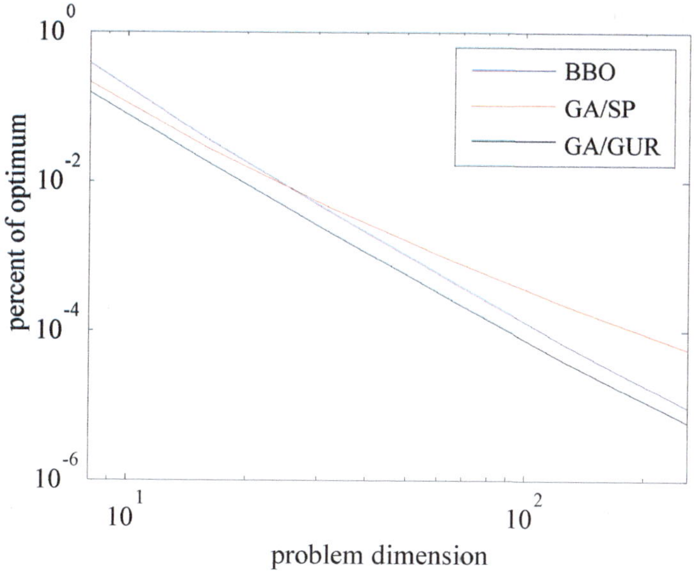

Figure 6.10. *Dynamic system model results for mutation rate = 10% per bit showing the steady-state proportion of optimal solutions. Reprinted from [SIM 11a] with permission from Elsevier. For a color version of this figure, see www.iste.co.uk/ma-simon/evolutionary.zip*

6.3. Applications to benchmark problems

Next, we apply the dynamic system model results of BBO, GA/GUR and GA/SP to standard benchmark functions to show how the theory can be used to study the performance of algorithms and the effect of parameter settings. These benchmark functions include the needle function which is given in equation [6.38], the one-max function which has a fitness proportional to the number of 1-bits in each bit string, the deceptive function which is the same as the one-max function except that the bit string with all zeros has the highest fitness [BAC 96, YAO 99], and some continuous-domain functions that we list in the left column in Table 6.1 and which are described briefly in Table 3.2 in Chapter 3. A more detailed description of these functions can be found in Appendix A. We implemented the continuous functions as two-dimensional functions whose independent variables are coded with 3 or 4 bits per independent variable. This gives an optimization problem with either 6 or 8 bits in total, which results in a search space cardinality of either 64 or 256. We initialized the population with a uniform distribution over all of the non-optimal solutions. The initial population did not include any optima. We recorded the percent of optimal solutions in the population after 10 generations, which gives an idea of how fast each algorithm converges. Tables 6.1 and 6.2 show the results. Note that these are not simulation results, but exact dynamic system model results.

Function	Cardinality = 64		
	BBO	GA/GUR	GA/SP
Needle	**6.97E-03**	2.32E-10	2.17E-06
One-max	46.04	**47.67**	42.64
Deceptive	**3.89E-03**	2.25E-04	2.33E-04
Sphere Function	**69.63**	47.69	49.21
Schwefel's Problem 2.22	**69.14**	25.80	51.27
Schwefel's Problem 1.2	**69.37**	25.29	49.21
Schwefel's Problem 2.21	**75.65**	66.84	62.89
Generalized Rosenbrock's Function	**10.37**	3.56	9.13
Step Function	**50.04**	25.04	37.56
Quartic Function	**39.09**	26.13	34.12
Generalized Schwefel's Problem 2.26	**15.93**	0.47	1.30
Generalized Rastrigin's Function	43.93	**50.05**	41.79
Ackley's Function	58.92	25.40	**71.38**
Generalized Griewank's Function	**67.82**	44.18	49.00
Generalized Penalized Function 1	**29.69**	24.98	27.77
Generalized Penalized Function 2	**32.83**	24.96	28.98

Table 6.1. *Dynamic system model results on benchmark functions with a cardinality of 64. The number in each cell indicates the percentage of optimal solutions in the population after 10 generations. The best result for each benchmark is shown in bold*

Function	Cardinality = 256		
	BBO	GA/GUR	GA/SP
Needle	**6.72E-03**	9.95E-06	6.48E-06
One-max	22.44	**23.84**	19.43
Deceptive	**1.03E-03**	5.36E-05	5.13E-05
Sphere Function	**35.95**	12.50	24.94
Schwefel's Problem 2.22	**38.10**	13.36	26.84
Schwefel's Problem 1.2	**35.91**	14.02	25.07
Schwefel's Problem 2.21	**49.03**	39.82	38.60
Generalized Rosenbrock's Function	**6.60**	4.18	6.49
Step Function	**25.51**	12.52	19.45
Quartic Function	**18.64**	12.49	17.01
Generalized Schwefel's Problem 2.26	**38.07**	15.69	8.70
Generalized Rastrigin's Function	**46.85**	14.64	42.43
Ackley's Function	43.40	54.17	**70.00**
Generalized Griewank's Function	**36.12**	15.07	25.61
Generalized Penalized Function 1	**26.22**	25.02	25.68
Generalized Penalized Function 2	13.03	12.51	**14.08**

Table 6.2. *Dynamic system results on benchmark functions with a cardinality of 256. The number in each cell indicates the percentage of optimal solutions in the population after 10 generations. The best result for each benchmark is shown in bold*

It can be seen from Tables 6.1 and 6.2 that for both the 64- and 256-bit problems, BBO performs the best in 13 out of 16 benchmarks. More importantly, for difficult problems (the needle and deceptive functions), BBO performs better than GA/GUR and GA/SP by orders of magnitude.

These results are not intended to give a comprehensive comparison between BBO and GAs; extensive comparisons between BBO and other EAs using standard benchmark functions are shown in [MA 10a]. The theory and results here are instead intended to show how dynamic system models can be used to compare EAs in situations where probabilities are extremely small and where Monte Carlo simulations are therefore not reliable. Dynamic system models can also be used to study the effect of parameter settings and learning approaches. Dynamic system models can also aid in the development of adaptive algorithms or parameter update schemes that work well on many different types of problems. These models can also be used to help understand the behavior of BBO; for example, how and why it works well, or does not work well, on certain types of problems.

6.4. Conclusions

In this chapter, we outlined dynamic system models for BBO based on the BBO Markov models, and in the process we obtained a dynamic system model for GA/GUR. These models give theoretically exact results, just like Markov models, and allow analytical comparisons between different types of BBO algorithms rather than a reliance on possibly unpredictable simulation results. Simulation results are important and necessary, but they can be misleading if used separately from theory. The Markov models and dynamic system models that we obtained provide a lot of room for additional application to BBO and other EAs.

7

Statistical Mechanics Approximations of BBO

In the previous two chapters, we set up the Markov model and dynamic system model of BBO to analyze the behavior of BBO. Those models tell us the exact probability distribution over all possible populations, but in practice, those models rapidly become impractical as the size of the search space grows, which limits their application to very small problems. This is the curse of dimensionality for analytical EA models. It is therefore natural to investigate other methods to obtain practical models. The statistical mechanics approximation is used in this chapter to analyze the behavior of BBO. The idea of this modeling approach comes from the field of statistical mechanics, which involves averaging the behavior of many molecular particles to model the behavior of a group of molecules. We can use this idea to model BBO behavior with large populations and to better understand the evolution of BBO in realistic, high-dimensional optimization problems.

Overview of the chapter

In this chapter, we will see that statistical mechanics approximation theory provides insight into BBO behavior. Section 7.1 provides the preliminary foundation for statistical mechanics approximations. Section 7.2 derives a statistical mechanics model for the basic BBO algorithm. Section 7.3 discusses extensions of the BBO statistical mechanics model.

7.1. Preliminary foundation

Statistical mechanics is a branch of physics that applies probability theory to the study of the thermodynamic behavior of systems consisting of a huge number of interacting entities. It provides methods to calculate the gross or average properties of systems by treating the small-scale motions as random. It is often applied to the computation of the thermodynamic properties of particles interacting with a heat bath, and has recently been applied to EAs, including mathematical modeling for genetic algorithms [PRÜ 94, SHA 94, RAT 95, RAT 96, RAT 96] and simulated annealing [VAN 87]. The advantage of statistical mechanics for describing the dynamics of EAs is that it only assumes knowledge of the cumulants of the fitness distribution and the average correlation within the population.

The cumulant is defined by the generating function, which is given by the natural logarithm of the Fourier transform of the probability distribution function. Cumulants of a probability distribution describe the shape of the distribution, and they contain the same information as the distribution function. The first and second cumulants are, respectively, called the mean and variance of the distribution, and the third cumulant is called the skewness and measures the degree of asymmetry of the distribution. All cumulants except the first two are zero for a Gaussian distribution, so the higher-order cumulants measure the non-Gaussian content of the distribution. Since EA populations are often initialized randomly, and fitness is a function of many variables, and the central limit theorem states that the accumulated distribution of random variables becomes more Gaussian as the number of variables increases, many real-world EA applications have a Gaussian fitness distribution during the early generations.

Consider a scalar random variable X which can take non-negative integer values. For each $k \geq 0$, there is a certain probability P_k that X takes the value k. So the *characteristic function* of the probability distribution of X can be described as:

$$\varphi(z) = \sum_{k \geq 0} \Pr(X = k) e^{kz} \quad [7.1]$$

If we expand the characteristic function as a Taylor series in z, we obtain:

$$\varphi(z) = 1 + \frac{\omega_1}{1!} z + \frac{\omega_2}{2!} z^2 + \frac{\omega_3}{3!} z^3 + \cdots \quad [7.2]$$

where the coefficients ω_k are called the moments of the probability distribution, which are given as:

$$\omega_k = E\left[X^k\right] \quad [7.3]$$

Expanding the natural logarithm of the characteristic function $\varphi(z)$, called the *cumulant generating function*, in a Taylor series gives:

$$
\begin{aligned}
H(z) &= \log\varphi(z) = \log(P_0 + P_1 e^{z} + P_2 e^{2z} + P_3 e^{3z} + \cdots) \\
&= \frac{\kappa_1}{1!} z + \frac{\kappa_2}{2!} z^2 + \frac{\kappa_3}{3!} z^3 + \cdots
\end{aligned}
\qquad [7.4]
$$

where the coefficients κ_k for k = 1, 2, 3..., are called the *cumulants* of the probability distribution. The relationships between the first three cumulants and the moments of the probability distribution have simple expressions:

$$
\begin{aligned}
\kappa_1 &= \omega_1 = E[X] \\
\kappa_2 &= \omega_2 - \omega_1^2 = E\left[X^2\right] - \left(E[X]\right)^2 \\
\kappa_3 &= \omega_3 - 3\omega_1\omega_2 + 2\omega_1^3 = E\left[X^3\right] - 3 \cdot E[X] \cdot E\left[X^2\right] + 2 \cdot \left(E[X]\right)^3
\end{aligned}
\qquad [7.5]
$$

Equation [7.5] shows that the first cumulant κ_1 is the mean of the distribution, and the second cumulant κ_2 is the variance. Since the Gaussian distribution is completely defined by its mean and variance and all the higher cumulants are zero, any distribution with very small values for the higher cumulants looks like a Gaussian. If a population is distributed approximately normally, we can truncate the series expansion of equation [7.4] and use just the first few cumulants to approximate the population distribution.

The cumulants of a probability distribution have a linearity property which can be formulated as follows [GRI 97].

THEOREM 7.1.– *If X_i is an independent random variable for i = 1, 2, 3...n, then the cumulants of the probability distribution can be found with the sum*:

$$
\kappa_k^{X_1 + X_2 + \cdots + X_n} = \kappa_k^{X_1} + \kappa_k^{X_2} + \cdots + \kappa_k^{X_n} = \sum_{i=1}^{n} \kappa_k^{X_i} \qquad [7.6]
$$

Later in this chapter, we will use this very important theorem to derive the statistical mechanics model of BBO.

EXAMPLE 7.1.–

This example serves to clarify the notations of cumulants, and is based on [REE 03] and [MA 16a]. The one-max problem is considered, which has only one optimum. The fitness of a solution is the number of ones in the binary string. Given

n as the length of a binary string, and X as the possible fitness values, the one-max problem is defined as follows:

$$\text{Maximize } X\text{, where } X=\sum_{j=1}^{n} x_j \qquad [7.7]$$

where $x_j \in I=\{0,1\}$ is bit j of a binary string. In this example, the length of the binary string n is set to 5, which is the maximum fitness, and the size of the population N is set to 24. If the initial population is random, it might be the following:

$$\begin{array}{cccccccc} 00011 & 01001 & 10000 & 01110 & 10101 & 11110 & 11010 & 00101 \\ 00001 & 01000 & 11011 & 00111 & 01100 & 11111 & 00000 & 01000 \\ 10111 & 10100 & 11010 & 00101 & 01110 & 01001 & 01101 & 00101 \end{array} \qquad [7.8]$$

Before calculating the cumulants, the independence of the bits is checked to confirm the linearity property of equation [7.6]. Considering bit 1 and bit 2, we obtain:

$$\begin{aligned} E[x_1]\cdot E[x_2] &= \left(\sum_{i=0,1} i\Pr[x_1=i]\right)\cdot\left(\sum_{i=0,1} i\Pr[x_2=i]\right) \\ &= (\Pr(x_1=1))\cdot(\Pr(x_2=1)) = (9/24)\cdot(13/24) = 0.204 \end{aligned} \qquad [7.9]$$

and:

$$\begin{aligned} E[x_1\cdot x_2] &= \sum_{i,j=0,1} i\cdot j\cdot \Pr[x_1=i, x_2=j] \\ &= \Pr(x_1=1, x_2=1) = 5/24 = 0.208 \end{aligned} \qquad [7.10]$$

We see that:

$$E[x_1\cdot x_2] \approx E[x_1].E[x_2] \qquad [7.11]$$

which is consistent with the fact that bit 1 and bit 2 are independent. Similarly, we can numerically confirm that all of the other bits in the population are independent, which confirms that the cumulants have the linearity property shown in equation [7.6] in the initial population.

The population distribution over the set of possible fitness values is obtained as:

$$\begin{aligned} &\Pr[X=0]=0.041,\quad \Pr[X=1]=0.167,\quad \Pr[X=2]=0.333 \\ &\Pr[X=3]=0.292,\quad \Pr[X=4]=0.125,\quad \Pr[X=5]=0.041 \end{aligned} \qquad [7.12]$$

Based on equation [7.2], the characteristic function of the probability distribution and its Taylor series are:

$$\begin{aligned}\varphi(z) &= 0.041 + 0.167e^{z} + 0.333e^{2z} + 0.292e^{3z} + 0.125e^{4z} + 0.041e^{5z} \\ &= 1 + \frac{2.414}{1!}z + \frac{7.152}{2!}z^2 + \frac{23.840}{3!}z^3 + \frac{86.772}{4!}z^4 + \cdots\end{aligned} \qquad [7.13]$$

Based on equation [7.5], the cumulants of the probability distribution are obtained as:

$$\begin{aligned}\kappa_1 &= \omega_1 = 2.414 \\ \kappa_2 &= \omega_2 - \omega_1^2 = 7.152 - (2.414)^2 = 1.324 \\ \kappa_3 &= \omega_3 - 3\omega_1\omega_2 + 2\omega_1^3 = 23.840 - 3\times 2.414 \times 7.152 + 2\times(2.414)^3 = 0.180\end{aligned} \qquad [7.14]$$

We find that the mean is 2.414 and the variance is 1.324. This distribution is approximately Gaussian as shown in Figure 7.1. From this example, it can be seen that cumulants can approximate the population distribution by truncating its series expansion.

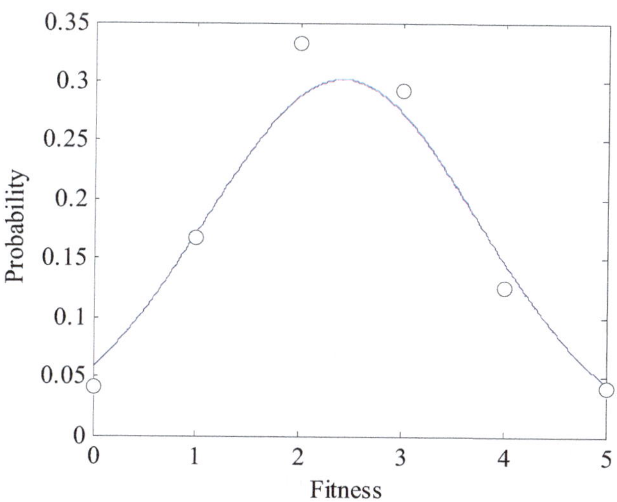

Figure 7.1. *Probability distribution of the fitness X. The solid line denotes a Gaussian distribution, and the dots denote the simulated probabilities of the population*

7.2. Statistical mechanics model of BBO

This section derives a statistical mechanics model for the basic BBO algorithm [MA 16a]. Before the formal derivation, we make some clarifications in the statistical mechanics model development.

First, we assume that the immigration rate λ and the emigration rate μ are independent of the population distribution. That is, the best fitness is the best fitness for the problem, not the best fitness in the current population. Similarly, the worst fitness is the worst fitness for the problem. Alternatives to this assumption are commonly used in practice, but this assumption is needed for the statistical mechanics model development of BBO.

Second, we assume that each bit in a given solution is independent because migration to each bit is independent of migration to the other bits, as shown in Figure 3.5 in Chapter 3. This is a simplistic assumption. Because of selection, BBO will introduce correlations between bits and will not remain independent for many generations. Our hypothesis here is that the bits will remain mostly independent, at least during the early generations. Similarly, mutation of each solution bit is independent, which helps maintain the independence of each bit in a given solution. The purpose of this assumption is to ensure that BBO migration and mutation satisfy Theorem 7.1.

7.2.1. *Migration*

The cumulants of a BBO fitness probability distribution provide certain average properties. The cumulants describe the fitness as a whole each generation. The cumulant dynamics tell us the trajectory of the distribution of the population fitness during the evolutionary process. In the following text, the equations of the dynamic evolution of BBO are derived based on the cumulants of the fitness distribution, and the specific one-max problem described in equation [7.7] is used to study the properties of BBO.

Consider a single bit position x_j, suppose that it has a randomly distributed population, and consider the effect of migration on the population values in this bit position. There are two ways in which a bit can be equal to 1 after an immigration trial (that is, after one iteration of the "for each decision variable s" loop in Figure 3.5). First, it can immigrate 1 from another solution in the population (that is, another solution emigrates a 1 to the bit). Second, it can be 1 before the immigration trial, and not immigrate. So the expected value of this bit after an immigration trial is computed as:

$$\begin{aligned} E\left[x_j\right] &= \sum_{i=0,1} i \Pr\left[x_j = i\right] = \Pr\left(x_j = 1\right) \\ &= \Pr\left(x_j = 1 \mid \text{immigration}\right)\Pr\left(\text{immigration}\right) \\ &\quad + \Pr(x_j = 1 \mid \text{no immigration})\Pr(\text{no immigration}) \\ &= \Pr(\text{emigrating bit} = 1)\Pr(\text{immigration}) \\ &\quad + \Pr(x_j = 1 \mid \text{no immigration})\Pr(\text{no immigration}) \end{aligned} \qquad [7.15]$$

We first consider the first probability on the right side of equation [7.15], which is the probability that an emigrating bit is 1. This probability is proportional to the sum of the emigration rates of all solutions whose jth bit is equal to 1:

$$\Pr(\text{emigrating bit} = 1) = \frac{\sum_{k \in \rho_j(1)} \mu_k}{\sum_{k=1}^{N} \mu_k} \qquad [7.16]$$

where $\rho_j(1)$ = {population indices of solutions whose jth bit is equal to 1} and N denotes the size of the population.

Next, we consider the third probability on the right side of equation [7.15]. If immigration does not occur, the probability that the value of bit j is 1 after an immigration trial is equal to the probability that the value of bit j is 1 before the immigration trial, which can be written as:

$$\Pr(x_j = 1 \mid \text{no immigration}) = \Pr(x'_j = 1) \qquad [7.17]$$

where x'_j denotes the value of bit j before an immigration trial.

The probability that the value of bit j is 1 before the immigration trial is denoted as:

$$\Pr(x'_j = 1) = E\left[x'_j\right] \qquad [7.18]$$

Note that the mean value of bit j before the immigration trial is:

$$E\left[x'_j\right] = \frac{\left|\rho_j(1)\right|}{N} \qquad [7.19]$$

where $|\rho_j(1)|$ denotes the number of solutions in the population whose jth bit is equal to 1. Combining equation [7.19] with equation [7.5], the first cumulant of the jth bit before the immigration trial is:

$$\kappa_1^{(j)} = \omega_1^{(j)} = \frac{|\rho_j(1)|}{N} \quad [7.20]$$

Next, we consider the probabilities Pr(immigration) and Pr(no immigration) in equation [7.15], which are the average values of the immigration curve; that is, they are independent of any information about the fitness value of any solution. Assuming that the migration curve is linear, as shown as Figure 3.1 in Chapter 3, the emigration rate μ_k is normalized as:

$$\mu_k = \frac{\text{fitness}}{n+1} = \frac{\text{number of 1 bits}}{n+1} \quad [7.21]$$

So the probability Pr(immigration) can be written as:

$$\begin{aligned} \Pr(\text{immigration}) &= 1 - \Pr(\text{emigration}) \\ &= 1 - \mu = 1 - E[\mu_k] = 1 - \frac{E[\text{number of 1 bits}]}{n+1} \end{aligned} \quad [7.22]$$

Furthermore, the average number of 1 bits in any given solution is $E[X]$, which denotes the mean of the fitness, and which is equal to the cumulant κ_1. The probabilities Pr(immigration) and Pr(no immigration) can thus be written as:

$$\begin{aligned} \Pr(\text{immigration}) &= 1 - \frac{\kappa_1}{n+1} \\ \Pr(\text{no immigration}) &= 1 - \Pr(\text{immigration}) = \frac{\kappa_1}{n+1} \end{aligned} \quad [7.23]$$

Note that if we used an immigration curve other than the one shown in Figure 3.1, we would need to re-compute the average of the immigration curve to find Pr(immigration) and Pr(no immigration).

We substitute equations [7.16], [7.17] and [7.23] into [7.15] to obtain the expected bit value after an immigration trial as:

$$E\left[x_j\right]=\frac{\sum_{k\in\rho_j(1)}\mu_k}{\sum_{k=1}^{N}\mu_k}\left(1-\frac{\kappa_1}{n+1}\right)+\Pr(x_j'=1)\left(\frac{\kappa_1}{n+1}\right) \qquad [7.24]$$

Note that we have normalized the emigration rate so that it has a minimum of 0 and a maximum of $n/(n+1)$. Other normalizations are possible as long as the emigration rate is between 0 and 1, but this is a typical approach. We use equation [7.21] to obtain:

$$\begin{aligned}\sum_{k=1}^{N}\mu_k &= \sum_{k=1}^{N}\frac{\text{number of 1 bits in the }k\text{th solution}}{n+1}\\ &= \frac{\text{number of 1 bits in entire population}}{n+1}\\ &= \frac{N\cdot E\left[X\right]}{n+1}=\frac{N\kappa_1}{n+1}\end{aligned} \qquad [7.25]$$

Now we note that:

$$\begin{aligned}\sum_{k\in\rho_j(1)}\mu_k &= \sum_{k\in\rho_j(1)}\frac{\text{number of 1 bits in the }k\text{th solution whose }j\text{th bit is 1}}{n+1}\\ &= \frac{1}{n+1}\left|\rho_j(1)\right|\cdot E\left[\text{number of 1 bits in any solution whose }j\text{th bit is 1}\right]\end{aligned} \qquad [7.26]$$

Further note that:

$$\begin{aligned}&E\left[\text{number of 1 bits in any solution whose }j\text{th bit is 1}\right]\\ &\quad=1+E\left[\text{number of 1 bits in remaining }(n-1)\text{ bits except the }j\text{th bit}\right]\\ &\quad=1+\sum_{\substack{k=1\\k\neq j}}^{n}\frac{\left|\rho_k(1)\right|}{N}=1+\sum_{k=1}^{n}\frac{\left|\rho_k(1)\right|}{N}-\frac{\left|\rho_j(1)\right|}{N}\\ &\quad=1+\sum_{k=1}^{n}\kappa_1^{(k)}-\kappa_1^{(j)}\end{aligned} \qquad [7.27]$$

Assuming that the cumulants have the linearity property described in equation [7.6], the cumulant of the binary string is equal to the sum of cumulants of each bit of the binary string:

$$\kappa_1 = \sum_{k=1}^{n} \kappa_1^{(k)} \qquad [7.28]$$

We combine equations [7.20], [7.26], [7.27] and [7.28] to obtain:

$$\begin{aligned}\sum_{k \in \rho_j(1)} \mu_k &= \frac{1}{n+1} |\rho_j(1)| \left(1 + \kappa_1 - \kappa_1^{(j)}\right) \\ &= \frac{N\kappa_1^{(j)}}{n+1} \left(1 + \kappa_1 - \kappa_1^{(j)}\right)\end{aligned} \qquad [7.29]$$

The expected value of bit x_j after an immigration trial can be written as:

$$\begin{aligned}\kappa_{1,t+1}^{(j)} = E\left[x_j\right] &= \frac{\sum_{k \in \rho_j(1)} \mu_k}{\sum_{k=1}^{N} \mu_k} \left(1 - \frac{\kappa_{1,t}}{n+1}\right) + \Pr(x_j' = 1) \left(\frac{\kappa_{1,t}}{n+1}\right) \\ &= \left(\frac{\kappa_{1,t}^{(j)}}{\kappa_{1,t}} \left(1 + \kappa_{1,t} - \kappa_{1,t}^{(j)}\right) \left(1 - \frac{\kappa_{1,t}}{n+1}\right) + \kappa_{1,t}^{(j)} \left(\frac{\kappa_{1,t}}{n+1}\right) \right) \\ &= \kappa_{1,t}^{(j)} + \left(1 - \frac{\kappa_{1,t}}{n+1}\right) \frac{\kappa_{1,t}^{(j)}}{\kappa_{1,t}} - \left(1 - \frac{\kappa_{1,t}}{n+1}\right) \frac{\left(\kappa_{1,t}^{(j)}\right)^2}{\kappa_{1,t}}\end{aligned} \qquad [7.30]$$

Note that the subscripts $t + 1$ and t denote quantities after and before migration, respectively. For the other cumulants of bit x_j, we note that bits can only take the values 0 and 1, so:

$$x_j^k = x_j \qquad [7.31]$$

for all values of k. So the moments of all orders are equal and are given as:

$$\omega_k^{(j)} = E\left[x_j^k\right] = E\left[x_j\right] \qquad [7.32]$$

Based on the relationships of moments and cumulants described in equation [7.5], we obtain the equations of cumulants $\kappa_2^{(j)}$ and $\kappa_3^{(j)}$ as follows:

$$\kappa_{2,t+1}^{(j)} = \kappa_{2,t}^{(j)} - \left(1-\frac{\kappa_{1,t}}{n+1}\right)^2 \left(\frac{\kappa_{2,t}^{(j)}}{\kappa_{1,t}}\right)^2 + \left(1-\frac{\kappa_{1,t}}{n+1}\right)\frac{\kappa_{3,t}^{(j)}}{2\kappa_{1,t}}$$
$$\kappa_{3,t+1}^{(j)} = \kappa_{3,t}^{(j)} + 2\left(1-\frac{\kappa_{1,t}}{n+1}\right)^3 \left(\frac{\kappa_{2,t}^{(j)}}{\kappa_{1,t}}\right)^3 - 3\left(1-\frac{\kappa_{1,t}}{n+1}\right)^2 \frac{\kappa_{2,t}^{(j)}\kappa_{3,t}^{(j)}}{\left(\kappa_{1,t}\right)^2} + \left(1-\frac{\kappa_{1,t}}{n+1}\right)\frac{\kappa_{4,t}^{(j)}}{\kappa_{1,t}} \quad [7.33]$$

The cumulant approximations of the population after migration are:

$$\kappa_{1,t+1} = \sum_{j=1}^{n} \kappa_{1,t+1}^{(j)} = \kappa_{1,t} + \left(1-\frac{\kappa_{1,t}}{n+1}\right)\frac{\kappa_{2,t}}{\kappa_{1,t}}$$
$$\kappa_{2,t+1} = \sum_{j=1}^{n} \kappa_{2,t+1}^{(j)} \approx \kappa_{2,t} - \left(1-\frac{\kappa_{1,t}}{n+1}\right)^2 \left(\frac{\kappa_{2,t}}{\kappa_{1,t}}\right)^2 + \left(1-\frac{\kappa_{1,t}}{n+1}\right)\frac{\kappa_{3,t}}{\kappa_{1,t}}$$
$$\kappa_{3,t+1} = \sum_{j=1}^{n} \kappa_{3,t+1}^{(j)} \approx \kappa_{3,t} + 2\left(1-\frac{\kappa_{1,t}}{n+1}\right)^3 \left(\frac{\kappa_{2,t}}{\kappa_{1,t}}\right)^3 - 3\left(1-\frac{\kappa_{1,t}}{n+1}\right)^2 \frac{\kappa_{2,t}\kappa_{3,t}}{\left(\kappa_{1,t}\right)^2} + \left(1-\frac{\kappa_{1,t}}{n+1}\right)\frac{\kappa_{4,t}}{\kappa_{1,t}} \quad [7.34]$$

Equations [7.33] and [7.34] are derived in detail in [MA 16a]; note that $\kappa_{2,t+1}$ and $\kappa_{3,t+1}$ are approximations. Cumulants κ_1 and κ_2 denote the mean and the variance of the population fitness distribution, and cumulant κ_3 denotes the skewness, which measures the degree of asymmetry of the distribution.

Cumulant κ_4 and other higher-order cumulants are not further discussed here. κ_4 denotes the kurtosis, which measures the peakedness of the distribution. EA researchers are typically more interested in the mean and the variance of fitness distributions, namely κ_1 and κ_2, and κ_4 only affects κ_3 based on equation [7.34]. In fact, it is apparent from equation [7.34] that the influence of κ_4 on κ_3 is relatively small. So we approximate κ_4 and other higher-order cumulants as zero, which gives us an approximation of the migration dynamics using just three variables.

7.2.2. *Mutation*

The derivation above considers only migration. Mutation is now added to the statistical mechanics model. BBO mutation is the operator that modifies a decision variable of a candidate solution with a small probability, as in other EAs. The view of mutation as a statistical mechanics model has been explained in [REE 03], and the same theory can apply to BBO.

For the one-max problem described in equation [7.7], suppose that we have a randomly distributed population. Consider the effect of mutation on the values in a single bit position. Similar to migration, there are two ways in which a bit can be equal to 1 after mutation. First, it can start as 0 and then mutate. Second, it can start as 1 and not mutate. If we suppose that the mutation probability is u, the expected value of this bit after mutation is computed as:

$$\begin{aligned} E\left[x_j\right] &= \sum_{i=0,1} i \Pr\left[x_j = i\right] = \Pr\left(x_j = 1\right) \\ &= \Pr\left(x_j = 1 \middle| \text{mutation}\right)\Pr\left(\text{mutation}\right) \\ &\quad + \Pr(x_j = 1 \mid \text{no mutation})\Pr(\text{no mutation}) \\ &= u \Pr\left[x'_j = 0\right] + (1-u)\Pr\left[x'_j = 1\right] \\ &= u(1-\Pr\left[x'_j = 1\right]) + (1-u)\Pr\left[x'_j = 1\right] \\ &= u + (1-2u)\Pr\left[x'_j = 1\right] \end{aligned} \qquad [7.35]$$

where x'_j denotes the value of bit j before mutation.

Substituting equation [7.18] into equation [7.35], we obtain:

$$E\left[x_j\right] = u + (1-2u)E\left[x'_j\right] \qquad [7.36]$$

Equation [7.32] indicates that the moments of all orders are equal, so we have:

$$\omega_{k,t+1}^{(j)} = u + (1-2u)\omega_{k,t}^{(j)} \qquad [7.37]$$

We now need to convert from moments to cumulants. Based on the relationship of moments and cumulants described in equation [7.5], we directly obtain the first cumulant $\kappa_1^{(j)}$ for one bit as:

$$\kappa_{1,t+1}^{(j)} = u + (1-2u)\kappa_{1,t}^{(j)} \qquad [7.38]$$

Thus, the first cumulant of the population after mutation is:

$$\begin{aligned}\kappa_{1,t+1} &= \sum_{j=1}^{n} \kappa_{1,t+1}^{(j)} \\ &= \sum_{j=1}^{n} \left(u + (1-2u)\kappa_{1,t}^{(j)}\right) \\ &= un + (1-2u)\kappa_{1,t}\end{aligned} \quad [7.39]$$

Next, we calculate the second cumulant $\kappa_2^{(j)}$ as:

$$\begin{aligned}\kappa_{2,t+1}^{(j)} &= \left(u + (1-2u)\kappa_{1,t}^{(j)}\right) - \left[u + (1-2u)\kappa_{1,t}^{(j)}\right]^2 \\ &= u(1-u) + (1-2u)^2\left(\kappa_{1,t}^{(j)} - \left(\kappa_{1,t}^{(j)}\right)^2\right) \\ &= u(1-u) + (1-2u)^2 \kappa_{2,t}^{(j)}\end{aligned} \quad [7.40]$$

Summing over all the bits gives us the second cumulant of the population after mutation as:

$$\kappa_{2,t+1} = u(1-u)n + (1-2u)^2 \kappa_{2,t} \quad [7.41]$$

Similarly, we derive the third cumulant of the population after mutation as:

$$\kappa_{3,t+1} = u(1-u)(1-2u)(n-2\kappa_{1,t}) + (1-2u)^3 \kappa_{3,t} \quad [7.42]$$

Finally, we combine equations [7.39], [7.41] and [7.42] with equation [7.34] for BBO migration to obtain the following model for BBO with both migration and mutation:

$$\begin{aligned}\kappa_{1,t+1} &= un + (1-2u)\left(\kappa_{1,t} + \left(1 - \frac{\kappa_{1,t}}{n+1}\right)\frac{\kappa_{2,t}}{\kappa_{1,t}}\right) \\ \kappa_{2,t+1} &= u(1-u)n + (1-2u)^2\left(\kappa_{2,t} - \left(1-\frac{\kappa_{1,t}}{n+1}\right)^2\left(\frac{\kappa_{2,t}}{\kappa_{1,t}}\right)^2 + \left(1-\frac{\kappa_{1,t}}{n+1}\right)\frac{\kappa_{3,t}}{\kappa_{1,t}}\right) \\ \kappa_{3,t+1} &= u(1-u)(1-2u)\left(n - 2\left(\kappa_{1,t} + \left(1-\frac{\kappa_{1,t}}{n+1}\right)\frac{\kappa_{2,t}}{\kappa_{1,t}}\right)\right) \\ &\quad + (1-2u)^3\left(\kappa_{3,t} + 2\left(1-\frac{\kappa_{1,t}}{n+1}\right)^3\left(\frac{\kappa_{2,t}}{\kappa_{1,t}}\right)^3 - 3\left(1-\frac{\kappa_{1,t}}{n+1}\right)^2\frac{\kappa_{2,t}\kappa_{3,t}}{\left(\kappa_{1,t}\right)^2} + \left(1-\frac{\kappa_{1,t}}{n+1}\right)\frac{\kappa_{4,t}}{\kappa_{1,t}}\right)\end{aligned} \quad [7.43]$$

The above equations give the statistical properties of BBO population fitness values after migration and mutation in terms of those at the previous generation. We can iterate these equations to obtain predictions for the evolution of BBO populations.

EXAMPLE 7.2.–

This example simulates the 100-bit one-max problem to verify the statistical mechanics model of BBO. The statistical approximations of BBO are obtained by iterating equation [7.34] if considering only migration, and by iterating equation [7.43] if considering both migration and mutation. We set $\kappa_{4,t} = 0$ in the iterations of equations [7.34] and [7.43]. The simulation parameters of BBO are as follows: population size of 50, maximum immigration rate and maximum emigration rate of 1, and 200 Monte Carlo simulations. Figures 7.2–7.4 show comparisons between theoretical (statistical approximation) and simulated BBO results with mutation rate $u = 0.01$.

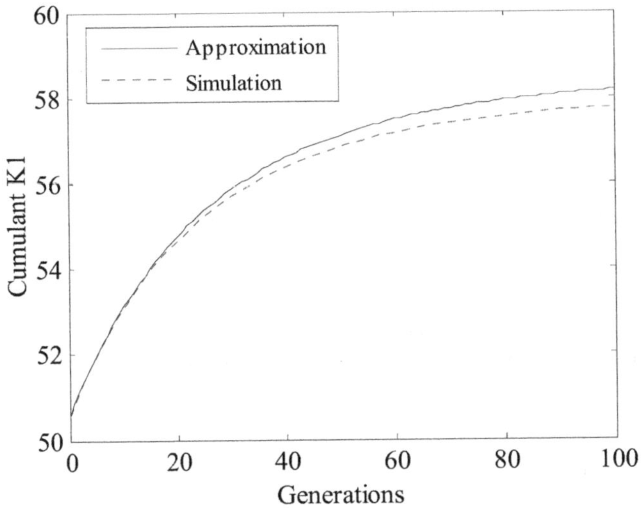

Figure 7.2. *BBO approximation and simulation results of cumulant κ_1 for the 100-bit one-max problem, where the solid line denotes approximation values and the dashed line denotes simulation values*

Several things are notable about Figures 7.2–7.4. First, the approximation results of BBO show that as the generation count increases, the first cumulant κ_1 (mean) increases and the second cumulant κ_2 (variance) decreases with the generation count. This is consistent with the expected behavior of BBO: as the generation count increases, the mean fitness of the population increases and the population becomes more uniform.

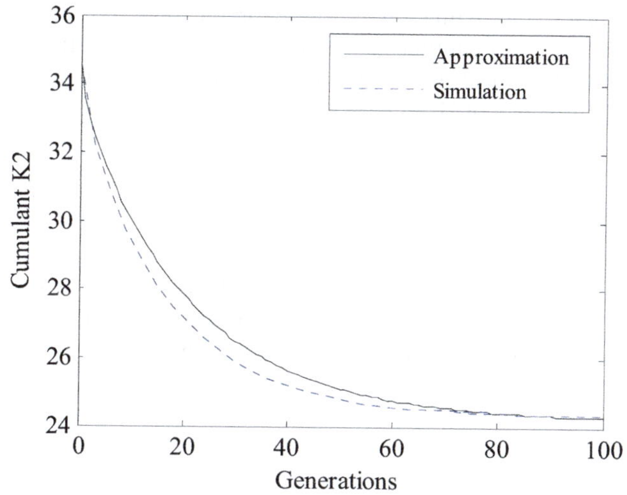

Figure 7.3. *BBO approximation and simulation results of cumulant κ_2 for the 100-bit one-max problem, where the solid line denotes approximation values and the dashed line denotes simulation values*

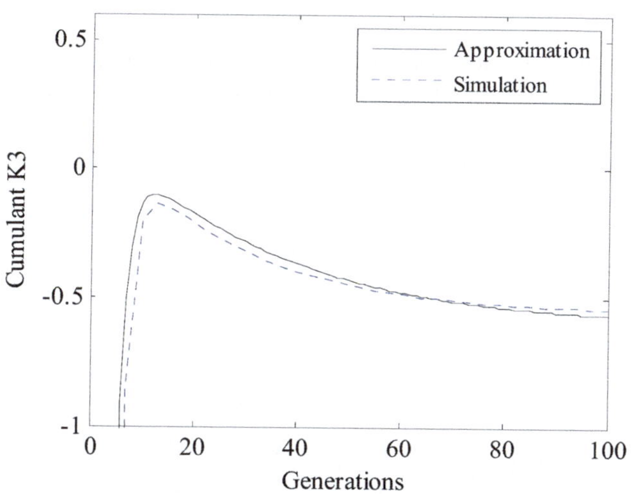

Figure 7.4. *BBO approximation and simulation results of cumulant κ_3 for the 100-bit one-max problem, where the solid line denotes approximation values and the dashed line denotes simulation values*

Second, the cumulant approximations and the simulation results match well. This indicates that the statistical mechanics model of BBO is reasonable and that the statistical mechanics approximation is valid for the one-max problem. This fact allows us to make quantitative conclusions from statistical mechanics approximations and to use those approximations to obtain valid conclusions regarding the performance of BBO with various tuning parameters.

EXAMPLE 7.3.–

In this example, we compare the statistical mechanics model of BBO with that of a simple GA using proportional selection and mutation [REE 03, PRÜ 94]. The statistical mechanics model of a GA with selection only is given as:

$$\begin{aligned}
\kappa_{1,t+1} &= \kappa_{1,t} + \frac{\kappa_{2,t}}{\kappa_{1,t}} \\
\kappa_{2,t+1} &= \kappa_{2,t} - \left(\frac{\kappa_{2,t}}{\kappa_{1,t}}\right)^2 + \frac{\kappa_{3,t}}{\kappa_{1,,t}} \\
\kappa_{3,t+1} &= \kappa_{3,t} + 2\left(\frac{\kappa_{2,t}}{\kappa_{1,t}}\right)^3 - 3\frac{\kappa_{2,t}\kappa_{3,t}}{\left(\kappa_{1,t}\right)^2} + \frac{\kappa_{4,t}}{\kappa_{1,t}}
\end{aligned} \tag{7.44}$$

Recall that the mutation model is given in equation [7.43]. So the three-parameter approximation equations for a simple GA with selection and mutation are:

$$\begin{aligned}
\kappa_{1,t+1} &= un + (1-2u)\left(\kappa_{1,t} + \frac{\kappa_{2,t}}{\kappa_{1,t}}\right) \\
\kappa_{2,t+1} &= u(1-u)n + (1-2u)^2\left(\kappa_{2,t} - \left(\frac{\kappa_{2,t}}{\kappa_{1,t}}\right)^2 + \frac{\kappa_{3,t}}{\kappa_{1,,t}}\right) \\
\kappa_{3,t+1} &= u(1-u)(1-2u)\left(n - 2\left(\kappa_{1,t} + \frac{\kappa_{2,t}}{\kappa_{1,t}}\right)\right) \\
&\quad + (1-2u)^3\left(\kappa_{3,t} + 2\left(\frac{\kappa_{2,t}}{\kappa_{1,t}}\right)^3 - 3\frac{\kappa_{2,t}\kappa_{3,t}}{\left(\kappa_{1,t}\right)^2} + \frac{\kappa_{4,t}}{\kappa_{1,t}}\right)
\end{aligned} \tag{7.45}$$

We see that the approximation equations of BBO in equation [7.43] are similar to those of the GA in equation [7.45], and the difference is the additional coefficient $1-\kappa_1/n+1$ in some of the terms of the BBO model. As the problem size n becomes large, the BBO approximation approaches the GA approximation. These equations can be used to compare the statistical mechanics models of BBO and GA for the 100-bit one-max problem with mutation rate $u = 0.01$, as shown in Figures 7.5–7.7.

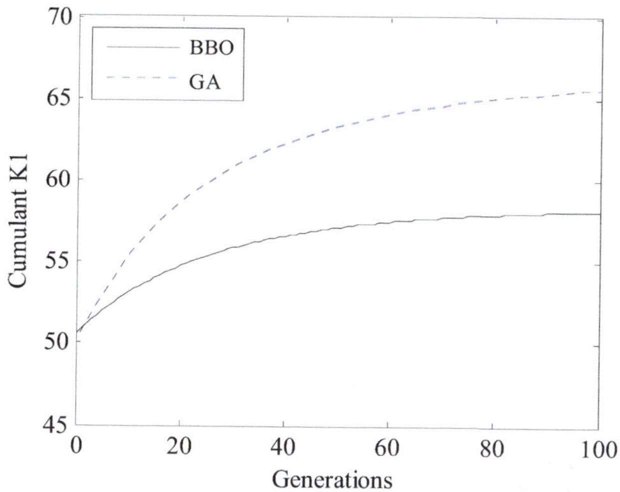

Figure 7.5. *Comparison of the approximation values of cumulant κ_1 between BBO and GA for the 100-bit one-max problem, where the solid line denotes BBO and the dashed line denotes GA*

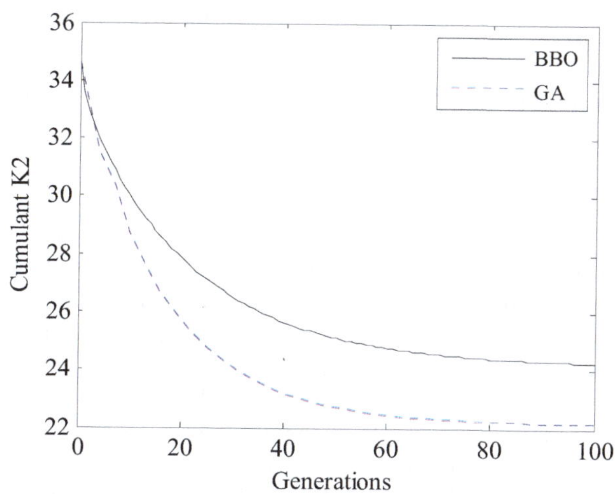

Figure 7.6. *Comparison of the approximation values of cumulant κ_2 between BBO and GA for the 100-bit one-max problem, where the solid line denotes BBO and the dashed line denotes GA*

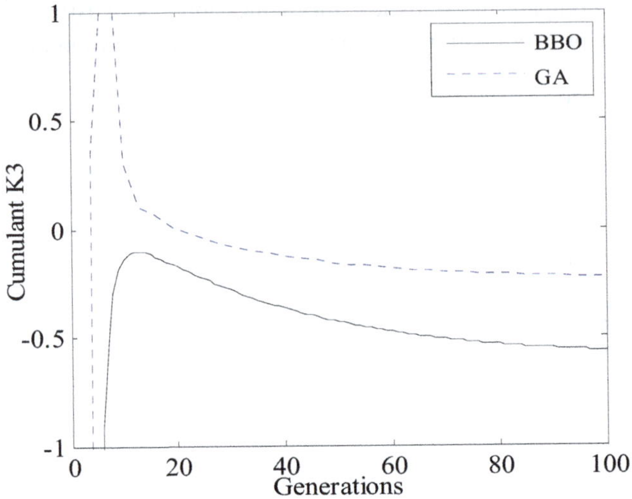

Figure 7.7. *Comparison of the approximation values of cumulant κ_3 between BBO and GA for the 100-bit one-max problem, where the solid line denotes BBO and the dashed line denotes GA*

From Figures 7.5–7.7, several things are notable about the approximation comparisons between BBO and GA. First, the trends of the cumulants are similar for BBO and GA. As the generation count increases, the mean (the first cumulant κ_1) increases and the variance (the second cumulant κ_2) decreases. The only qualitative difference we see is that the skewness (the third cumulant κ_3) of the GA suddenly changes slope during the first few generations, while the skewness of BBO changes more smoothly. These comparisons indicate that the expected behavior of BBO is similar to that of GA. This point has already been implied by equations [7.43] and [7.45]. It is not too surprising that BBO is similar to GA, because many EAs can be expressed in terms of each other. Chapters 5 and 6 compared the optimization performance of BBO and GA and have shown similar results. On the other hand, the figures show that differences do exist between BBO and GA. This indicates that BBO has its own particular characteristics that are distinct from GA. A BBO solution uses its own fitness before deciding how likely it is to accept features from other candidate solutions. This simple and intuitive idea does not have an analogy in genetics, but is motivated by biogeography. This implies that it is useful to retain the distinction between BBO and GA, and that the biogeographical foundation of BBO can motivate many algorithmic extensions.

Second, Figure 7.5 shows that the mean fitness of BBO is smaller than that of GA, and Figure 7.6 shows that the fitness variance of BBO is larger than that of GA.

This indicates that GA might have better convergence properties than BBO for the one-max problem.

Finally, we should be wary of drawing general conclusions about BBO versus GA comparisons from these results because the BBO and GA algorithms used here are basic versions, and the results in this experiment are limited to the 100-bit one-max problem.

7.3. Further discussion

Next, we will discuss some questions related to the statistical mechanics model of BBO which affect the approximation performance.

7.3.1. *Finite population effects*

We have obtained the statistical mechanics model of BBO based on an infinite population, and have used the average properties of the population to determine the dynamic trajectory of the population as a whole. However, in actual simulations of BBO, we have a finite population and can no longer ignore the effect of population size. In statistical mechanics, if a balloon contained only two molecules, the chance of both of them being on one side of the balloon is quite high. Similarly, we must consider fluctuations caused by the finite population, which we view as sampling effects. In BBO, we effectively sample the fitness N times, where N is the population size. To account for finite population effects, we need to know the relationships of mean, variance and higher-order cumulants of the probability distribution, along with their expected values from this distribution.

In general, the expected values of the cumulants of a finite sample are not the same as the cumulants of the distribution from which the sample is taken, and the size of the sample affects these values. In [REE 03], the expected values of mean, variance and the third cumulant are derived as:

$$\begin{aligned} E[\kappa_1] &= \kappa_1 \\ E[\kappa_2] &= \left(1-\frac{1}{N}\right)\kappa_2 \\ E[\kappa_3] &= \left(1-\frac{1}{N}\right)\left(1-\frac{2}{N}\right)\kappa_3 \end{aligned} \qquad [7.46]$$

where the terms on the left-hand sides represent the expected values of the cumulants of a sample of size N. Equation [7.46] shows that only the mean is unaffected by sample size. Note that as $N \to \infty$, the expected values of these cumulants are equal to the cumulants of the distribution. We can directly modify equation [7.43] to account for population size by multiplying by the appropriate factors.

7.3.2. Separable fitness functions

We have been treating the fitness of a population as a random variable with some probability distribution. We have estimated the cumulants of the distribution for the one-max problem, and have studied the way they change over time. One notable characteristic of the one-max problem is that its variables are separable. We can extend the derivation of the statistical mechanics model, which was limited to the one-max problem, to other functions if we assume that each bit in a given solution is separable. In Ma *et al.* [MA 16a, MA 16b, MA 16c], the derivation of a statistical mechanics model of BBO for a general separable function is given. Here we provide the approximation equations. A general separable function is written as:

$$f(x) = x_1 \qquad [7.47]$$

where:

$$x = [x_1 \quad \cdots \quad x_n] \qquad [7.48]$$

$x_j \in I = \{0,1\}$ is bit j of a binary string, and n is the length of this binary string as well as the number of features in each candidate solution.

The first two cumulants of the statistical mechanics model of BBO for a general separable function are:

$$\begin{aligned} \kappa_{1,t+1} &= \kappa_{1,t} + \left(1 - \frac{\kappa_{1,t}}{n+1}\right)\frac{\kappa_{2,t}}{\kappa_{1,t}} \\ \kappa_{2,t+1} &\approx \kappa_{2,t} - \left(1 - \frac{\kappa_{1,t}}{n+1}\right)^2 \left(\frac{\kappa_{2,t}}{\kappa_{1,t}}\right)^2 + \left(1 - \frac{\kappa_{1,t}}{n+1}\right)\frac{\kappa_{3,t}}{\kappa_{1,t}} \end{aligned} \qquad [7.49]$$

Comparing equation [7.49] with the first two equations of [7.34], we see that the cumulants $\kappa_{1,t+1}$ and $\kappa_{2,t+1}$ are the same. That is, the dynamics of the mean and the variance are identical for all separable functions. Note that higher-order cumulants quantify the non-Gaussian content of the distribution. These results indicate that approximations of BBO dynamics based on statistical mechanics can model how BBO works not only for the simple one-max problem, but also for general but separable fitness functions. Furthermore, for many complex, non-separable functions, if we can find a method to approximate them as separable functions, we could successfully derive the statistical mechanics model of BBO for these functions. From this point of view, statistical mechanics terminology in evolutionary computation is a methodology that can approximate BBO behavior for functions other than separable functions. We can also apply this methodology to the analysis of other EAs.

7.4. Conclusions

In this chapter, we outlined a computational model for BBO based on ideas from statistical mechanics, and studied the dynamics of BBO for the one-max problem and other separable functions. The statistical mechanics model describes the evolution of the statistical properties of the BBO population fitness. Note that the statistical mechanics model is completely different from the Markov model described earlier in this book. The Markov model gives the probability of arriving at any given population as the generation count approaches infinity. But the Markov model may be of little practical use for real-world problems because transition probabilities are not known in real applications. The statistical mechanics model gives the average behavior of population fitness, which can lead to a better understanding of the evolution of BBO populations in realistic, high-dimensional optimization problems.

We have shown three mathematical models of the dynamics of BBO in the past three chapters: the Markov model, the dynamic systems model and the statistical mechanics model. These models can provide insight into which problems are suitable for BBO, and why, from different viewpoints. The efficiency of BBO depends on the problem representation and the BBO tuning parameters. These parameters include the population size, the immigration curve shape, the emigration curve shape and the mutation rate. For many problems, BBO can be effective when a good representation is chosen and the parameters are set to appropriate values. When poor choices are made, BBO can perform similarly to a random search. The three mathematical models, each of which are functions of BBO tuning parameters, can successfully predict the improvement in fitness from one generation to the next, and can assist in finding optimal values of the BBO tuning parameters. For example, consider a problem with very expensive fitness function evaluations; we may even need to do physical experiments to evaluate fitness. However, we could also use one of our three models to tune BBO parameters during early generations to quickly improve fitness, or to predict whether BBO will perform better than other EAs. These mathematical models of BBO could be useful for developing effective BBO modifications. More generally, models of BBO dynamics can be useful in producing insights into how the algorithm behaves and when it is likely to be effective.

8

BBO for Combinatorial Optimization

Up until this point, this book has emphasized the framework and mathematical theory of BBO algorithms. In this chapter, we begin to discuss the applications of BBO. This chapter uses BBO to solve combinatorial optimization problems, which comprise a subset of mathematical optimization. Combinatorial optimization has important applications in many fields, including artificial intelligence, machine learning, auction theory and software engineering. Combinatorial optimization can be thought of as finding the optimal object from among a finite set of candidate objects:

$$\min_{x} f(x) \quad \text{where } x \in \{x_1, x_2, \cdots, x_{N_x}\} \qquad [8.1]$$

where N_x is the cardinality of the search space. Theoretically, we can solve equation [8.1] by evaluating $f(x)$ for all N_x possible solutions. But it is not feasible to check every possible solution when the combinatorial problem has a large search space.

Some classic combinatorial optimization problems include the traveling salesman problem, the knapsack problem, the minimum spanning tree problem and many others. Figure 8.1 shows an example of a one-dimensional knapsack problem: which boxes should be chosen to maximize the amount of money while still keeping the overall weight less than or equal to 15 kg? A combinatorial optimization problem could consider both the weight and volume of the boxes.

Figure 8.1. *An example of a one-dimensional knapsack problem. For a color version of this figure, see www.iste.co.uk/ma-simon/evolutionary.zip*

Another combinatorial optimization example is the minimum spanning tree problem, which is a spanning tree of a connected, undirected graph. The goal here is to connect all the vertices with the minimum total edge weights. The minimum spanning tree problem is a common combinatorial optimization problem, as shown in Figure 8.2.

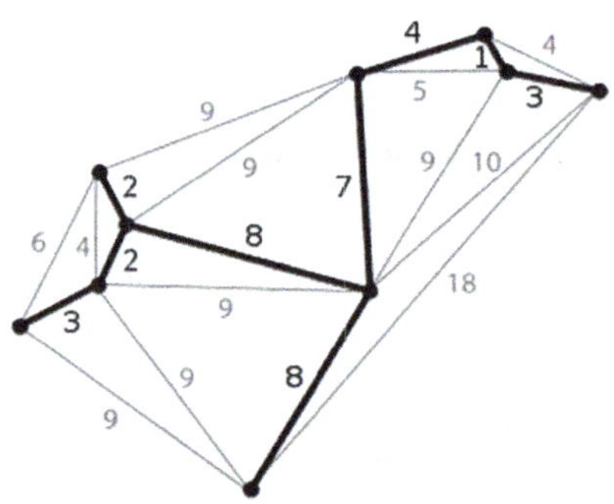

Figure 8.2. *A planar graph and its minimum spanning tree, where each edge is labeled with its weight, which is roughly proportional to its length*

Overview of the chapter

Most of this chapter uses BBO algorithms to solve the TSP, which is perhaps the most famous, applicable and widely studied combinatorial optimization problem. Section 8.1 gives an overview of the TSP, and section 8.2 discusses and presents the application of BBO for the solution of TSPs. Sections 8.3 and 8.4 discuss the graph coloring problem and the knapsack problem, which are two other popular combinatorial optimization problems.

8.1. Traveling salesman problem

The TSP can be described by the following question: given a list of cities and the distances between each pair of cities, what is the shortest possible route that visits each city exactly once and returns to the original city? It is a classic problem in combinatorial optimization that has intrigued mathematicians and computer scientists for years. Most important, it has many applications in science and engineering, including robotics, circuit board drilling, welding, manufacturing, transportation and many other areas.

Many people have studied this problem. The easiest and most expensive solution is to simply try all possibilities. The problem with this is that for n cities, there are $(n-1)!$ possible combinations. This means that for a 50-city TSP, there are over $49! = 6.1\times10^{62}$ combinations, which is not feasible to solve using exhaustive search. For a circuit board with tens of thousands of holes, it is required to determine the best order in which a laser will drill. An efficient solution to this problem can reduce production costs for the manufacturer.

In general, the TSP describes a salesman who must travel between n cities. The order in which he goes is something he does not care about, as long as he visits each city once during his trip, and finishes where he began. We assume that there is a known distance $D(i, j)$ between cities i and j for all $i\in[1, n]$ and $j\in[1, n]$, and that $D(i, j) = D(j, i)$. This is called the symmetric TSP because the distance from city i to city j is the same as the distance from city j to city i. Another possibility is $D(i, j) \neq D(j, i)$, which is called the asymmetric TSP.

In this chapter, we mainly discuss the symmetric TSP. An example of a valid tour in a 4-city TSP is as follows:

$$\text{Valid 4-city tour: } \quad 3 \rightarrow 2 \rightarrow 4 \rightarrow 1 \rightarrow 3 \qquad [8.2]$$

where the number denotes the city index and "→" denotes an edge or segment of a tour. Equation [8.2] consists of four edges, which indicates that an n-city tour includes n edges.

In the TSP, we try to minimize the total distance. Suppose that n cities in a TSP are listed in the order $x_1 \to x_2 \to \cdots \to x_n \to x_1$. Then, the total distance is:

$$D_T = D(x_n, x_1) + \sum_{i=1}^{n-1} D(x_i, x_{i+1}) \qquad [8.3]$$

Note that we use the term "distance" in a general sense. It might refer to physical distance, financial cost or any other quantity that we want to minimize in a combinatorial problem.

8.2. BBO for the TSP

This section discusses the application of BBO for solving the TSP [DU 13]. In order to use BBO to solve the TSP, we need to code candidate solutions in BBO differently than we have up to this point in the book. An element in a solution is a city, so each element in the solution contains no information by itself. It is only the order of elements that determines the goodness of a solution. In order to determine the distance of the tour, we need to know the order of all the cities in the tour. Since the original BBO algorithm is not designed for combinatorial problems, we need to modify BBO to apply it to the TSP.

8.2.1. *Population initialization*

Population initialization is usually the first step for BBO. For most problems, we do not have a good understanding of the effect of each independent variable. That means we cannot create an initial population based on our expertise, so we instead randomly create the initial population. However, randomization is an inefficient way to create an initial population. If we could initialize BBO intelligently rather than randomly, we could greatly increase our chances of finding a good solution.

One simple and common way to initialize a candidate TSP solution is with a nearest-neighbor strategy [COV 67]. The detailed procedure for an n-city TSP is described as follows:

1) Initialize $i = 1$, and randomly select a city $x_1 \in [1, n]$ as the starting city.

2) $x_{i+1} \leftarrow \arg\min_\sigma \{D(x_i,\sigma): \sigma \notin x_k \text{ for } k \in [1,i]\}$. That is, find the city that is closest to x_i that has not yet been assigned to an element of the tour, and assign it to x_{i+1}.

3) Increment i by one, and if $i = n$, terminate; otherwise, go to step 3.

At the end of this process, we have an open tour $x_1 \to x_2 \to \cdots \to x_n$ that gives us a reasonable guess for a good TSP solution. Note that here and in the below text, we use an open tour as a candidate solution for convenience. If we need a closed tour as in equation [8.2], we simply add the starting city to the end of the open tour.

EXAMPLE 8.1.–

Consider the following distance matrix:

$$D = \begin{bmatrix} \times & 3 & 2 & 9 & 3 \\ 3 & \times & 5 & 8 & 11 \\ 2 & 5 & \times & 4 & 6 \\ 9 & 8 & 4 & \times & 10 \\ 3 & 11 & 6 & 10 & \times \end{bmatrix} \qquad [8.4]$$

where D_{ij}, which we can also write as $D(i,j)$, represents the distance between city i and city j. If we start at city 1, the nearest-neighbor algorithm gives the tour $1 \to 3 \to 4 \to 2 \to 5$, which has a total cost of 25. If we start at city 2, the algorithm gives the tour $2 \to 1 \to 3 \to 4 \to 5$, which has a total cost of 19.

For a general $n \times n$ symmetric distance matrix D, since the distance between city i and city j is the same as that between city j and city i, there are $n(n-1)/2$ terms above the diagonal. Therefore, a symmetric n-city TSP has $n(n-1)/2$ unique edges.

To intelligently initialize BBO for solving the TSP, we can use nearest-neighbor initialization, as described above, for just one solution in the population, or for a few solutions in the population, or for the entire population. But if we initialize too many solutions this way, then we will probably obtain many duplicate solutions. We could also use a stochastic nearest-neighbor initialization algorithm. In this case, the probability of assigning a given city to x_{i+1} at each iteration would be inversely proportional to its distance from x_i. We could also take the nearest-neighbor algorithm to another level by performing a "nearest two-neighbor" algorithm. In this approach, given x_i, we would assign a city to x_{i+1} that results in the smallest combined distance $D(x_i, x_{i+1}) + D(x_{i+1}, \sigma)$, where σ is allowed to be equal to any city $\neq x_k$ for $k \leq i+1$.

8.2.2. *Migration in the TSP*

BBO migration is a method for combining or modifying features based on parent solutions to create offspring, or new solutions. It is the most important component in BBO. Four types of migration methods for solving the TSP are introduced here: order migration, cycle migration, inver-over migration and matrix migration. The first three migrations use path representation, and the last migration uses matrix representation.

Path representation is the most natural way of representing a TSP tour. In path representation, the vector:

$$x = \begin{bmatrix} x_1 & x_2 & \cdots & x_n \end{bmatrix} \quad [8.5]$$

represents the n-city tour $x_1 \to x_2 \to \cdots \to x_n$.

A. Order migration

Order migration (OM) in BBO is similar to order crossover in GAs [DAV 85]. OM in BBO first chooses the immigrating and emigrating solutions based on the immigration rate and the emigration rate, respectively. Then, a section of the tour from the immigrating solution remains unchanged, which results in a child solution that has a partial tour. OM completes the child solution by copying the remaining required cities from the emigrating solution to the child solution, while maintaining the relative order of those cities from the emigrating solution.

EXAMPLE 8.2.–

Figure 8.3 shows BBO order migration for a TSP. We randomly select a subtour from the immigrating solution; suppose that we select subtour [4 5 6 7] from the immigrating solution, which gives a partial child solution. Now we see that the child solution still needs cities 1, 2, 3, 8, 9. Those cities occur in the order [2 1 8 9 3] in the emigrating solution. We therefore copy those cities in that order to the child solution to obtain the complete child solution [2 1 8 4 5 6 7 9 3].

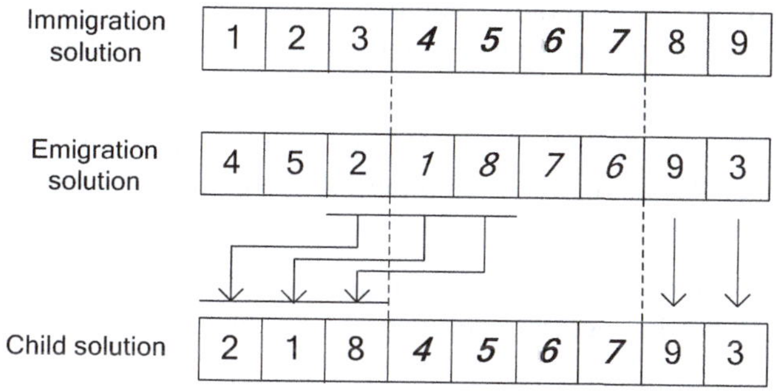

Figure 8.3. *Order migration in BBO*

B. Cycle migration

Cycle migration (CM) in BBO is similar to cycle crossover (CX) in GAs [OLI 87]. CM in BBO first chooses the immigrating and emigrating solutions based on the immigration rate and the emigration rate, respectively. Then, we create a child solution from two parent solutions in a way that preserves as much sequence information as possible from the immigrating solution, while completing the child solution with information from the emigrating solution. The basic procedure of CM is as follows:

1) Randomly select a city as the starting point in the immigrating solution and record its position.

2) In the emigrating solution, find the city in the position recorded above in the immigrating solution, and record this city. Go back to the immigrating solution, search for the city we found in the emigrating solution, and record its position in the immigrating solution.

3) Repeat step 2 until we obtain a complete cycle, which means that we have returned to the starting city. Then, copy the remaining cities from the emigrating solution to the child solution.

EXAMPLE 8.3.–

Figure 8.4 shows BBO cycle migration for a TSP. We create a child solution as follows:

1) Select a random index between 1 and n. Suppose that we select 1, and position 1 in the immigrating solution is city 1, so the child solution is initialized as [1 − − − − − − − −].

2) Position 1 in the emigrating solution is city 4, and city 4 occurs in the fourth position of the immigrating solution, so the child is augmented to become [1 – – 4 – – – – –].

3) Position 4 in the emigrating solution is city 8, and city 8 occurs in the eighth position of the immigrating solution, so the child is augmented to become [1 – – 4 – – – 8 –].

4) Position 8 in the emigrating solution is city 3, and city 3 occurs in the third position of the immigrating solution, so the child is augmented to become [1 – 3 4 – – – 8 –].

5) Position 3 in the emigrating solution is city 2, and city 2 occurs in the second position of the immigrating solution, so the child is augmented to become [1 2 3 4 – – – 8 –].

6) Position 2 in the emigrating solution is city 1, but the child solution already includes city 1. Therefore, we copy the remaining cities from the emigrating solution to the child solution, which gives [1 2 3 4 7 6 9 8 5].

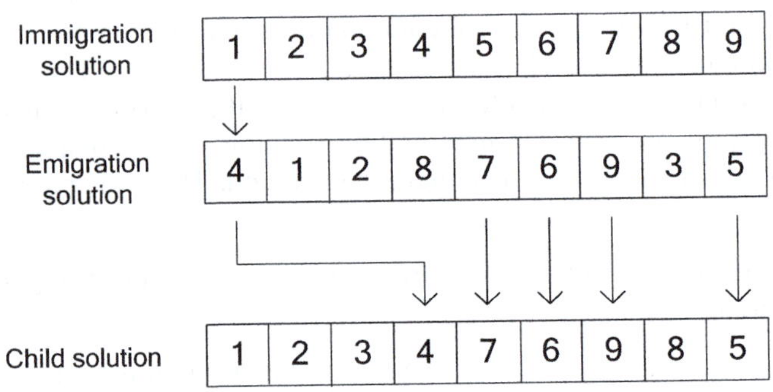

Figure 8.4. *Cycle migration in BBO*

C. Inver-over migration

Inver-over migration in BBO is similar to inver-over crossover in GAs [TAO 98]. Inver-over migration in BBO does not require expression transformation, and can guarantee that the generated child represents a valid and complete tour. Like other migration options, inver-over migration first chooses the immigrating and emigrating solutions based on the immigration rate and the emigration rate, respectively. Then, we perform the following procedure:

1) Randomly select a position s in the immigrating solution. Suppose that position s in the immigrating solution is city r.

2) Suppose that city r is in the kth position in the emigrating solution. Set the $(k + 1)$st position in the emigrating solution as the end-point city e.

3) Reverse the order of the cities between the city in the $(s + 1)$st position of the immigrating solution and city e to obtain the child solution.

EXAMPLE 8.4.–

Figure 8.5 illustrates inver-over migration in BBO for a TSP. We randomly select position $s = 4$ from the immigrating solution. We see that the fourth position in the immigrating solution is city 4. We see that city 4 occurs in the first position in the emigrating solution. So we set $e = 1$ as the end-point city. We then reverse the order of the cities between the city in the fifth position and city 1 of the immigrating solution to obtain the child solution.

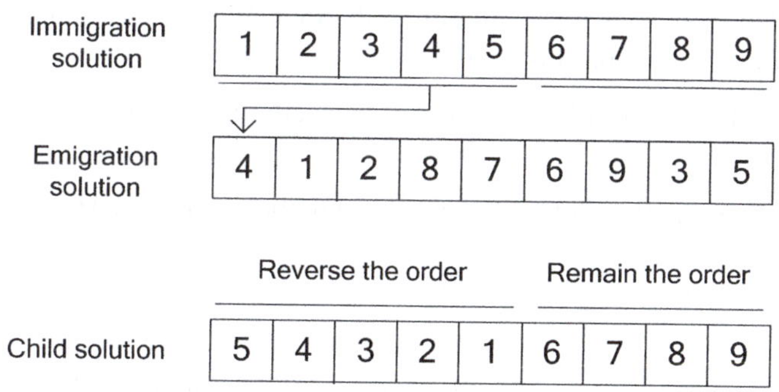

Figure 8.5. *Inver-over migration in BBO*

D. Matrix migration

The next migration method that we discuss is called matrix migration, which is similar to matrix crossover in GAs [FOX 91]. In matrix migration, the tour representation is different from the above three path representations. An n-city tour is represented by an $n \times n$ matrix M containing only zeros and ones. $M_{ik} = 1$ if and only if city i occurs before city k in the tour. For instance, consider the matrix:

$$M = \begin{bmatrix} 0 & 1 & 0 & 1 & 1 \\ 0 & 0 & 0 & 1 & 1 \\ 1 & 1 & 0 & 1 & 1 \\ 0 & 0 & 0 & 0 & 1 \\ 0 & 0 & 0 & 0 & 0 \end{bmatrix} \qquad [8.6]$$

The ones in the first row indicate that city 1 is before cities 2, 4 and 5. The ones in the second row indicate that city 2 is before cities 4 and 5. The ones in the third row indicate that city 3 is before cities 1, 2, 4 and 5. The one in the fourth row indicate that city 4 is before city 5. Finally, the fact that the fifth row is comprised of all zeros indicates that city 5 is the last city in the tour. Therefore, equation [8.6] represents the tour $3 \rightarrow 1 \rightarrow 2 \rightarrow 4 \rightarrow 5$.

Another way to interpret equation [8.6] is to note that the row with the most ones is the first city, the row with the second most ones is the second city, and so on; the row with the kth most ones is the kth city in the tour. We note several properties for an $n \times n$ matrix M that represents a valid tour.

1) Exactly one row of M has $(n - 1)$ ones, exactly one row of M has $(n-2)$ ones, and so on.

2) The above property allows us to find the number of ones in M:

$$\text{Number of ones} = \sum_{\mathrm{i}=1}^{\mathrm{n}} (n-i) = n(n-1)/2 \qquad [8.7]$$

3) No city occurs before itself in the tour, so $M_{\mathrm{ii}} = 0$ for all $\mathrm{i} \in [1, n]$.

4) If city i occurs before city j, and city j occurs before city k, then city i occurs before city k. That is,

$$\left(M_{ij} = 1 \quad \text{and} \quad M_{jk} = 1\right) \Rightarrow M_{ik} = 1 \qquad [8.8]$$

The advantage of matrix migration is that it is straightforward and easy to operate, as we shall soon see. With matrix migration, a child solution can inherit partial information from both immigrating and emigrating solutions, but it will also contain unique information. The drawback of matrix migration is that all sequence information is represented by matrices, and it requires a high computational effort when converting between tour information in a vector expression and tour information in a matrix expression.

Now we discuss the details of matrix migration in BBO. First, we use roulette-wheel selection to select two parent solutions. Once the parents are selected, we perform matrix migration operation on the two parent matrices to obtain the child matrix. We have a couple of ways that we can combine parent matrices to obtain children.

The first way is the intersection method. We use an example to illustrate intersection. Suppose that equation [8.6] represents parent M_1, and that the second parent is given as:

$$M_2 = \begin{bmatrix} 0 & 1 & 1 & 0 & 1 \\ 0 & 0 & 1 & 0 & 1 \\ 0 & 0 & 0 & 0 & 0 \\ 1 & 1 & 1 & 0 & 1 \\ 0 & 0 & 1 & 0 & 0 \end{bmatrix} \qquad [8.9]$$

where M_2 represents the tour $4 \to 1 \to 2 \to 5 \to 3$. We obtain the intersection of M_1 and M_2 by performing an element-by-element logical AND operation on the two matrices. This gives the partially defined child as:

$$M = M_1 \wedge M_2 = \begin{bmatrix} 0 & 1 & 0 & 0 & 1 \\ 0 & 0 & 0 & 0 & 1 \\ 0 & 0 & 0 & 0 & 0 \\ 0 & 0 & 0 & 0 & 1 \\ 0 & 0 & 0 & 0 & 0 \end{bmatrix} \qquad [8.10]$$

This does not represent a valid tour, but it does indicate that city 1 is before cities 2 and 5, city 2 is before city 5 and city 4 is before city 5. This ordering occurs in the partially defined child because the same ordering occurs in both parents. In fact, this is the only ordering that is common in both parents. At this point, we can pseudo-randomly add ones to M until it is a valid tour (that is, until it satisfies all of the properties enumerated above). For example, we might choose to add ones to M to obtain:

$$M = \begin{bmatrix} 0 & 1 & \mathbf{1} & \mathbf{1} & 1 \\ 0 & 0 & 0 & 0 & 1 \\ 0 & 0 & 0 & 0 & 0 \\ 0 & \mathbf{1} & \mathbf{1} & 0 & 1 \\ 0 & \mathbf{1} & \mathbf{1} & 0 & 0 \end{bmatrix} \qquad [8.11]$$

where the added ones are denoted in bold. M now satisfies all the properties of a valid tour, and so we can transform M from its matrix expression to a sequential representation to obtain the tour $1 \to 4 \to 5 \to 2 \to 3$.

The second way to combine parent matrices is with the union method. We use the same example parent matrices as above to illustrate union. Suppose that equations [8.6] and [8.9] represent parents M_1 and M_2. We obtain the union of M_1 and M_2 by performing an element-by-element logical OR operation on the two matrices to obtain the partially defined child:

$$M = M_1 \vee M_2 = \begin{bmatrix} 0 & 1 & 1 & 1 & 1 \\ 0 & 0 & 1 & 1 & 1 \\ 1 & 1 & 0 & 1 & 1 \\ 1 & 1 & 1 & 0 & 1 \\ 0 & 0 & 1 & 0 & 0 \end{bmatrix} \qquad [8.12]$$

We next select a random cut point that divides M into four quadrants (not necessarily of equal size). Suppose that we generate a random cut point at the second row and the second column. We write M with the upper-left and lower-right quadrants unchanged, but with the lower-left and upper-right quadrants replaced with undefined terms:

$$M = \begin{bmatrix} 0 & 1 & \times & \times & \times \\ 0 & 0 & \times & \times & \times \\ \times & \times & 0 & 1 & 1 \\ \times & \times & 1 & 0 & 1 \\ \times & \times & 1 & 0 & 0 \end{bmatrix} \qquad [8.13]$$

We next make necessary changes to M to remove contradictions. For example, $M_{34} = 1$ and $M_{43} = 1$, so one of those elements needs to be changed to a 0. This gives the corrected but still partially defined child as:

$$M = \begin{bmatrix} 0 & 1 & \times & \times & \times \\ 0 & 0 & \times & \times & \times \\ \times & \times & 0 & 0 & 0 \\ \times & \times & 1 & 0 & 1 \\ \times & \times & 1 & 0 & 0 \end{bmatrix} \quad [8.14]$$

Finally, we pseudo-randomly add ones to the off-diagonal blocks in M until it is a valid tour (that is, until it satisfies all of the properties enumerated earlier). For example, we might choose to add ones to M to obtain:

$$M = \begin{bmatrix} 0 & 1 & 1 & 1 & 1 \\ 0 & 0 & 1 & 1 & 1 \\ 0 & 0 & 0 & 0 & 0 \\ 0 & 0 & 1 & 0 & 1 \\ 0 & 0 & 1 & 0 & 0 \end{bmatrix} \quad [8.15]$$

M now satisfies all of the properties of a valid tour, and we can transform M from its matrix expression to a sequential representation to obtain the tour $1 \to 4 \to 5 \to 2 \to 3$.

8.2.3. *Mutation in the TSP*

This section discusses a few ways to mutate TSP solutions. We restrict our discussion to path representations.

A. Inversion

Inversion reverses the order of the tour between two randomly selected indices [FOG 90]. For example, tour x could be mutated to become x_m as follows:

$$\begin{aligned} x &= 1 \to \underbrace{2 \to 3 \to 4 \to 5} \to 6 \to 7 \to 8 \to 9 \\ x_m &= 1 \to \overbrace{5 \to 4 \to 3 \to 2} \to 6 \to 7 \to 8 \to 9 \end{aligned} \quad [8.16]$$

where we randomly selected the start and the end point of the mutated segment. Inversion is also called 2-opt mutation [BEY 02]. There are $n(n-1)/2$ unique ways to implement inversion in an n-city TSP tour. The lowest-cost solution that results from all possible inversions of an n-city TSP tour always results in a tour without any crossed edges.

2-opt is a simple but effective mutation method. Instead of replacing two points in the solution, kopt, which is a mutation method, adaptively chooses the number of points to break and reconnect [JOH 97]. According to simulation results, as the number of replaced points increases, the performance of k-opt increases. But the computational burden also increases. We need to find a balance between the expected performance and the computational burden. For the first few optimization generations, the population is still diverse and there is a lot of information to exploit. In this case, we do not need k-opt to act aggressively, so k should be a small number. But as the optimization algorithm progresses, it begins to converge. In this situation, we need to increase the effectiveness of k-opt to increase the rate of improvement, so we should use a larger value of k. We conclude that k should increase as the generation count increases. One way of doing this is shown as follows:

$$k = \left\lfloor \frac{g_c c}{2 g_m} \right\rfloor \quad [8.17]$$

where c is the number of cities, g_m is the generation limit and g_c is the current generation number.

B. Insertion

Insertion moves the city in position i to position k, where i and k are randomly selected [FOG 88]. For example, suppose that we have the tour $x = [1\,2\,3\,4\,5\,6\,7\,8\,9]$. Suppose that we randomly select i = 4 and k = 2. We then move city 7, which is in position 4, to position 2 to obtain the mutated tour

$$x_m = 1 \to 4 \to 2 \to 3 \to 5 \to 6 \to 7 \to 8 \to 9 \quad [8.18]$$

C. Displacement

Displacement is a generalization of insertion [MIC 96a, Chapter 10]. Displacement takes the sequence of q cities beginning at the ith position and moves them to the kth position in the tour, where q, i and k are randomly selected. For example, suppose that we have the tour $x = [1\,2\,3\,4\,5\,6\,7\,8\,9]$ and randomly select

$q = 2$, $i = 4$ and $k = 2$. We then take the two-city sequence beginning at position 4 (cities 4 and 5) and move it to position 2 to obtain the mutated tour:

$$x_m = 1 \to 4 \to 5 \to 2 \to 3 \to 6 \to 7 \to 8 \to 9 \quad [8.19]$$

We could also combine displacement with inversion by reversing the order of the selected cities before we move them to their new position.

D. Reciprocal exchange

Reciprocal exchange swaps the cities in the ith and kth positions, where i and k are randomly selected [BAN 90]. For example, suppose we have the tour $x = [1\ 2\ 3\ 4\ 5\ 6\ 7\ 8\ 9]$. Suppose that we randomly select $i = 5$ and $k = 1$. We swap the cities in the first and fifth positions to obtain the mutated tour:

$$x_m = 5 \to 2 \to 3 \to 4 \to 1 \to 6 \to 7 \to 8 \to 9 \quad [8.20]$$

We could generalize this method by swapping sequences of cities rather than single cities. We could then combine this generalization with inversion by reversing the order of one or more of the swapped sequences.

8.2.4. *Implementation framework*

Given the background of the preceding sections, we now present a BBO algorithm to solve the TSP, which is shown in Figure 8.6. We have two options in the implementation of Figure 8.6.

1) We have several options for migration, as discussed in section 8.2.2. We could also use more than one migration method, probabilistically switching between various methods from one generation to the next. Furthermore, we could keep track of which migration method gives the best results and adapt the frequency of the migration methods depending on the fitness of their offspring.

2) We have several options for mutation, as discussed in section 8.2.3. As with migration, we could use more than one mutation method, probabilistically switching between various methods from one generation to next. We could keep track of which mutation method gives the best results and adapt the frequency of the mutation methods depending on the fitness of their results.

Initialize a population of candidate solutions $\{x_k\}$ for $k \in [1, N]$ (see section 8.2.1)
Represent the candidate solutions using either path or matrix representations
Calculate the tour distance for each candidate solution
While not (termination criterion)
    For each x_k, set emigration rate μ_k proportional to the fitness of x_k, where μ_k is normalized to [0, 1]
    For each x_k, set immigration rate $\lambda_k = 1 - \mu_k$
    $\{z_k\} \leftarrow \{x_k\}$
    For each solution z_k
        Use λ_k to probabilistically decide whether to immigrate to z_k
        If immigrating then
            Use $\{\mu_i\}$ to probabilistically select the emigrating solution x_j
            Create a child c_k using one of the migration methods discussed in section 8.2.2:
                If using path representation then
                    Use one of the first three migration methods to create c_k
                Else if using matrix representation then
                    Use the fourth migration method to create c_k
                End if
        End if
        Probabilistically decide whether to mutate z_k using one of the mutation methods described in section 8.2.3
    Next solution
    $\{x_k\} \leftarrow \{z_k\}$
Next generation

Figure 8.6. *Outline of BBO for solving the TSP.* $\{x_k\}$ *is the population of solutions,* $\{z_k\}$ *is a temporary population of solutions and* x_k *is the kth candidate solution*

EXAMPLE 8.5.–

We investigate the Att-48 TSP, a data set of 48 capitals of the US. This problem is available on the TSPLIB website at http://www.iwr.uni-heidelberg.de/groups/comopt/software/TSPLIB95/. We implement BBO for the TSP with the following parameters: we use a population size $N = 50$; we initialize the population using nearest-neighbor strategy as described in section 8.2.1; we use path representation and inver-over migration as described in section 8.2.2; we use a mutation rate of 0.01 and inversion mutation as described in section 8.2.3; we define the fitness of a tour as shown in equation [8.3]; we run 20 Monte Carlo simulations, each for 300 generations. Figure 8.7 shows a plot of the 48 cities and the best minimum-distance closed tour of the 20 simulations. The average minimum-distance tour length is 48,958.1.

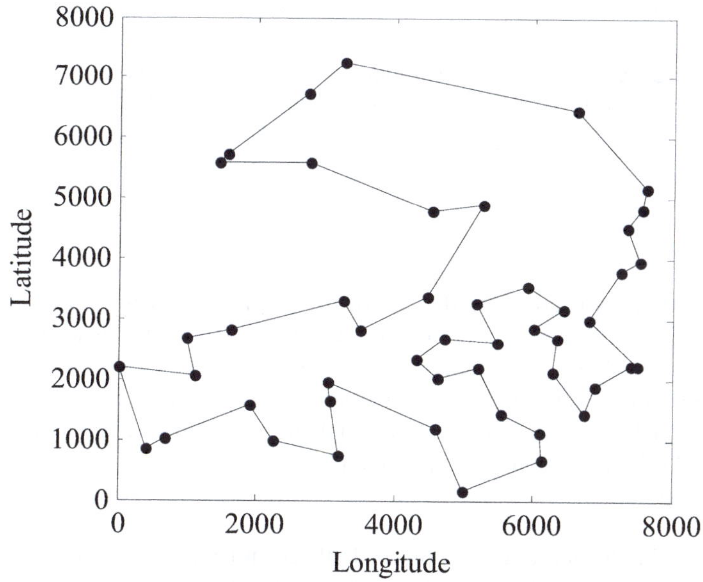

Figure 8.7. *The Att-48 TSP cities and the best minimum-distance tour*

EXAMPLE 8.6.–

In this example, we investigate the effect of different migration and mutation options for the TSP. We use the same TSP problem and parameters as described in Example 8.5. Table 8.1 gives the average minimum-distance tour for different migration and mutation options. Note that we try three different migration methods with inversion mutation, and we try three different mutation methods with OM. From Table 8.1, we see that OM works best when we use inversion mutation, and inversion mutation works best when we use inver-over migration.

Inversion mutation		Inver-over migration	
Order migration	44,598.6	Inversion mutation	48,958.1
Cycle migration	51,049.2	Insertion mutation	55,661.8
Inver-over migration	48,958.1	Reciprocal exchange	53,160.4

Table 8.1. *Average minimum-distance tours for different migration and mutation options for the Att-48 TSP*

EXAMPLE 8.7.–

In this example, we compare the performance of BBO with GA/GUR for the TSP. The difference between a GA and a BBO implementation lies in the selection of the parents. With BBO, the parents are selected based on the immigration rate and the emigration rate. With GA/GUR, parents are selected using roulette-wheel selection, but each variable in a solution is taken from a different parent. Here for BBO, we use OM and inversion mutation, and for GA/GUR, we use order crossover and inversion mutation.

We evaluated BBO and GA/GUR on six TSP benchmarks, all of which are available in [REI 08]. Att-48 is a data set of 48 capitals of the US. Berlin-52 is a data set of 52 locations in Berlin, Germany. ST-70 is a 70-city problem, CH-130 is a 130-city problem, GR-202 is a 202-city problem and RAT-575 is a 575-city problem.

Table 8.2 shows the average minimum-distance tour for BBO and GA/GUR after 300 generations over 20 Monte Carlo simulations. The algorithms are still converging after 300 generations, so the numbers in Table 8.2 should not be compared with the best published solutions, but only with each other to measure the effect of BBO and GA/GUR for the TSP. Table 8.2 shows that BBO is significantly better than GA/GUR for all six of the benchmarks for the first 300 generations.

TSP problem	BBO	GA/GUR
Att-48	44,598.6	50,321.7
Berlin-52	15,816.3	17,074.6
ST-70	2,078.2	2,255.9
CH-130	33,954.1	35,762.3
GR-202	24,762.5	2,674.2
RAT-595	10,175.3	10,408.5

Table 8.2. *Average minimum-distance tours between BBO and GA/GUR for TSP benchmarks*

8.3. Graph coloring

Graph coloring is a special case of graph labeling; it is an assignment of labels traditionally called "colors" to elements of a graph subject to certain constraints. In its simplest form, it is a way of coloring the vertices of a graph such that no two adjacent vertices share the same color; this is called vertex coloring. Similarly, edge coloring assigns a color to each edge so that no two adjacent edges share the same color, and face (or map) coloring in a planar graph assigns a color to each face or region so that no two faces that share a boundary have the same color.

There are many related but distinct graph coloring problem. The classical graph coloring problem is defined as follows: determine the smallest number of colors n such that each node of a connected graph can be colored with one of these n colors under the constraint that linked nodes are not assigned the same color. This problem is also called the n-coloring problem. The smallest number of colors needed to color a graph G is called its chromatic number and is often denoted $\chi(G)$. A graph that can be assigned an n-coloring is n-colorable, and it is called n-chromatic if its chromatic number is exactly n.

Figure 8.8 shows a graph coloring problem with 7 vertices and 13 edges, along with its solution. The minimum number of colors needed to solve the problem is $\chi(G)=4$. We have used different shapes instead of different colors in Figure 8.8 for the sake of illustration. Note from the right side of Figure 8.8 that none of the circles are connected to each other, none of the squares are connected to each other and there is only one triangle and one oval.

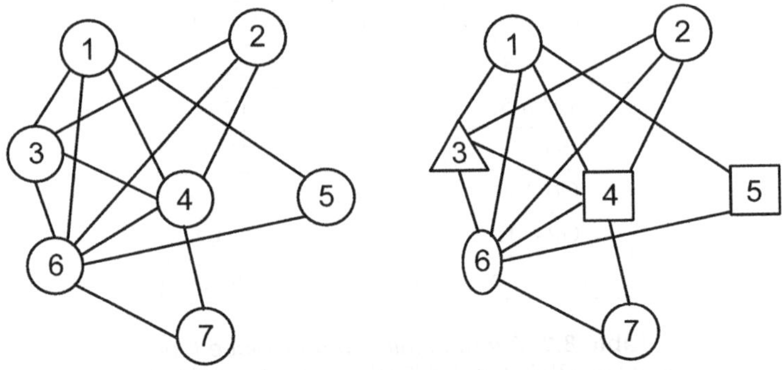

Figure 8.8. *Illustration of a graph coloring problem with 7 vertices and 13 edges. The left figure shows the unsolved graph and the right figure shows the solved graph, where we have used shapes to indicate colors*

In this section, we combine BBO with a greedy algorithm to solve the graph coloring problem. With this approach, each candidate solution in the population stores an ordered list of vertices as its solution features. Information about the order of the vertices is shared among solutions using inver-over migration. The greedy algorithm of Figure 8.9 assigns colors to the vertices. The number of colors assigned by the greedy algorithm to a given solution is the cost of that solution, and is used to assign emigration and immigration rates in BBO.

C = indices of available colors

For each vertex v_i

  $N_i \leftarrow$ {neighbors of v_i }

  $C(N_i) \leftarrow$ {colors of x : $x \in N_i$ }

  $c_i \leftarrow \min \{k \in C : k \notin C(N_i)\}$

  Assign c_i as the color of v_i

Next vertex

Figure 8.9. *A greedy algorithm for the graph coloring problem. For each vertex, this algorithm finds the next available color that is not used by a neighbor*

To compare performance, we use GA/GUR with the same greedy algorithm. We evaluated three graph coloring benchmarks from the website http://mat.gsia.cmu.edu/COLOR/instances.html [TRI 95]. The first benchmark is the Leighton graph [LEI 79] and includes 450 vertices and 17,343 edges. The second benchmark is based on the Mycielski transformation and includes 191 vertices and 2,360 edges. The third benchmark is called Flat and includes 300 vertices and 21,695 edges.

Table 8.3 shows the average minimum number of colors for BBO and GA/GUR after 300 generations over 20 Monte Carlo simulations. We see from Table 8.3 that BBO is significantly better than GA/GUR for all three of the benchmarks.

Graph coloring problem	BBO	GA/GUR
Leighton	91.6	92.8
Mycielski	25.7	26.1
Flat	133.2	136.5

Table 8.3. *Average minimum number of colors for BBO and GA/GUR for graph coloring benchmarks*

8.4. Knapsack problem

The knapsack problem is another combinatorial optimization problem: given a set of items, each with a weight and a value, determine the quantity of each item to include in a collection so that the total weight is less than or equal to a given limit and the total value is as large as possible. The knapsack problem often arises in resource allocation with financial constraints, and it has been studied for more than a century. The name "knapsack problem" dates back to the early work on the problem and refers to the commonplace problem of packing your most valuable items without overloading your luggage.

The most common knapsack problem is the 0/1 knapsack problem, which can be described as follows. Consider a set of n items where the ith item has weight w_i and profit p_i. The problem is to select a subset of the n items to maximize overall profit without exceeding the weight constraint b. The problem can be mathematically modeled as follows:

$$\begin{aligned} &\text{Maximize} \quad \sum_{i=1}^{n} w_i x_i \\ &\text{Subject to} \quad \sum_{i=1}^{n} p_i x_i \leq b, \quad \text{where } x_i \in \{0,1\} \quad \forall \quad i \in \{1,2,\cdots,n\} \end{aligned} \qquad [8.21]$$

x_i takes the value 1 or 0, which represents the selection or rejection of the ith item. Ten benchmark knapsack problems are studied here as summarized in Table 8.4. The parameters used in BBO and GA/GUR are the same as those in Example 8.5.

Table 8.5 shows comparisons of the best performance of BBO and GA/GUR after 300 generations over 20 Monte Carlo simulations. The results show that both algorithms perform the same for F03, F04 and F09, and BBO performs better than GA/GUR for the other seven benchmarks, which indicates that BBO is a good tool for solving knapsack problems.

Fun.	Dim.	Parameters (w, p, b)
F01	10	w = {95, 4, 60, 32, 23, 72, 80, 62, 65, 46}; p = {55, 10, 47, 5, 4, 50, 8, 61, 85, 87}; b = 269
F02	20	w = {92, 4, 43, 83, 84, 68, 92, 82, 6, 44, 32, 18, 56, 83, 25, 96, 70, 48, 14, 58}; p = {44, 46, 90, 72, 91, 40, 75, 35, 8, 54, 78, 40, 77, 15, 61, 17, 75, 29, 75, 63}; b = 878
F03	4	w = {9, 11, 13, 15}; p = {6, 5, 9, 7}; b = 20
F04	4	w = {6, 10, 12, 13}; p = {2, 4, 6, 7}; b = 11
F05	15	w = {56.358531, 80.87405, 47.987304, 89.59624, 74.660482, 85.894345, 51.353496, 1.498459, 36.445204, 16.589862, 44.569231, 0.466933, 37.788018, 57.118442, 60.716575}; p = {0.125126, 19.330424, 58.500931, 35.029145, 82.284005, 17.41081, 71.050142, 30.399487, 9.140294, 14.731258, 98.852504, 11.908322, 0.89114, 53.166295, 60.176397}; b = 375
F06	10	w = {30, 25, 20, 18, 17, 11, 5, 2, 1, 1}; p = {20, 18, 17, 15, 15, 10, 3, 1, 1}; b = 60
F07	7	w = {70, 20, 39, 37, 7, 5, 10}; p = {31, 10, 20, 19, 4, 3, 6}; b = 50
F08	23	w = {983, 982, 981, 980, 979, 978, 488, 976, 972, 486, 486, 972, 972, 485, 485, 969, 966, 483, 964, 963, 961, 958, 959}; p = {81, 980, 979, 978, 977, 976, 487, 974, 970, 485, 485, 970, 970, 484, 484, 976, 974, 482, 962, 961, 959, 958, 857}; b = 10000
F09	5	w = {33, 24, 36, 37, 12}; p = {15, 20, 17, 8, 31}; b = 80
F10	20	w = {84, 83, 43, 4, 44, 6, 82, 92, 25, 83, 56, 18, 58, 14, 48, 70, 96, 32, 68, 92}; p = {91, 72, 90, 46, 55, 8, 35, 75, 61, 15, 77, 40, 63, 75, 29, 75, 17, 78, 40, 44}; b = 879

Table 8.4. *The dimensions and parameters of the 10 benchmark knapsack problems*

Fun.	BBO	GA/GUR	Fun.	BBO	GA/GUR
F01	290	265	F06	54	49
F02	950	860	F07	112	105
F03	35	35	F08	9,820	9,724
F04	23	23	F09	130	130
F05	470	395	F10	1,020	870

Table 8.5. *Performance comparison between BBO and GA/GUR for the 10 benchmark knapsack problems*

8.5. Conclusion

In this chapter, we have summarized the TSP and have discussed some of the most commonly used TSP representations and operators. We have shown how BBO combines these representations and operators to solve the TSP and obtained good results. In addition, we have discussed BBO for the solution of other combinational optimization problems including the graph coloring problem and the knapsack problem.

There are many other popular and practical combinatorial optimization problems, including the minimum spanning tree problem, the job shop scheduling problem and the bin packing problem. Evolutionary algorithms have been applied to all of these problems, but there is plenty of room for BBO research on these problems. There are several BBO variations that have yet to be applied to some of these combinatorial optimization problems. Finally, we note that we need more theoretical results to quantify the performance of BBO (and other EAs) on combinatorial problems to provide guidance for practical applications.

9

Constrained BBO

All real-world optimization problems are constrained, at least implicitly if not explicitly. Constrained optimization is the optimization of an objective function in the presence of constraints on the solution. Without loss of generality, a constrained optimization problem can be written as:

$$\min_x f(x) \quad \text{such that} \quad g_i(x) \leq 0 \;\; \text{for} \;\; i \in [1, m] \qquad \text{and} \quad h_j(x) = 0 \;\; \text{for} \;\; j \in [1, p] \qquad [9.1]$$

This problem includes $(m+p)$ constraints, m of which are inequality constraints and p of which are equality constraints. The set of x that satisfies all $(m+p)$ constraints is called the feasible set, and the set of x that violates one or more constraints is called the infeasible set:

$$\text{feasible set} \quad \Gamma = \{x : g_i(x) \leq 0 \text{ for } i \in [1, m] \text{ and } h_j(x) = 0 \text{ for } j \in [1, p]\} \qquad \text{infeasible set} \quad \overline{\Gamma} = \{x : x \notin \Gamma\} \qquad [9.2]$$

Constraints could be linear or nonlinear, and the objective of a constrained evolutionary algorithm is to minimize $f(x)$ while at the same time satisfying the constraints $g_i(x)$ and $h_j(x)$.

Overview of the chapter

In this chapter, we discuss how to modify BBO for constrained optimization problems. Section 9.1 provides notation and concepts that often arise in constrained optimization. Section 9.2 introduces several popular constraint-handling approaches used in EAs, which are also suitable for BBO. Section 9.3 shows how we can combine BBO with these popular constraint-handling methods to obtain constrained BBO algorithms, and presents a comparative study of these constrained BBO algorithms on standard constrained benchmark problems. The concluding section of this chapter provides references to additional resources, and suggests several important topics for future research for constrained BBO.

9.1. Constrained optimization

Various techniques have been proposed to handle constrained optimization problems. The methods that are based on penalty functions are the most popular approaches, thanks to their simplicity and ease of application. In this chapter, our constrained BBO algorithms will also use penalty function methods. So it is first necessary to review penalty function methods.

Penalty function methods usually penalize candidate solutions that violate constraints. Penalty function approaches for general constrained optimization problems were first proposed by Courant [COU 43]. They are often cited as being the most popular algorithms for constrained optimization, but other approaches for constrained EAs are rapidly gaining in popularity. We could allow infeasible solutions in the EA population but penalize them in terms of cost, or in terms of selection for contributing to the next generation. This approach generally does not penalize feasible solutions, no matter how close they are to the constraint boundary.

Penalty function methods transform the standard constrained optimization problem of equation [9.1] into the following unconstrained problem:

$$\begin{aligned} &\min_x \phi(x), \text{ where } \phi(x) = f(x) + \sum_{i=1}^{m} r_i G_i(x) + \sum_{j=1}^{p} c_j L_j(x) \\ &G_i(x) = [\max(0, g_i(x))]^\beta \\ &L_j(x) = |h_j(x)|^\gamma \end{aligned} \quad [9.3]$$

where r_i and c_j are positive constants that are called penalty factors, and β and γ are positive constants that are often set equal to 1 or 2. $\phi(x)$ is called the penalized cost function, and we obtain $\phi(x)$ as a weighted sum of the original cost function $f(x)$ and the constraint violation magnitudes $\{G_i(x)\}$ and $\{L_j(x)\}$. We see that if $x \in \Gamma$, then $\phi(x) = f(x)$. However, if $x \notin \Gamma$, then $\phi(x) > f(x)$ by an amount that increases with the amount of constraint violation.

Now that we have a penalized cost function $\phi(x)$, we can run an EA that uses $\phi(x)$ as the cost function to select candidate solutions for the next generation. We can therefore extend the unconstrained BBO algorithms discussed in this book to constrained optimization problems; we simply use $\phi(x)$ instead of $f(x)$ as the cost function.

The constraints $h_j(x) = 0$ are very unforgiving. If we randomly generate an initial population in a continuous search domain, we have essentially zero chance of obtaining candidate solutions that satisfy equality constraints. Therefore, we often change the hard equality constraints to soft constraints that require $h_j(x)$ to be approximately zero rather than exactly zero; that is:

$$|h_j(x)| \leq \varepsilon \qquad [9.4]$$

where ε is a small positive constant.

Depending on the value of ε, we may have a reasonable chance of obtaining solutions that satisfy the soft constraint of equation [9.4]. One of the ways of assigning ε is to use relatively large values of ε early in the EA so that we can obtain some feasible solutions, and then gradually decrease ε as the generation count increases [BRE 09, ZAV 09]. Many research papers that compare constrained optimization algorithms on benchmark functions use $\varepsilon = 0.0001$ [LIA 06].

The conversion of equality constraints to inequality constraints transforms equation [9.3] to:

$$\min_x \phi(x), \text{ where } \phi(x) = f(x) + \sum_{i=1}^{m+p} r_i G_i(x)$$
$$G_i(x) = \begin{cases} [\max(0, g_i(x))]^\beta & \text{for } i \in [1, m] \\ [\max(0, |h_i(x)| - \varepsilon)]^\beta & \text{for } i \in [m+1, m+p] \end{cases} \qquad [9.5]$$

where we have simplified the problem by setting $\gamma = \beta$.

Problems like equation [9.5] can be solved with static or dynamic methods. Static methods use values of r_i, β and ε that are independent of the EA generation number. In contrast, dynamic methods use values of r_i, β and ε that depend on the EA generation number. Static methods are simpler to implement but dynamic methods may perform better because of their flexibility. Dynamic methods may be able to intelligently adapt their weights, based on the population distribution or the problem characteristics, to improve performance. Dynamic methods often increase r_i and β, and decrease ε, as the generation count increases. This increases the weight given to constraint violation, which results in a gradual attraction of more and more infeasible solutions toward the feasible region.

9.2. Constraint-handling methods

This section discusses several popular constraint-handling approaches used in EAs, which are variations of the basic penalty function method described in the previous section. These constraint-handling approaches are suitable for implementation in BBO.

9.2.1. *Static penalty methods*

The first constraint-handling method is a static penalty method. Homaifar *et al.* [HOM 94] set β=2 and r_i as a function of the constraint violation magnitude in equation [9.5]; that is, r_i is a non-decreasing function of $G_i(x)$. Sometimes the penalty factor r_i is set equal to one of a set of discrete values depending on the amount of the constraint violation:

$$r_i = \begin{cases} R_{i1} & \text{if } G_i(x) \in (0, T_{i1}] \\ R_{i2} & \text{if } G_i(x) \in (T_{i1}, T_{i2}] \\ \vdots & \\ R_{iq} & \text{if } G_i(x) \in (T_{i,q-1}, \infty] \end{cases} \qquad [9.6]$$

where q is the user-specified number of constraint levels, the R_{ij} values are user-defined weights and the T_{ij} values are user-defined constraint thresholds. This is a static approach because the penalty on the constraints is not a function of the generation count. The disadvantage of this approach is that it requires $(2q-1)(m+p)$ tuning parameters, although we can reduce this number by combining some of the weight levels and thresholds to simplify the algorithm.

9.2.2. *Superiority of feasible points*

The method of the superiority of feasible points modifies the penalized cost function of equation [9.5] as follows [POW 93]:

$$\begin{aligned}\min_x \phi'(x), \text{ where } \quad \phi'(x) &= \phi(x)+\theta(x) \\ &= f(x)+\sum_{i=1}^{m+p} r_i G_i(x)+\theta(x)\end{aligned} \qquad [9.7]$$

where $\theta(x)$ is an additional term that is designed to guarantee that $\phi'(x) \le \phi'(\overline{x})$ for all $x \in \Gamma$ and for all $\overline{x} \notin \Gamma$. That is, $\phi'(x) \le \phi'(\overline{x})$ for all feasible x and for all infeasible $\overline{x}$. This requirement can be satisfied by setting $\theta(x)$ as follows:

$$\theta(x) = \begin{cases} 0 & \text{if } x \in \Gamma \\ \max f(y) : y \in \Gamma & \text{if } x \notin \Gamma \end{cases} \qquad [9.8]$$

assuming that $f(x) \ge 0$ for all x. A less conservative way to implement this method, which does not assume that $f(x) \ge 0$, is to set $\theta(x)$ as follows [MIC 96b]:

$$\theta(x) = \begin{cases} 0 & \text{if} x \in \Gamma \\ \max\left\{0, \max_{y \in \Gamma} f(y) - \min_{y \notin \Gamma} \phi(y)\right\} & \text{if} x \notin \Gamma \end{cases} \qquad [9.9]$$

This definition of $\theta(x)$ gives $\phi'(x) = \phi(x)$ for all x if $\phi(x) \le \phi(\overline{x})$ for all $x \in \Gamma$ and for all $\overline{x} \notin \Gamma$. That is, if the penalized cost function of equation [9.5] results in all feasible solutions being ranked better than all infeasible solutions, then we do not make any changes to equation [9.5]. However, if equation [9.5] results in $\phi(x) > \phi(\overline{x})$ for some $x \in \Gamma$ and for some $\overline{x} \notin \Gamma$, then equation [9.9] shifts the penalized cost function values of all the infeasible solutions so that $\min_{\overline{x}} \phi'(\overline{x}) = \max_x \phi'(x)$; that is, the best infeasible penalized cost is equal to the worst feasible penalized cost.

The method of the superiority of feasible points may be an attractive approach if the optimization problem includes difficult constraints. If the constraints are hard to satisfy, then this method provides a lot of selection pressure for feasible points to remain in the population, which allows their information to continue to the next generation.

9.2.3. *The eclectic evolutionary algorithm*

The eclectic EA provides another approach to enforcing the superiority of feasible points [MOR 98]. It defines the penalized cost function as:

$$\phi(x)=\begin{cases} f(x) & \text{if } x\in\Gamma \\ K\cdot\left(1-\dfrac{s(x)}{m+p}\right) & \text{if } x\notin\Gamma \end{cases} \qquad [9.10]$$

where K is a large constant, $m+p$ is the total number of constraints and $s(x)$ is the number of constraints that are satisfied by x. The user-defined constant K needs to be large enough to ensure that $\phi(x)>\phi(\overline{x})$ for all $x\in\Gamma$ and for all $\overline{x}\notin\Gamma$. If we use a ranking method to select solutions for recombination, then there is no upper bound for K. However, if we use roulette-wheel selection, or some other method that uses absolute values of $\phi(\cdot)$ for selection, then we need to be careful not to set K too large; we want to make sure that although infeasible solutions are ranked worse than feasible solutions, infeasible solutions still have a reasonable chance of being selected for recombination.

The eclectic EA differs from equation [9.7] because the eclectic EA does not evaluate the cost $f(x)$ for infeasible solutions. This could result in significant computational savings. Also, the eclectic EA only considers the number of constraint violations in determining the penalized cost function, and it does not consider the magnitude of the constraint violations. Equation [9.5], on the other hand, considers only the magnitude of constraint violation, and it does not consider the number of constraint violations. This could provide another computational advantage to the eclectic EA because in real-world problems it is often much easier to count the number of constraint violations than to quantify the exact level of those violations.

9.2.4. *Dynamic penalty methods*

If the penalized cost function of equation [9.5] uses $\beta=1$ or 2, and $r_i=(ct)^\alpha$, where c and α are constants, and where t is the generation count, then we obtain:

$$\begin{aligned}\phi(x)&=f(x)+(ct)^\alpha M(x)\\ M(x)&=\sum_{i=1}^{m+p}G_i(x)\end{aligned} \qquad [9.11]$$

This approach is called a dynamic penalty method because the penalty on the constraints increases with the generation count [JOI 94]. However, in order to be successful with this approach, the cost $f(\cdot)$ and the constraint violation magnitude $M(\cdot)$ should be normalized so that the penalized cost function $\phi(\cdot)$ is written as follows:

$$\phi(x) = f'(x) + (ct)^{\alpha} M'(x)$$
$$M'(x) = \begin{cases} M(x)\big/\max\limits_x M(x) & \text{if } \max\limits_x M(x) > 0 \\ 0 & \text{if } \max\limits_x M(x) = 0 \end{cases} \qquad [9.12]$$
$$f'(x) = f(x)\big/\max\limits_x |f(x)|$$

assuming that $f(x) > 0$ for all x. This ensures that the components of the penalized cost $\phi(x)$ have approximately the same magnitude. Another option is to combine a dynamic penalty method with the superiority of feasible points method from section 9.2.2. With this approach, the penalized cost is written as:

$$\phi(x) \begin{cases} f'(x) & \text{if } x \in \Gamma \\ f'(x) + (ct)^{\alpha} M'(x) + \theta(x) & \text{if } x \notin \Gamma \end{cases} \qquad [9.13]$$

where $\theta(x)$ is defined such that all feasible points have a lower penalized cost than all infeasible points. Joines and Houck [JOI 94] report typical constant values of $c = 1/2$ and $\alpha = 1$ or 2, but appropriate values of c depend on the maximum generation count. For shorter EA simulations, c should be larger than $1/2$ by one or two orders of magnitude. If c is too small, then the constraint violation penalty will be too small and the EA will assign too high a selection probability on solutions with low costs but large constraint violations.

Next, we introduce a popular dynamic penalty method called the exponential dynamic penalty method, which is proposed in Carlson and Shonkwiler [CAR 98] as:

$$\phi(x) = f(x) \exp(M(x)/T) \qquad [9.14]$$

where $M(x)$ is the constraint violation magnitude defined in equation [9.11], and T is a monotonically non-increasing function of the generation count t. $T = 1/\sqrt{t}$ is proposed in Carlson and Shonkwiler [CAR 98], which gives $\lim_{t\to\infty} T = 0$, so the penalized cost of infeasible solutions tends to infinity as the generation count tends to infinity.

Equation [9.14] assumes that $f(x) \geq 0$ for all x; otherwise, the constraint penalty would serve to decrease the cost. If this assumption is not satisfied, then we should shift the cost function values before we penalize them. We can also add a tuning parameter to the penalty part of $\phi(x)$:

$$\begin{aligned} \phi(x) &= f'(x)\exp(\alpha M'(x)/T) \\ f'(x) &= f(x) - \min_x f(x) \end{aligned} \qquad [9.15]$$

where the normalized constraint violation magnitude $M'(x)$ is defined in equation [9.12], and α is a tuning parameter to adjust the relative importance of constraint violations. We find that values of α around 10 usually work pretty well.

As with the additive penalty method described in equation [9.12], we could combine the exponential dynamic penalty method with the superiority of feasible points method in section 9.2.2. With this approach, the penalized cost is written as:

$$\phi(x) = \begin{cases} f'(x) & \text{if } x \in \Gamma \\ f'(x)\exp(M(x)/T) + \theta(x) & \text{if } x \notin \Gamma \end{cases} \qquad [9.16]$$

or:

$$\phi(x) = \begin{cases} f'(x) & \text{if } x \in \Gamma \\ f'(x)\exp(\alpha M'(x)/T) + \theta(x) & \text{if } x \notin \Gamma \end{cases} \qquad [9.17]$$

where $\theta(x)$ is large enough to ensure that all feasible points have a lower cost than all infeasible points.

9.2.5. *Adaptive penalty methods*

Dynamic penalty methods often work better than static methods, but they require additional tuning, which is problem-dependent. Penalties that are too high discourage exploration of the infeasible set, but sometimes we need to use infeasible solutions to find good solutions that satisfy the constraints. However, penalties that are too low result in too much exploration of the infeasible set, and poor convergence to feasible solutions. These considerations motivate adaptive penalty methods. Adaptive methods use feedback from the population to adjust the penalty weights. One adaptive approach is proposed in Carlson and Shonkwiler [CAR 98], which sets the penalty weights of equation [9.5] as follows:

$$r_i(t+1) = \begin{cases} r_i(t)/\beta_1 & \text{if case 1} \\ \beta_2 r_i(t) & \text{if case 2} \\ r_i(t) & \text{otherwise} \end{cases} \qquad [9.18]$$

where t is the generation number, β_1 and β_2 are constants satisfying $\beta_1 > \beta_2 > 1$, case 1 means that the best solution was feasible for each of the past k generations, and case 2 means that there were no feasible solutions in any of the past k generations. The generation window k is a tuning parameter that affects the speed of adaptation. We see that if the best solution in the population is feasible, we decrease the constraint weight to allow more infeasible solutions in the population. If there are no feasible solutions in the population, we increase the constraint weight to try to obtain some feasible solutions. The goal is to obtain a balanced mix of feasible and infeasible solutions to thoroughly explore the search space and the constraints. Typical constant values for this method are $r_i(1) = 1$, $\beta_1 = 4$, $\beta_2 = 3$ and $k = n$, where n is the problem dimension [HAD 93].

9.2.6. *The niched-penalty approach*

The niched-penalty approach is motivated by the difficulty of tuning the parameters of penalty methods [DEB 99, DEB 00a]. It uses tournament selection to select solutions for recombination according to the following rules:

– given two feasible solutions, the one with the lower cost wins the tournament;

– given one feasible solution and one infeasible solution, the feasible solution wins the tournament;

– given two infeasible solutions, the one with the smaller constraint violation wins the tournament.

This method is attractive because of its simplicity, and it does not require any tuning of penalty parameters. A comparison of two infeasible solutions does not require any cost function evaluations, which can reduce computational effort. The niched-penalty approach often obtains good results on constrained optimization problems. However, its simplicity may also be a disadvantage because it considers a feasible solution with a very high cost to be better than a slightly infeasible solution with a very low cost. Therefore, it may not work well for problems whose solutions are on the constraint boundary, which is the case for many real-world optimization problems [LEG 09, RAY 09].

The "niched" part of this approach is not integral to its constraint-handling ability, but is intended to preserve diversity in the population, and is described as follows. We do not allow solutions to participate in a tournament with each other if they are far from each other in domain space. After we randomly choose solutions for tournament selection, we then compute their distance from each other. If the solutions are too far apart from each other, then we randomly choose different solutions for the tournament. This helps prevent distant clusters of solutions from disappearing from the population and thus maintains diversity.

9.2.7. *Stochastic ranking*

Stochastic ranking adds a random component to constrained EAs [RUN 00]. Since randomness is such an important component of EAs, it makes sense to include randomness in their constraint-handling approach. Stochastic ranking sometimes ranks candidate solutions according to their cost $f(\cdot)$, and sometimes ranks them according to their constraint violation magnitude. The decision of how to rank solutions is stochastic. When we compare two solutions x_1 and x_2, we consider solution x_1 to be better than x_2 if one of the following applies:

– both solutions are feasible and $f(x_1) < f(x_2)$;

– a randomly generated number $r \sim U[0,1]$ is less than a user-defined probability P_f, and $f(x_1) < f(x_2)$;

– neither of the above conditions are satisfied, and x_1 has a smaller constraint violation than x_2.

Otherwise, we consider x_2 to be better than x_1. We see that we might compare x_1 and x_2 on the basis of their costs, or we might compare them on the basis of their constraint violations, depending on the outcome of a random number generator. After we have compared and sorted all of the solutions in the population, we then perform selection and recombination for the next generation. Probability values $P_f \in (0.4, 0.5)$ give good results for many benchmark problems.

9.2.8. ε*-level comparisons*

ε-level comparisons are similar to the static penalty approach of section 9.2.1 in which different penalty function weights are used depending on the level of constraint violation. However, ε-level comparisons use only two levels of constraint violations for ranking [TAK 09].

First, we quantify the constraint violation $M(x)$ of each solution x by either combining all constraint violations or by finding the maximum constraint violation:

$$M(x) = \begin{cases} \sum_{i=1}^{m+p} G_i(x) & \text{constraint sum method} \\ \max_i G(x) & \text{maximum constraint method} \end{cases} \qquad [9.19]$$

Second, we rank two solutions x and y as follows:

$$x \text{ is better than } y \text{ if}: \begin{cases} f(x) < f(y) & \text{and } M(x) \leq \varepsilon \text{ and } M(y) \leq \varepsilon, \text{ or} \\ f(x) < f(y) & \text{and } M(x) = M(y), \text{ or} \\ M(x) < M(y) & \text{and } M(y) > \varepsilon \end{cases} \qquad [9.20]$$

where $\varepsilon \geq 0$ is a user-defined constraint violation threshold. We see that a constraint violation that is less than ε is considered to be a feasible solution for the purpose of ranking. Note that if $\varepsilon = \infty$, then solutions are ranked solely on the basis of cost. If $\varepsilon = 0$, then feasible solutions are ranked on the basis of cost, infeasible solutions are ranked solely on the basis of their constraint violation and feasible solutions are always ranked better than infeasible solutions. We typically decrease ε as the generation count increases, which gradually increases the importance of constraint satisfaction:

$$\varepsilon(0) = M(x_p)$$
$$\varepsilon(t) = \begin{cases} \varepsilon(0)(1-t/T_c)^c & \text{if } 0 < t < T_c \\ 0 & \text{if } t \geq T_c \end{cases} \qquad [9.21]$$

where $\varepsilon(t)$ is the value of ε during the tth generation, x_p is the solution with the pth smallest constraint violation, $p = N/5$, N is the population size, and c and T_c are tuning parameters that are often set to values of about $c = 100$ and $T_c = t_{\max}/5$ [TAK 09]. We could also try other tuning parameters and other profiles for decreasing ε as a function of t.

9.3. BBO for constrained optimization

Since the constraint-handling methods discussed above are solely concerned with how to rank candidate solutions, we can use any EA in conjunction with any of these constraint-handling methods. So the combination of BBO with these constraint-handling methods is conceptually simple. In this section, we use the BBO algorithm of Figure 3.5 in Chapter 3, while calculating the fitness of each candidate solution using equation [9.5] with one of the eight constraint-handling methods discussed above. Then, we use the new fitness values to calculate the immigration rate and the emigration rate in the BBO algorithm. This results in constrained BBO algorithms.

Although the implementations of constrained BBO algorithms are easy, their optimization performances may be completely different. In this section, we present comparisons between constrained BBO algorithms with different constraint-handling methods. We use equation [9.11] to measure the constraint violation magnitude of a candidate solution, where $G_i(x)$ is given by equation [9.5] with $\beta = 1$. The constraint-handling methods that we test, and their tuning parameters, include the following:

– EE: the eclectic EA of section 9.2.3;

– DP: the dynamic penalty method of equation [9.11] in section 9.2.4 with $c = 10$ and $\alpha = 1$;

– DS: the dynamic penalty method combined with the superiority of feasible points method, as defined by equation [9.13] in section 9.2.4, with $c = 10$ and $\alpha = 2$;

– EP: the exponential dynamic penalty method of equation [9.15] in section 9.2.4 with $\alpha = 10$;

– AP: the adaptive penalty method of equation [9.18] in section 9.2.5 with $\beta_1 = 4, \beta_2 = 3$ and $k = n$, where n is the problem dimension;

– NP: the niched-penalty approach of section 9.2.6;

– SR: the stochastic ranking method of section 9.2.7 with $P_f = 0.45$;

– LC: the ε-level comparison method of section 9.2.8 with $c = 100$, $T_c = 200$ and $p = N/5$, where N is the population size.

We test constrained BBO algorithms with the above constraint-handling methods on two sets of 2006 and 2010 IEEE Congress on Evolutionary Computation (CEC) benchmark functions listed in Appendix B. The 2006 CEC benchmarks are briefly summarized in Table 9.1 [LIA 06]. The 2010 benchmarks are briefly summarized in Table 9.2 [MAL 10]. We use the dimension $D = 10$ for each benchmark, and randomly generated the offset values $\{o_i\}$ and rotation matrix M for each benchmark.

Function	D	Function type	L_i	N_i	L_e	N_e	A
G01	13	Quadratic	9	0	0	0	6
G02	20	Nonlinear	0	2	0	0	1
G03	10	Polynomial	0	0	0	1	1
G04	5	Quadratic	0	6	0	0	2
G05	4	Cubic	2	0	0	3	3
G06	2	Cubic	0	2	0	0	2
G07	10	Quadratic	3	5	0	0	6
G08	2	Nonlinear	0	2	0	0	0
G09	7	Polynomial	0	4	0	0	2
G10	8	Linear	3	3	0	0	6
G11	2	Quadratic	0	0	0	1	1
G12	3	Quadratic	0	1	0	0	0
G13	5	Nonlinear	0	0	0	3	3
G14	10	Nonlinear	0	0	3	0	3
G15	3	Quadratic	0	0	1	1	2
G16	5	Nonlinear	4	34	0	0	4
G17	6	Nonlinear	0	0	0	4	4
G18	9	Quadratic	0	13	0	0	6
G19	15	Nonlinear	0	5	0	0	0
G20	24	Linear	0	6	2	12	16
G21	7	Linear	0	1	0	5	6
G22	22	Linear	0	1	8	11	19
G23	9	Linear	0	2	3	1	6
G24	2	Linear	0	2	0	0	2

Table 9.1. *2006 CEC benchmark functions. L_i and N_i are the number of linear and nonlinear inequality constraints, L_e and N_e are the number of linear and nonlinear equality constraints, D is the number of dimensions and A is the number of active constraints at the solution*

The parameters that we use in the BBO algorithm are: population size 50, maximum immigration rate and maximum emigration rate 1, and maximum mutation rate 0.01 with a mutated value randomly chosen from a uniform distribution in the search domain. We use linear migration curves. We allow BBO to run for 50,000 fitness function evaluations. We run 25 Monte Carlo simulations on each benchmark to obtain representative performances. We use an elitism parameter of two, which means that we keep the two best solutions from each generation to the next.

Function	Function type	N_i	N_e	ρ
C01	Non-separable	2	0	0.997689
C02	Separable	2	1	0.000000
C03	Non-separable	0	1	0.000000
C04	Separable	0	4	0.000000
C05	Separable	0	2	0.000000
C06	Separable	0	2	0.000000
C07	Non-separable	1	0	0.505123
C08	Non-separable	1	0	0.379512
C09	Non-separable	0	1	0.000000
C10	Non-separable	0	1	0.000000
C11	Rotated	0	1	0.000000
C12	Separable	1	1	0.000000
C13	Separable	3	0	0.000000
C14	Non-separable	3	0	0.003112
C15	Non-separable	3	0	0.003210
C16	Non-separable	2	2	0.000000
C17	Non-separable	2	1	0.000000
C18	Non-separable	1	1	0.000000

Table 9.2. *2010 CEC benchmark functions. N_i and N_e are the number of inequality and equality constraints and ρ is the ratio of the size of the feasible set to the size of the search space*

For the purpose of elitism, we define the best solution as the feasible solution with the lowest cost. If there are not any feasible solutions, then we define the best solution as the one with the lowest penalized cost, where we obtain penalized cost using one of the constraint-handling methods discussed above which ranks infeasible solutions better than feasible solutions. We can afford to use this approach when there are a relatively large number of solutions and the ranking is used for selection for recombination. But, for saving two elite solutions from one generation to the next, we need to make sure that feasible solutions are always preferred above infeasible ones for the purpose of defining elite solutions. This ensures that once BBO finds a feasible solution, it will always have at least one feasible solution for the rest of simulation.

Tables 9.3 and 9.4 summarize the performance of the constraint-handling BBO algorithms on the 2006 and 2010 CEC benchmarks, respectively. For the 2006 CEC benchmarks, we notice that for most of the functions, the BBO algorithms with EP, AP and NP give the best performance, and the other algorithms cannot find feasible solutions (except for functions G20, G21 and G22, for which none of the algorithms could find feasible solutions). This indicates that some functions, for example G04, are very easy, meaning that any method works well, and other functions, for example G20, are very hard, meaning that no method works well. The results also indicate that different constraint-handling methods have significantly different performance levels. EP, AP and NP combined with the BBO are better than other constraint-handling methods for the 2006 CEC benchmarks.

Fun.	Best known solution	EE	DP	DS	EP	AP	NP	SR	LC
G01	– 15.000	– 14.998	– 14.982	– 14.791	**– 15.000**	**– 15.000**	**– 15.000**	– 14.991	– 14.842
G02	– 0.80361	– 0.80215	– 0.79246	– 0.78346	**– 0.80361**	**– 0.80361**	**– 0.80361**	– 0.80192	– 0.78118
G03	– 1.0005	– 1.0004	– 1.0000	– 0.9992	**– 1.0005**	**– 1.0005**	**– 1.0005**	– 1.0001	– 0.9997
G04	– 30665.5	**– 30665.5**	– 30665.4	– 30665.1	**– 30665.5**	**– 30665.5**	**– 30665.5**	**– 30665.5**	**– 30665.5**
G05	5126.497	5127.046	5127.048	5129.993	**5126.497**	**5126.497**	**5126.497**	5129.913	5129.391
G06	– 6961.81	– 6961.00	– 6957.07	– 6942.65	**– 6961.81**	**– 6961.81**	**– 6961.81**	– 6960.87	– 6957.39
G07	24.306	27.051	30.547	34.119	24.272	26.272	**24.306**	35.574	28.562
G08	– 0.09582	**– 0.09582**	– 0.09425	– 0.09006	**– 0.09582**	**– 0.09582**	**– 0.09582**	**– 0.09582**	– 0.08917
G09	680.630	680.677	688.637	699.758	**680.630**	**680.630**	**680.630**	681.328	698.027
G10	7049.248	7057.369	7236.683	7377.902	**7049.248**	**7049.248**	**7049.248**	7139.645	7147.586
G11	0.74990	**0.74990**	0.74992	0.74994	**0.74990**	**0.74990**	**0.74990**	**0.74990**	0.74997
G12	– 1.00000	– 0.99994	– 0.99514	– 0.98725	**– 1.00000**	**– 1.00000**	**– 1.00000**	– 0.99991	**– 1.00000**
G13	0.053942	0.055046	0.054264	0.054931	**0.053942**	**0.053942**	**0.053942**	0.056891	0.055284
G14	– 47.765	**– 47.765**	**– 47.765**	**– 47.765**	**– 47.765**	**– 47.765**	**– 47.765**	**– 47.765**	**– 47.765**
G15	961.715	962.627	968.648	972.114	**961.715**	**961.715**	**961.715**	965.736	967.728
G16	– 1.90516	– 1.90511	– 1.90264	– 1.90017	**– 1.90516**	**– 1.90516**	**– 1.90516**	– 1.90508	– 1.90382
G17	8853.540	8853.600	8854.231	8855.703	**8853.540**	**8853.540**	**8853.540**	8853.692	8855.376
G18	– 0.86602	– 0.86552	– 0.86206	– 0.85742	**– 0.86602**	**– 0.86602**	**– 0.86602**	– 0.86303	– 0.85954
G19	32.656	32.659	32.661	32.678	**32.656**	**32.656**	**32.656**	32.665	**32.656**
G20	0.204979	–	–	–	–	–	–	–	–
G21	193.725	–	–	–	–	–	–	–	–
G22	236.431	–	–	–	–	–	–	–	–
G23	– 400.055	– 391.319	– 381.794	– 362.830	**– 396.034**	– 389.748	– 391.653	– 384.183	– 348.958
G24	– 5.50801	**– 5.50801**	– 5.50800	– 5.50800	**– 5.50801**	**– 5.50801**	**– 5.50801**	**– 5.50801**	– 5.50801

Table 9.3. *Comparison of the best feasible cost function values found by constraint-handling BBO algorithms on the 2006 CEC benchmarks. The best value in each row is indicated in bold*

Fun.	EE	DP	DS	EP	AP	NP	SR	LC
C01	-3.41E-01	-1.98E-01	-2.16E-01	**-7.65E-01**	-7.43E-01	-7.12E-01	-6.65E-01	-4.43E-01
C02	-5.19E+00	-2.07E+00	-1.23E+00	-4.56E+00	**-7.01E+00**	-2.33E+00	-1.24E+00	-6.89E+01
C03	3.76E+02	2.50E+02	6.24E+02	**1.25E+01**	3.23E+01	7.48E+01	1.26E+02	4.35E+01
C04	1.23E+02	4.32E+02	2.31E+02	**1.27E+01**	4.41E+01	7.58E+01	3.22E+02	1.78E+02
C05	-4.36E+01	-4.78E+01	-2.19E+01	**-7.78E+02**	-1.25E+02	-1.47E+02	-7.69E+02	-1.12E+02
C06	-4.45E+01	-5.11E+01	-1.63E+01	-4.87E+01	-1.46E+02	**-2.34E+02**	-4.55E+01	-1.39E+02
C07	2.58E+03	4.80E+03	1.47E+03	2.69E+02	4.58E+02	**1.14E+02**	7.78E+02	4.65E+02
C08	2.36E+06	5.82E+06	1.17E+06	7.84E+05	2.36E+05	**1.46E+05**	7.89E+05	3.65E+05
C09	1.24E+08	6.12E+08	7.72E+07	**1.36E+07**	4.21E+07	7.63E+08	5.36E+08	2.16E+07
C10	8.94E+07	7.93E+07	1.11E+07	**2.31E+06**	4.57E+06	3.26E+06	1.02E+07	3.50E+06
C11	-4.78E+00	-5.82E+00	**-7.85E+00**	-7.63E+00	-7.12E+00	-4.63E+00	-3.69E+00	**-7.85E+00**
C12	-4.56E+02	-7.90E+02	-2.36E+02	**-8.54E+02**	-2.24E+02	-2.13E+02	-2.17E+02	-1.48E+02
C13	-8.52E+01	-2.74E+01	-4.11E+02	-2.36E+02	-3.68E+03	**-7.58E+03**	-6.63E+02	-2.21E+02
C14	1.17E+09	2.91E+09	4.50E+09	4.45E+07	**1.29E+07**	4.51E+07	5.28E+08	7.58E+08
C15	2.35E+09	8.54E+10	3.58E+09	**1.21E+09**	3.54E+09	2.36E+09	1.27E+09	1.30E+09
C16	4.66E-01	1.51E-01	7.66E-01	3.36E-01	**1.28E-02**	3.17E-02	3.68E-01	4.45E-02
C17	1.20E+00	6.92E+00	2.36E+00	**7.85E-02**	8.96E-01	1.22E-01	1.08E-01	1.28E-01
C18	1.66E-01	2.42E-01	4.15E-01	4.10E-01	**1.54E-01**	8.57E-01	6.63E-01	3.07E-01

Table 9.4. *Comparison of the best feasible cost function values found by constraint-handling BBO algorithms on the 2010 CEC benchmarks. The best value in each row is indicated in bold*

For the 2010 CEC benchmarks, we notice that for most of the functions, BBO with EP, AP and NP give the best performance, while the other algorithms cannot find feasible solutions (except for function C11, for which DS and LS give the best performance). In particular, BBO with EP obtains the best feasible cost function values for functions C01, C03, C04, C05, C09, C10, C12, C15 and C17; BBO with AP obtains the best feasible cost function values for functions C02, C14, C16 and C18; and BBO with NP obtains the best feasible cost function values for functions C06, C07, C08 and C13. This indicates that for the 2010 CEC benchmarks, EP, AP and NP combined with BBO are better than the other constraint-handling methods.

Based on the above results, it appears that some constraint-handling methods provide similar performance levels, and some constraint-handling methods provide notably different constraint-handling abilities. In general, constrained BBO is a competitive algorithm for solving constrained optimization problems.

9.4. Conclusion

We see from this chapter that there are many constraint-handling methods that can be used with BBO. Some of these methods have similar performance levels, and others have significantly different performance levels. We showed how BBO can be combined with these constraint-handling methods to produce various constraint-handling BBO algorithms. These combinations of BBO with various constraint-handling methods can serve as a template for the extension of any other EA for constrained optimization. Rather than trying all possible constraint-handling BBO algorithms in search of the best combination, it is more instructive to remember the basic principles, summarized above, that we have uncovered.

Many other constraint-handling methods have been proposed, and new ones continually appear in the literature. Constrained optimization surveys can be found in [EIB 01, COE 02a, COE 02b, COE 11]. Any reader who is interested in further research should note that Carlos Coello Coello maintains a bibliography of papers related to constrained evolutionary optimization, which includes 1,357 references as of May 2016 [COE 16a, COE 16b].

Constrained evolutionary optimization is an active research area because: (1) it is a relatively new area; (2) it is lacking in theoretical results; (3) real-world optimization problems are almost always constrained. In this concluding section, we mention some important topics for future constrained evolutionary optimization research.

Incorporation of constraint-handling methods with newer EAs

Much current research includes the incorporation of standard constraint-handling methods, such as those discussed in this chapter, into newer EAs. The literature continually introduces new EAs. These new EAs are often nothing more than modifications of older EAs, but sometimes they have distinctive new features and capabilities. It is important to explore how well current constraint-handling methods perform when incorporated into different types of EAs. The relative performance levels of different EAs on unconstrained problems does not necessarily correlate with their relative performance levels on constrained problems.

Constrained optimization theory and mathematical models

Theoretical results would be a fruitful area for future research in constrained optimization. This book discusses Markov models, dynamic system models and statistical mechanics approximation models for BBO. Perhaps those tools, or others, could also be used to analyze constrained EAs.

Combinations of different constraint-handling methods

Just as EAs can be combined in various ways, constraint-handling methods can also be combined. For example, an ensemble of constraint-handling methods could all use the same cost function results, and the best method at each generation would dominate the next generation [MAL 10]. As a further level of abstraction beyond ensembles, hyper-heuristics combine multiple EAs and multiple constraint-handling methods into a single algorithm. Recall that a heuristic is a family of algorithms (for example, a family of BBO variations). A hyper-heuristic is a family of families of algorithms (for example, a family containing a BBO heuristic and other heuristics). Hyper-heuristics can be used for any type of optimization problem, but we mention them here because of their promise for constrained optimization problems [TIN 13].

10

BBO in Noisy Environments

Many optimization problems in science and engineering suffer from the effects of noise, which poses a challenge for EAs. Noise corrupts the calculation of objective functions via imperfect sensors, measurement devices and approximate numerical simulators [BEY 07, HAN 09, MA 15a, SCH 93, YU 08]. Noise results in two types of undesirable effects in EAs: (1) a superior candidate solution may be erroneously believed to be inferior, and (2) an inferior candidate solution may be erroneously believed to be superior. These effects result in false optima and reduced EA optimization performance, including reduced convergence rates and non-monotonic fitness improvement.

Noise-handling methods in EAs can be classified into two categories: methods which require an increase in computational cost, including explicit averaging methods and implicit averaging methods, and methods which perform hypotheses testing on the noise, including the use of approximate fitness models and modification of selection schemes.

Explicit averaging methods include re-sampling, which is the most common approach to deal with noise [PIE 04]. Re-sampling of the fitness values involves several noisy fitness value measurements, followed by averaging to obtain an improved fitness estimate. Averaging an increased number of samples reduces the variance of the estimated fitness. As the number of samples increases to infinity, the uncertainty in the fitness estimate decreases to zero, which transforms the noisy problem into a noiseless one.

Implicit averaging methods increase the population size so that solutions can be re-evaluated during the normal course of evolution, and so that neighboring solutions can be evaluated, which gives fitness estimates in neighboring regions of the search space. It has been shown that a large population size reduces the influence of noise on the optimization process [FIT 88].

The main idea of approximated model methods is that measured fitness values of neighboring individuals can give good fitness estimates without extra evaluations [NER 08].

Overview of the chapter

In this chapter, BBO is applied to the optimization of noisy problems. A noisy problem is one in which the fitness function is corrupted by random noise. Noise interferes with the BBO immigration rate and emigration rate, and adversely affects optimization performance. Section 10.1 introduces the notation of noisy fitness functions, and section 10.2 analyzes the effect of noise on BBO using the Markov model derived in Chapter 5. Section 10.3 incorporates the re-sampling approach into BBO to alleviate the effects of noise, and section 10.4 discusses the Kalman BBO algorithm, which uses a Kalman filter to estimate fitness function values. Section 10.5 demonstrates the performance of BBO with re-sampling, along with Kalman BBO, on a set of standard benchmarks.

10.1. Noisy fitness functions

Fitness function evaluations in EAs are often accompanied by noise [JIN 05, SIM 13]. For example, sensor inaccuracies can cause noise in experimental fitness function evaluations. Also, if we measure fitness function values with simulation software, then approximation errors in our software could cause noise in fitness function evaluations.

A noisy fitness function evaluation could result in a high fitness being mistakenly assigned to a low-fitness solution. Conversely, it could result in a low fitness being mistakenly assigned to a high-fitness solution. Figure 10.1 illustrates the PDFs of two noisy but unbiased fitness functions $f(x_1)$ and $f(x_2)$. We see that the true value of $f(x_1)$ is 0 and the true value of $f(x_2)$ is 5, but the evaluations are noisy. Therefore, x_1 might have an evaluated fitness that is greater than that of x_2. This situation would result in an inaccurate assessment of the relative fitness values of x_1 and x_2, which could result in an EA selecting the wrong solution for recombination. That is, noise can deceive an EA.

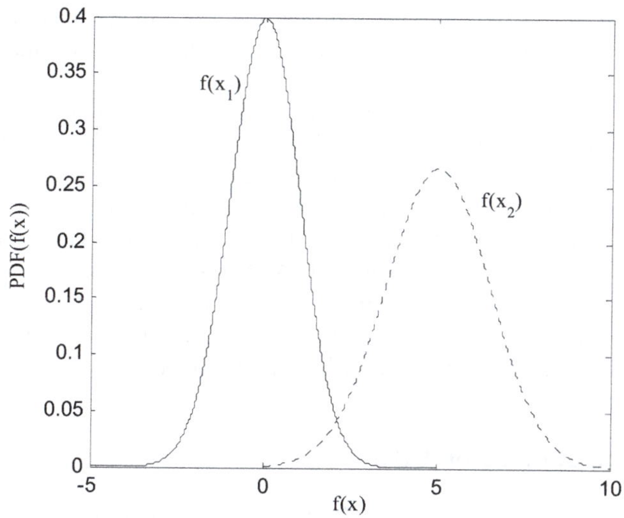

Figure 10.1. *The PDFs of two fitness functions. x_2, which has a true value of 5, is more fit than x_1, which has a true value of 0. But depending on the noise that is realized during fitness function evaluation, the EA might think that x_1 is more fit than x_2. This could result in incorrect selection for the next generation*

When we have noisy fitness function evaluations, we cannot be sure which solution is best. Consider two solutions, x_1 and x_2. Their true fitness values are denoted by $f_t\,(x_1)$ and $f_t\,(x_2)$, respectively, and the fitness function evaluations are denoted by $f\,(x_1)$ and $f\,(x_2)$, respectively. Assume that the true fitness of x_1 is better than that of x_2, that is,

$$f_t\,(x_1) > f_t\,(x_2) \qquad [10.1]$$

But the evaluated fitness of x_1 may be less than that of x_2 because of noise:

$$f\,(x_1) < f\,(x_2) \qquad [10.2]$$

That is, $f\,(x_1) < f\,(x_2)$ does not necessarily imply that $f_t\,(x_1) < f_t\,(x_2)$. However, if we know the PDFs of $f\,(x_1)$ and $f\,(x_2)$, then we can calculate the probability that $f_t\,(x_1) > f_t\,(x_2)$ given specific values of $f\,(x_1)$ and $f\,(x_2)$. We will not go through the mathematics here, but Du [DU 09] has computed the probability of relative fitness changes due to noise.

During EA execution, we do not have the PDF of the noisy fitness function since we do not know the true fitness function. However, we might know the PDF of the true fitness function. This situation is analogous to that shown in Figure 10.1, except that instead of treating the noisy fitness function as a random variable with a mean equal to the true fitness function, we can treat the true fitness function as a random variable with a mean equal to the noisy evaluated fitness function value.

Fitness noise can be represented in a very general form, but here we assume the most simple and most common type of noise, which is additive and Gaussian, because it is the predominant noise model due to its frequent occurrence in measurement systems. Additive noise is often assumed to be Gaussian due to its wide prevalence in both natural and engineering systems. Non-Gaussian noise, such as Cauchy noise, has also been considered [ARN 03]. It is plausible to assume that the noise cannot exceed certain limits due to the characteristics of the fitness measurement instrument. These assumptions have theoretical and practical impacts on noisy EAs, but are not considered further in this chapter.

10.2. Influence of noise on BBO

BBO applications (along with other EA applications) are typically implemented on deterministic problems. That means that the fitness evaluation of each solution is noise-free. But in the real-world, noiseless environments do not exist. In a noisy environment, the evaluated fitness is not equal to the true fitness, the immigration and emigration rates in BBO will be calculated incorrectly, and BBO migration may not accomplish its intended purpose.

In BBO, the immigration and emigration rates are assigned to different solutions according to the fitness of each solution. If the noise has a strong effect on the fitnesses of the solutions, the numerical order of the evaluated fitnesses could be much different from the numerical order of the true fitnesses. A solution with good fitness may be assigned a low emigration rate and a high immigration rate, and a solution with poor fitness may be assigned a high emigration rate and a low immigration rate. This is the opposite of what the true immigration and emigration rates ought to be. Therefore, the immigration and emigration rates may not accurately reflect the fitnesses of the solutions. This means that a solution with poor fitness may have a greater chance to emigrate its SIVs to other solutions compared to a solution with good fitness. If this happens, the migration mechanism of BBO is corrupted, and BBO will not perform as well as in a noiseless environment.

Now we give an example of the effect of noise on BBO performance using the Markov transition probabilities derived in Chapter 5 [MA 15a].

EXAMPLE 10.1.–

Suppose that we have a 2-bit problem ($q = 2$, $n = 4$) with a population size $N = 3$. The search space consists of bit strings $x = \{x_1, x_2, x_3, x_4\} = \{00, 01, 10, 11\}$

with corresponding fitness values $f=\{f_t(x_1), f_t(x_2), f_t(x_3), f_t(x_4)\}$ $=\{0.2, 0.4, 0.6, 0.8\}$. Suppose that the three candidate solutions in the current population are $y=\{x_1, x_2, x_3\}=\{00, 01, 10\}$. In the noise-free case, the fitness of x_1 is $f_t(x_1)=0.2$, and its corresponding immigration rate and emigration rate are $\lambda_1=0.8$ and $\mu_1=0.2$, as indicated by a linear migration curve. The fitness of x_2 is $f_t(x_2)=0.4$, with corresponding immigration rate and emigration rate $\lambda_2=0.6$ and $\mu_2=0.4$. We perform probabilistic migration to see whether x_1 and x_2 can transition to the optimal solution $x_4=11$ at the next generation. Based on equation [5.45] in Chapter 5, the probability of x_1 transitioning to the optimal solution due to migration only (no mutation) is

$$\begin{aligned}\Pr(x_1 \to x_4) &= \prod_{s=1}^{2}\left[\left((1-\lambda_1)\mathbf{1_0}(x_1(s)-x_4(s))\right)+\left(\lambda_1\frac{\sum_{j\in\varsigma_4(s)} v_j\mu_j}{\sum_{j=1}^{4} v_j\mu_j}\right)\right] \\ &= \left[(1-\lambda_1)\mathbf{1_0}(x_1(1)-x_4(1))+\lambda_1\frac{\sum_{j\in\{3,4\}} v_j\mu_j}{\sum_{j=1}^{4} v_j\mu_j}\right] \\ &\times\left[(1-\lambda_1)\mathbf{1_0}(x_1(2)-x_4(2))+\lambda_1\frac{\sum_{j\in\{2,4\}} v_j\mu_j}{\sum_{j=1}^{4} v_j\mu_j}\right] \\ &= 0.107\end{aligned} \quad [10.3]$$

The probability of x_2 transitioning to the optimal solution due to migration only is

$$\begin{aligned}\Pr(x_2 \to x_4) &= \prod_{s=1}^{2}\left[\left((1-\lambda_2)\mathbf{1_0}(x_2(s)-x_4(s))\right)+\left(\lambda_2\frac{\sum_{j\in\varsigma_4(s)} v_j\mu_j}{\sum_{j=1}^{4} v_j\mu_j}\right)\right] \\ &= \left[(1-\lambda_2)\mathbf{1_0}(x_2(1)-x_4(1))+\lambda_2\frac{\sum_{j\in\{3,4\}} v_j\mu_j}{\sum_{j=1}^{4} v_j\mu_j}\right] \\ &\times\left[(1-\lambda_2)\mathbf{1_0}(x_2(2)-x_4(2))+\lambda_2\frac{\sum_{j\in\{2,4\}} v_j\mu_j}{\sum_{j=1}^{4} v_j\mu_j}\right] \\ &= 0.180\end{aligned} \quad [10.4]$$

Next suppose that noise corrupts the measured fitness of x_1 and x_2. Suppose that the measured fitness of x_1 is $f(x_1)=0.3$ and the measured fitness of x_2 is $f(x_2)=0.2$, so that $f(x_1)>f(x_2)$. In this case, the immigration rate and the emigration rate of x_1 are $\lambda_1'=0.7$ and $\mu_1'=0.3$, respectively, and the immigration rate and the emigration rate of x_2 are $\lambda_2'=0.8$ and $\mu_2'=0.2$, respectively. We perform a migration trial to see whether x_1 and x_2 can transition to the optimal solution $x_4=11$. Based on equation [5.45] in Chapter 5, the probability of x_1 transitioning to the optimal solution due to migration only is

$$\begin{aligned}&\Pr_{\text{noise}}(x_1 \to x_4)\\&=\prod_{s=1}^{2}\left[\left((1-\lambda_1)\mathbf{1}_0\left(x_1(s)-x_4(s)\right)\right)+\left(\lambda_1\frac{\sum_{j\in\varsigma_4(s)}v_j\mu_j}{\sum_{j=1}^{4}v_j\mu_j}\right)\right]\\&=0.049\end{aligned} \quad [10.5]$$

The probability of x_2 transitioning to the optimal solution due to migration only is

$$\begin{aligned}&\Pr_{\text{noise}}(x_2 \to x_4)\\&=\prod_{s=1}^{2}\left[\left((1-\lambda_2)\mathbf{1}_0\left(x_2(s)-x_4(s)\right)\right)+\left(\lambda_2\frac{\sum_{j\in\varsigma_4(s)}v_j\mu_j}{\sum_{j=1}^{4}v_j\mu_j}\right)\right]\\&=0.151\end{aligned} \quad [10.6]$$

We see that the probabilities that the two solutions x_1 and x_2 transition to the optimal solution change significantly. We further find that these two probabilities both decrease, with the probability of x_1 decreasing from 0.107 to 0.049, and the probability of x_2 decreasing from 0.180 to 0.151.

Now suppose that the mutation rate probability p_m is 0.1 per bit. We can combine equations [5.45] and [5.46] in Chapter 5 to find the following transition probabilities in the noise-free and noisy cases:

$$\begin{aligned}\Pr_m(x_1 \to x_4)&=0.132\\\Pr_m(x_2 \to x_4)&=0.197\\\Pr_{m,\text{noise}}(x_1 \to x_4)&=0.082\\\Pr_{m,\text{noise}}(x_2 \to x_4)&=0.169\end{aligned} \quad [10.7]$$

We see that even with mutation, the probability of transitioning to the optimal solution x_4 decreases when noise corrupts the fitness evaluations. However, mutation tends to even out the probabilities. Without mutation, we saw that the probability of x_1 transitioning to the optimal solution decreases from 0.107 to 0.049, a decrease of 54%; and the probability of x_2 transitioning to the optimal solution decreases from 0.180 to 0.151, a decrease of 16%. However, with a mutation rate of 0.1, we saw that the probability of x_1 transitioning to the optimal solution decreases from 0.132 to 0.082, a decrease of 38%; and the probability of x_2 transitioning to the optimal solution decreases from 0.197 to 0.169, a decrease of 14%. Noise damages the migration mechanism of BBO, but some of that damage can be mitigated with a high mutation rate.

10.3. BBO with re-sampling

In this section, we apply re-sampling to BBO to improve BBO performance in noisy environments [MA 15a]. In noisy problems, evaluated fitness values include noise. Therefore, as we showed in the previous section, the evaluated values are not perfectly accurate, and they do not perfectly reflect the true value of the fitness. BBO uses the fitness of each solution to determine its immigration and emigration rates. Because of noise, evaluated fitness is not equal to true fitness, the immigration and emigration rates in BBO are incorrect, and this negatively affects BBO migration.

Re-sampling is a simple approach that samples the fitness of each candidate solution several times and calculates the average as the evaluated fitness. If we evaluate a fitness function for a given candidate solution ι times, and the noise values of those ι samples are independent, then the variance of the average fitness function decreases by a factor of ι [GRI 97, MIT 05].

Suppose that the ith sample $g_i(x)$ of the fitness function of a candidate solution x is given by

$$g_i(x) = f_t(x) + w_i \qquad [10.8]$$

where $f_t(x)$ is the true fitness, and w_i is the zero-mean additive noise with a variance of σ^2 at the ith measurement. If we re-sample the evaluated fitness function l times, the best estimate of the true fitness is

$$\hat{f}(x)=\frac{1}{l}\sum_{i=1}^{l}g_i(x) \qquad [10.9]$$

and the variance of $\hat{f}(x)$ is σ^2/l. Figure 10.2 illustrates this idea. The average of a set of l noisy fitness function evaluations is l times as accurate as a single evaluation.

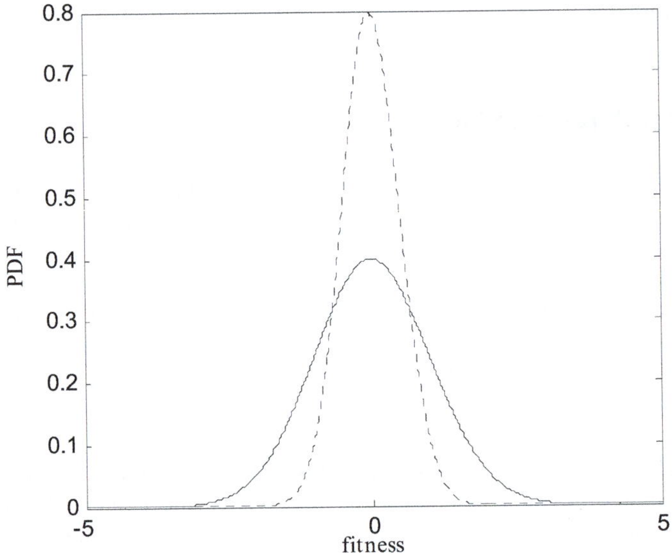

Figure 10.2. *The re-sampling strategy for noisy fitness function evaluation. The solid line shows the PDF of a noisy fitness function. The dashed line shows the PDF of the average of four fitness function evaluations. Both have a mean of zero, but the averaged evaluation has a variance that is 1/4 that of a single evaluation. The averaged evaluation is likely to be much closer to its mean than a single evaluation*

However, the re-sampling strategy is theoretically valid only if the fitness function evaluation noise is independent from one sample to the next. For instance, suppose that we measure the fitness of candidate solutions with noisy instrumentation. If the instrumentation noise is time-correlated with itself from one sample time to the next, then averaging l samples does *not* reduce the variance by a factor of l. In this case, the amount by which the variance is reduced depends on the noise correlation from one sample to the next.

If we have l fitness evaluations $\{g_i(x)\}$ of a candidate solution x, then we can find an estimate $\hat{\sigma}^2$ of the variance σ^2 of the fitness estimate as follows:

$$\hat{f}(x)=\frac{1}{l}\sum_{i=1}^{l}g_i(x)$$
$$\hat{\sigma}^2=\frac{1}{l-1}\sum_{i=1}^{l}\left(\hat{f}(x)-g_i(x)\right)^2 \qquad [10.10]$$

Intuitively, it seems that the equation for $\hat{\sigma}^2$ should have l instead of $(l-1)$ in the denominator, but the $(l - 1)$ term is preferred because it gives an unbiased estimate of the variance [SIM 06]. We can use equation [10.10] to see how many times we have to sample a noisy fitness function to achieve a desired variance in our fitness function evaluation. The desired variance is user-defined and depends on the particular problem. As $l \to \infty$, the variance goes to 0 and our fitness value estimate becomes error-free.

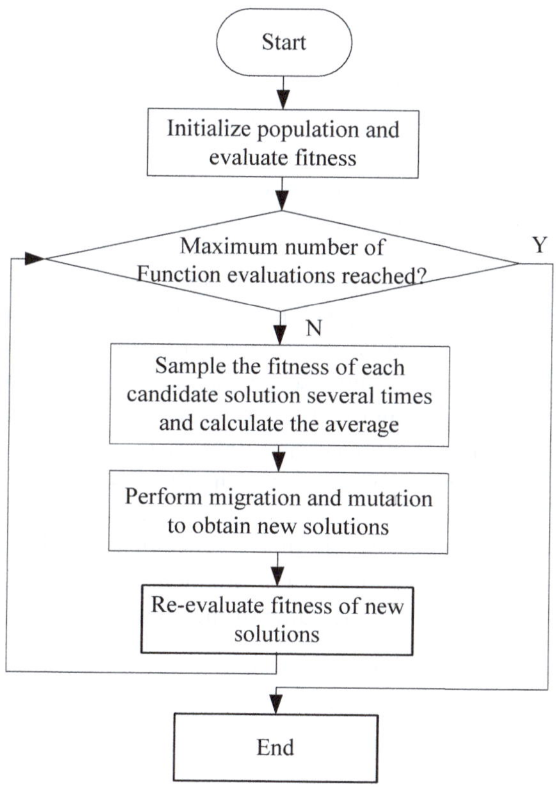

Figure 10.3. *Flowchart of BBO with re-sampling*

Re-sampling is a straightforward and effective way to handle noise in fitness functions, and one of the most important contributions of re-sampling is that it does not need any control parameters except for the number of re-samples. The flowchart of BBO with re-sampling is shown in Figure 10.3 [MA 15a]. It is worth pointing out that in Figure 10.3 we can use GA, PSO or any other EA instead of BBO to alleviate the effects of noise.

10.4. The Kalman BBO

The Kalman filter was invented by Kalman [KAL 60]. It is a recursive filter that can estimate states in a noisy environment. The contribution of the Kalman filter in noisy environments has been significant, and it is the theoretical foundation of many famous applications. One of the most important contributions of the Kalman filter is that it can provide an estimate of the true state of a dynamic system in a noisy environment.

The Kalman BBO is designed for problems with noisy fitness function evaluations. In noisy optimization problems, each fitness is the sum of the true fitness and a random noise. Therefore, the evaluated fitness is not equal to the true fitness. According to section 10.2, noise adversely affects the emigration and immigration rates of each solution. The Kalman filter provides a better estimate of the true fitness of the solution compared to the evaluated one.

The Kalman BBO assumes that the fitness of a given solution x is constant within a single Kalman filter estimation cycle. We further assume that fitness evaluation noise is not a function of x. Each fitness is a scalar, therefore the Kalman filter only needs to estimate a scalar for each solution, not a vector. In this case, the Kalman filter is simplified to its scalar version, keeps track of the uncertainty of each fitness estimate and thus reduces the complexity of the calculations compared to the vector version of the Kalman filter.

We denote the variance of a single fitness function evaluation as R. We denote the variance of the fitness estimate of a solution x after k fitness function evaluations as $P_k(x)$. We denote the value of the kth fitness function evaluation of x as $g_k(x)$. Finally, we denote our estimate of the fitness of x after k fitness function evaluations as $\hat{f}_k(x)$. With this notation, we can use Kalman filter theory to write:

$$\begin{aligned} \hat{f}_{k+1}(x) &= \hat{f}_k(x) + \frac{P_k(x)\left(g_{k+1}(x) - \hat{f}_k(x)\right)}{P_k(x) + R} \\ P_{k+1}(x) &= \frac{P_k(x) R}{P_k(x) + R} \end{aligned} \qquad [10.11]$$

For $k = 0, 1, 2, \cdots$. We initialize $P_0(x) = \infty$ for all x, which gives

$$\begin{aligned} \hat{f}_1(x) &= g_1(x) \\ P_1(x) &= R \end{aligned} \qquad [10.12]$$

That is, our estimate $\hat{f}_1(x)$ after the first fitness function evaluation $g_1(x)$ is simply equal to the first evaluation. Moreover, the uncertainty $p_1(x)$ in our fitness estimate after the first evaluation is simply equal to the uncertainty in the evaluation.

Equation [10.11] shows that each time we evaluate the fitness of x, we modify our estimate of $f(x)$ based on the previous estimate, its uncertainty and the most recent fitness function evaluation result $g(x)$. Equation [10.11] shows that:

$$\begin{aligned} \lim_{P_k(x) \to 0} \hat{f}_{k+1}(x) &= \hat{f}_k(x) \\ \lim_{P_k(x) \to \infty} \hat{f}_{k+1}(x) &= g_{k+1}(x) \end{aligned} \qquad [10.13]$$

In other words, if we are completely certain of the fitness of x (that is, $P_k(x) = 0$), then further evaluations of the fitness of x will not change our estimate of its fitness. On the other hand, if we are completely uncertain of the fitness of x ($P_k(x) = \infty$), then we will set our estimate of its fitness equal to the next fitness function evaluation result.

Equation [10.11] also shows that each time we evaluate the fitness of x, our uncertainty $P(x)$ in its value decreases; that is, our confidence in its estimated value increases. Equation [10.11] shows that:

$$\begin{aligned}
&\lim_{R\to 0} \hat{f}_{k+1}(x) = g_{k+1}(x) \\
&\lim_{R\to 0} P_{k+1}(x) = 0 \\
&\lim_{R\to \infty} \hat{f}_{k+1}(x) = \hat{f}_k(x) \\
&\lim_{R\to \infty} P_{k+1}(x) = P_k(x)
\end{aligned} \qquad [10.14]$$

These results agree with intuition. If the fitness function noise variance R is 0, then the fitness function evaluation is perfect, thus our estimate is simply equal to the fitness function evaluation result, and the uncertainty in our fitness function estimate is 0. On the other hand, if the fitness function noise variance R is infinite, then the noise is so large that fitness function evaluations do not provide us with any information. In this case, additional fitness function evaluations do not change our estimate of the fitness function value, nor do they reduce our uncertainty in its true value.

The Kalman BBO keeps track of the fitness function estimate $\hat{f}_k(x)$ and variance $P_k(x)$ for each solution x from one fitness function evaluation to the next $(k = 1, 2, \cdots)$. We allocate a user-defined fraction F of the available fitness function evaluations to generate and evaluate new solutions. We initialize our fitness estimate and variance for each new solution as described in equation [10.12]. We use the fraction $(1-F)$ of the available evaluations to re-evaluate existing solutions. After re-evaluations, we update our fitness estimate and variance as shown in equation [10.11]. Each time we have sufficient resources for a fitness function evaluation, we generate a random number r that is uniformly distributed on [0, 1]. If $r < F$, then we perform BBO migration and mutation to generate a new solution, and then we evaluate its fitness; otherwise, we re-evaluate an existing solution.

When it is time to re-evaluate an existing solution, we consider two guiding principles. First, we can generate more information by re-evaluating solutions whose fitness estimate variance is high. Second, we can generate more useful information by re-evaluating solutions whose estimated fitness is good. That is, we do not care too much about obtaining high precision in the estimate of low-fitness solutions because we are probably not interested in recombining them for future BBO generations. Stroud [STR 01] suggests the following strategy to select a solution x_s for re-evaluation:

$$\begin{aligned}&\bar{f} \leftarrow \text{mean of the population's estimated fitness values}\\&\sigma \leftarrow \text{standard deviation of the population's estimated fitness values}\\&x_s \leftarrow \arg\max\left\{P(x) : \hat{f}(x) > \bar{f} - \sigma\right\}\end{aligned} \qquad [10.15]$$

where we have omitted the subscript k on $\hat{f}(x)$ and $P(x)$ for convenience; we use the most recently updated values of $\hat{f}(x)$ and $P(x)$ for each solution x in equation [10.15]. The equation shows that among all solutions whose estimated fitness is greater than one standard deviation below the mean, we select the one with the largest uncertainty for re-evaluation. This strategy assumes that $f(x)$ is fitness, so large $f(x)$ is better than small $f(x)$.

10.5. Experimental results

In this section, we investigate the performance of BBO in noisy environments. A representative set of noiseless and noisy benchmark functions are used for performance testing. For the noiseless functions, we use the 13 benchmark functions briefly described in Table 3.2 in Chapter 3. A more detailed description of these functions can be found in Appendix A. The noisy benchmark functions are defined as

$$f_{Noisy}(x) = f_t(x) + |N(0,1)| \qquad [10.16]$$

where $|N(0,1)|$ is the absolute value of a Gaussian random variable with mean 0 and variance 1. Note that all benchmark functions are minimization problems.

We compare the performance of BBO in the noiseless case, BBO in the noisy case, BBO with re-sampling and Kalman BBO. The parameters used in each BBO algorithm are population size equal to 50, maximum immigration rate and maximum emigration rate equal to 1 and maximum mutation rate equal to 0.01 with a mutated value randomly chosen from a uniform distribution in the search domain. We use linear migration curves, l = 5 fitness re-samples for BBO with re-sampling and a fixed number of total fitness evaluations for each benchmark and each algorithm to provide fair performance comparisons. We terminate after

20,000 fitness function evaluations, and run 25 Monte Carlo simulations on each benchmark to obtain representative performances.

Table 10.1 summarizes the performance of the BBO algorithms on these benchmarks and shows the mean minimum values found by each algorithm. We observe that the performance of BBO is dramatically different for noiseless and noisy benchmark functions, and noise strongly affects BBO performance. However, when we incorporate re-sampling and Kalman filtering into BBO, we obtain good optimization results that are almost the same as those obtained for most of the noiseless benchmark functions. This result indicates that re-sampling and Kalman filtering can alleviate the effect of noise for these benchmark functions. Furthermore, we see that for the noisy benchmark functions, BBO with re-sampling performs better than Kalman BBO on six functions, and Kalman BBO performs better than BBO with re-sampling on the other seven functions. This result shows that BBO with re-sampling achieves almost the same performance as Kalman BBO for noisy optimization problems.

Function	BBO in noiseless case	BBO in noisy case	BBO with re-sampling	Kalman BBO
F01	2.17E–02	4.26E+01	6.57E–02	9.67E–02
F02	1.84E–03	1.78E+00	1.64E–02	7.65E–02
F03	6.33E–02	3.57E+01	7.52E–01	8.94E–02
F04	5.68E–14	4.28E+00	6.38E–14	7.83E–24
F05	9.24E–01	1.19E+01	9.65E–01	7.89E–01
F06	0.00E+00	7.80E+01	1.25E–10	7.82E–05
F07	1.37E–15	5.26E–01	4.58E–10	9.63E–10
F08	2.63E–06	3.45E+00	7.84E–06	3.41E–06
F09	1.55E–13	7.89E+02	3.25E–02	2.30E–02
F10	0.00E+00	1.31E+01	7.80E–05	1.17E–05
F11	7.49E–01	2.65E+02	9.65E–01	9.52E–01
F12	2.26E–30	5.74E+05	7.83E+00	7.84E+00
F13	1.28E–10	2.30E+02	4.57E–02	1.26E–02

Table 10.1. *Comparison of results for BBO in the noiseless case, BBO in the noisy case, BBO with re-sampling and Kalman BBO. The table shows the best solution achieved, averaged over 25 Monte Carlo simulations*

10.6. Conclusion

In this chapter, we investigated noisy fitness functions and noise effects on BBO performance using a Markov model. Analysis indicated that migration between candidate solutions, the most critical operation of BBO, can be corrupted by fitness function evaluation noise. The analysis was confirmed with an example using a BBO Markov model.

We used re-sampling and Kalman filtering in BBO to alleviate the effect of random noise on the fitness function evaluations of numerical benchmark functions. We also compared BBO in the noiseless case, BBO in the noisy case, BBO with re-sampling, and Kalman BBO. Our numerical simulations showed the following: (1) BBO is a powerful evolutionary algorithm for noiseless benchmark functions, but fitness function evaluation noise is indeed a problem for BBO; (2) BBO with re-sampling and Kalman BBO achieve almost the same optimization performance for noisy benchmark functions, and they can greatly improve the performance of BBO in noisy environments.

This chapter focused on the fitness of candidate solutions contaminated by additive and normally distributed noise. There are several important areas for further studies. First, in many real-world applications, different types of fitness function noise can be encountered, so it is of interest to combine BBO with re-sampling or Kalman filtering to address other types of fitness function noise. Also, other types of noise (besides fitness function noise) can arise in optimization. For example, in distributed optimization, some nodes might temporarily drop out due to communication glitches; or during experimental optimization, some parameters might be corrupted during fitness function evaluation. Future research could explore the effects of these and other types of noise on EA performance. Another important area for future work is to explore the optimization performance of BBO combined with other noise-handling methods; for example, dynamic re-sampling, which uses different re-sampling rates at different points in the search domain, or other types of filtering besides Kalman filtering.

11

Multi-objective BBO

Most real-world optimization problems are multi-objective, and therefore multi-objective optimization has been applied in many fields of science, engineering, economics and logistics. Multi-objective optimization typically includes multiple objectives which usually conflict. For example, minimizing vehicle cost while maximizing comfort, or maximizing vehicle performance while minimizing fuel consumption and pollutant emissions, involves two and three conflicting objectives, respectively.

Multi-objective optimization is also called multi-criteria optimization, multi-performance optimization and vector optimization. In this chapter, we assume that a candidate solution is n-dimensional, and that our multi-objective optimization problem (MOP) is a minimization problem for each objective. An MOP can then be written as follows:

$$\min_x f(x) = \min_x \{f_1(x), f_2(x), \cdots, f_k(x)\} \qquad [11.1]$$

That is, we want to minimize a vector $f(x)$ of functions. Of course, we cannot minimize a vector in the typical sense of the word *minimize*. Nevertheless, our goal in an MOP is to simultaneously minimize all k functions $f(x)$. MOPs were first solved by evolutionary algorithms in [ROS 67], and we call those implementations multi-objective evolutionary algorithms (MOEAs). MOEAs have been widely studied by the operations research community for many years [SCH 85, EHR 05, SIM 13a, SIM 13b, ZIT 04].

Overview of the chapter

In this chapter, we discuss how to modify biogeography-based optimization (BBO) for MOPs. Section 11.1 provides some notation and concepts that often occur in MOPs. Section 11.2 shows how we can combine BBO with some well-known MOEA approaches. The combination of BBO with various MOEA approaches results in several multi-objective biogeography-based optimization (MOBBO) algorithms. Section 11.3 presents an application of MOBBO to an automatic warehouse scheduling problem. The concluding section of this chapter provides references to additional resources and suggests several important topics for future MOBBO research.

11.1. Multi-objective optimization problems

This section outlines some basic notation and concepts that are related to MOPs. We first list some definitions that are often used in multi-objective optimization:

1) *Domination*: a solution x^* is said to dominate x if the following two conditions hold: (1) $f_i(x^*) \le f_i(x)$ for all $i \in [1, k]$ and (2) $f_j(x^*) < f_j(x)$ for at least one $j \in [1, k]$. That is, x^* is at least as good as x for all objective function values, and it is better than x for at least one objective function value. We use the notation:

$$x^* \succ x \qquad [11.2]$$

to indicate that x^* dominates x. This notation can be confusing because the symbol $\succ$ looks like a "greater than" symbol, but since we deal with minimization problems in this chapter, the symbol $\succ$ means that the function values of x^* are less than or equal to those of x. However, this notation is standard in the literature, so this is the notation that we use. The statement "x^* is superior to x" is identical to the statement "x^* dominates x".

2) *Weak domination*: a solution x^* is said to weakly dominate x if $f_i(x^*) \le f_i(x)$ for all $i \in [1, k]$. That is, x^* is at least as good as x for all objective function values. Note that if x^* dominates x, then it also weakly dominates x.

Also note that if $f_i(x^*) = f_i(x)$ for all $i \in [1, k]$, then x^* and x weakly dominate each other. We use the notation:

$$x^* \succeq x \quad [11.3]$$

to indicate that x^* weakly dominates x. Some authors use the equivalent terminology that x^* covers x.

3) *Non-dominated*: a solution x^* is said to be non-dominated if there is no x that dominates it. Non-inferior, admissible and efficient are synonyms for non-dominated.

4) *Pareto optimal solutions*: a Pareto optimal solution x^*, also called a Pareto solution, is one that is not dominated by any other x in the search space. That is,

$$\{x^* \text{ is Pareto optimal}\} \Leftrightarrow$$

$$\{\nexists\, x : (f_i(x) \le f_i(x^*) \text{ for all } i \in [1, k],$$

$$\text{and } f_j(x) < f_j(x^*) \text{ for some } j \in [1, k])\} \quad [11.4]$$

5) *Pareto optimal set*: the Pareto optimal set, also called the Pareto set and denoted as P_s, is the set of all x^* that are non-dominated:

$$P_s = \{x^* : [\nexists\, x : (f_i(x) \le f_i(x^*) \text{ for all } i \in [1, k],$$

$$\text{and } f_j(x) < f_j(x^*) \text{ for some } j \in [1, k])]\} \quad [11.5]$$

The Pareto set is also called the efficient set, and it is sometimes called the admissible set, although this latter term usually implies constraint satisfaction rather than Pareto optimality.

6) *Pareto front*: the Pareto front, also called the non-dominated set and denoted as P_f, is the set of all vector functions $f(x)$ corresponding to the Pareto set:

$$P_f = \{f(x^*) : x^* \in P_s\} \quad [11.6]$$

EXAMPLE 11.1.–

This example uses an airplane trip to illustrate Pareto optimal solutions and a Pareto front. Suppose that the two objectives in this problem are travel time and ticket price. We find the tickets shown in Table 11.1 for sale.

Ticket	Travel time (hrs)	Ticket price ($)
A	10	1,700
B	9	2,000
C	8	1,800
D	7.5	2,300
E	6	2,200

Table 11.1. *Travel time and ticket price options of an airplane trip*

If we compare tickets A and B, we cannot say that either is superior without knowing the relative importance of travel time versus ticket price. However, we can see that C is better than B in both objectives, so we say that C dominates B. As long as C is a feasible option, there is no reason we would choose B.

We also see that D is dominated by E. The rest of the options (A, C and E) have a trade-off associated with travel time versus ticket price, so none is clearly superior to the others. We call this the non-dominated set of solutions because none of the solutions are dominated. Usually, solutions of this type form a typical shape, as shown in Figure 11.1. Solutions that lie along the line are non-dominated solutions, while those that lie above and to the right of the line are dominated because there is always at least one solution on the line that has at least one objective that is better. The line is called the Pareto front and solutions on it are called Pareto optimal solutions. All Pareto optimal solutions are non-dominated. It is important in MOPs to find solutions as close as possible to the Pareto front.

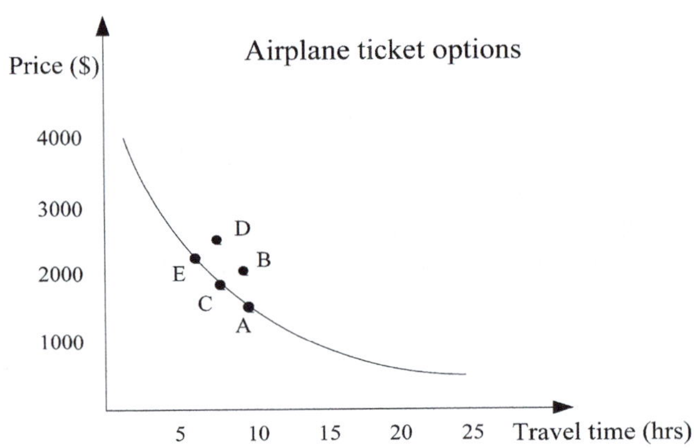

Figure 11.1. *Pareto optimal solutions and Pareto front example using an airplane trip*

It is important to determine performance metrics for multi-objective optimization algorithms; that is, how can we judge the performance of an MOEA? For a single-objective optimization algorithm, performance is usually straightforward: performance is quantified by the minimum value of the cost function that is determined. However, even in single-objective optimization, we might be interested in several different performance metrics for an EA. We might be interested not only in finding the minimum cost function value, but also in quickly finding a "good" solution that is not necessarily the best. We might also be interested in finding many good solutions in diverse regions of the search space. So even in the apparently straightforward problem of single-objective optimization, we may have several possible performance metrics. This complication increases with multiple-objective optimization. Some potential criteria for MOEA performance might be the following [ZIT 03]:

1) Maximize the number of solutions that we find within a certain distance of the true Pareto set.

2) Minimize the average distance between the MOEA-approximated Pareto set and the true Pareto set.

3) Maximize the diversity of the solutions that we find in the approximated Pareto set.

4) Minimize the distance from the MOEA-approximated Pareto front to an ideal point.

Criteria 1 and 2 are concerned with finding the "best" approximation of the true Pareto set. Criterion 3 is concerned with finding a diverse set of solutions so that the human decision maker has enough resources to make an informed decision among the possible trade-offs. In contrast to the other criteria, Criterion 4 is concerned with finding solution candidates that are as close as possible to the decision maker's ideal solution, which may or may not exist. However, most MOEAs are primarily concerned with finding the best approximation to the true Pareto set.

Criteria 1 and 2 assume that we know the true Pareto set in the first place, so those criteria might be useful when testing MOEAs on well-understood benchmarks, but the criteria are useless (unless modified) when running an MOEA on a real-world optimization problem with an unknown Pareto set. But if we know the true

Pareto set P_s, and an MOEA gives us an approximate Pareto set $\hat{P}_s$, the average distance $M(P_s, \hat{P}_s)$ between them can be computed as:

$$M_1(P_s, \hat{P}_s) = \frac{1}{|\hat{P}_s|} \sum_{x \in \hat{P}_s} \min_{x^* \in P_s} \|x^* - x\| \qquad [11.7]$$

where $\|\cdot\|$ is any user-specified distance metric.

Criterion 3 can be measured in a few different ways. First, we could measure the average distance of each solution to its nearest neighbor in the approximated Pareto set. Second, we could measure the distance between the two extreme solutions in the approximated Pareto set. Third, we could compute the normalized number of solutions that are beyond a certain threshold from each element in the approximate Pareto set [ZIT 00]:

$$M_2(\hat{P}_s) = \frac{1}{|\hat{P}_s|} \sum_{x \in \hat{P}_s} \left| x' \in \hat{P}_s : \|x' - x\| > \sigma \right| \qquad [11.8]$$

where σ is a user-specified distance threshold. In general, M_2 increases as the number of elements in $\hat{P}_s$ increases, and also as the diversity of the elements in $\hat{P}_s$ increases. Khare *et al.* [KHA 03] discusses some additional diversity metrics for MOPs.

Criterion 4 is called target vector optimization [WIE 92], goal attainment [WIL 93] or goal programming. It assumes that the user is thinking of some ideal point in objective function space, and it requires a definition of "distance". Usually, we use the Euclidean distance D_2, also called the two-norm distance, between an objective function vector f and an ideal point f^*. The squared distance between f and f^* is defined as follows:

$$D_2^2(f^*(x), f(x)) = \|f^*(x) - f(x)\|_2^2 = \sum_{i=1}^{k} (f_i^*(x) - f_i(x))^2 \qquad [11.9]$$

However, we can also use other distance measures, such as the weighted two-norm, the one-norm or the infinity-norm.

EXAMPLE 11.2.–

Figure 11.2 shows the performance of three different EAs on an MOP. Figure 11.2(a) shows a solution that is fairly diverse and reasonably close to the true Pareto front. Figure 11.2(b) shows a solution that is more diverse than Figure 11.2(a) in the sense that the distance between the extreme solutions is farther, but Figure 11.2(b) includes only three solutions while Figure 11.2(a) includes four solutions. Figure 11.2(c) shows solutions that are closer to the true Pareto front than Figure 11.2(a) or (b), but the diversity is not good. Which of the three solutions is "best"? It depends on the priorities of the decision maker.

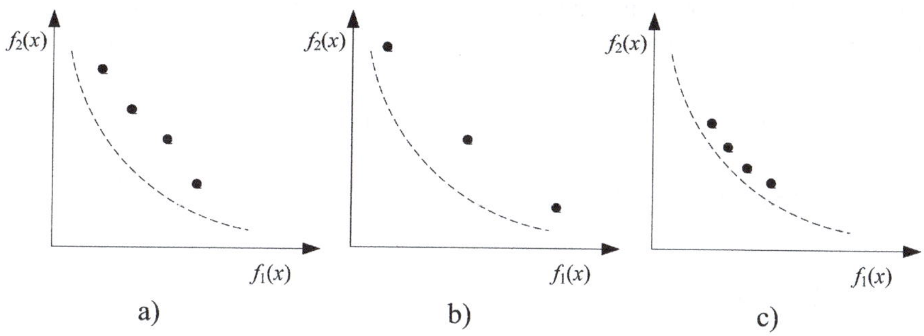

Figure 11.2. *Three potential EA solutions to a two-objective MOP, where the true Pareto front is the dotted line and the circles are the approximations that were found by each EA. Which solution is "best"? It depends on the priorities of the decision maker with respect to solution diversity and closeness to the true Pareto front*

Another metric that researchers often use to measure the quality of a Pareto front is its hypervolume. Suppose that an MOEA has found M points in an approximate Pareto front $\hat{P}_f = \{f(x_j)\}$ for $j = [1, M]$, where $f(x_j)$ is a k-dimensional function. The hypervolume can be computed as:

$$S(\hat{P}_f) = \sum_{j=1}^{M} \prod_{i=1}^{k} f_i(x_j) \qquad [11.10]$$

Given two MOEAs that compute two Pareto front approximations to a given MOP, we can use the hypervolume measure to quantify how good the two approximations are relative to each other. For a minimization problem, a smaller hypervolume indicates a better Pareto front approximation.

Hypervolume cannot be blindly used as an indicator of Pareto front quality. Equation [11.10] shows that an empty Pareto front approximation ($M = 0$) gives the smallest possible value of S. Therefore, a more accurate measure is the normalized hypervolume, which is given as:

$$S_n(\hat{P}_f) = \frac{S(\hat{P}_f)}{M} \qquad [11.11]$$

EXAMPLE 11.3.–

Suppose that we have two MOEAs, each of which is designed to approximate the Pareto front of a two-objective minimization problem. Figure 11.3 shows their Pareto front approximations. Figure 11.3(a) has the Pareto front approximation points:

$$\hat{P}_f(1) = \{[f_1(x_j), f_2(x_j)]\} = \{[1,5],[2,3],[5,1]\} \qquad [11.12]$$

which has the hypervolume $5+6+5=16$. Figure 11.3(b) has the Pareto front approximation points:

$$\hat{P}_f(2) = \{[f_1(x_j), f_2(x_j)]\} = \{[1,4],[3,3],[4,1]\} \qquad [11.13]$$

which has the hypervolume $4+9+4=17$. According to the hyper volume measure of equation [11.10], Figure 11.3(a) gives a slightly better $\hat{P}_f$ than Figure 11.3(b).

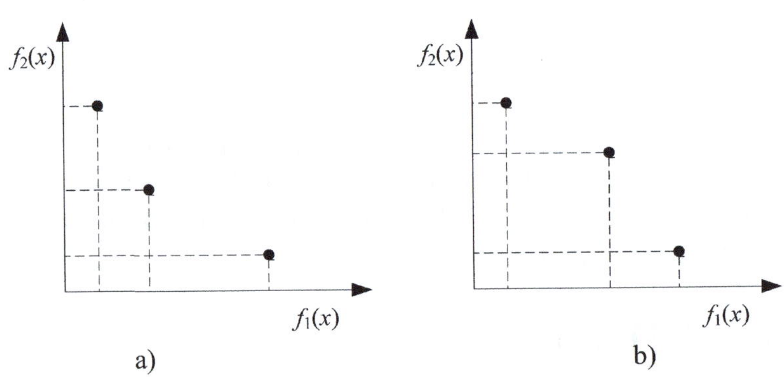

Figure 11.3. *Two Pareto front approximations to a two-objective MOP. A hypervolume measurement is used to quantify the goodness of each approximation. The approximation on the left has a hypervolume of 16, and the approximation on the right has a hypervolume of 17*

11.2. Multi-objective BBO

This section shows how BBO is combined with some of the popular MOEA approaches in the literature to obtain various MOBBO algorithms. This section could also serve as a template for the extension of any other EA to multi-objective optimization.

11.2.1. *Vector evaluated BBO*

Vector evaluated biogeography-based optimization (VEBBO) combines ideas from BBO with the vector evaluated genetic algorithm (VEGA). First, we recall the VEGA algorithm, which was one of the original MOEAs and operates by performing selection on the population using one objective function at a time [SCH 85]. This gives a set of subpopulations, one set for each objective function. We then select solutions from the subpopulations to obtain the parents for the next generation, and combine the parents using standard recombination methods to obtain children. Figure 11.4 gives an outline of VEGA.

Initialize a population of candidate solution $P=\{x_j\}$ for $j\in[1,N]$

$M\leftarrow\lceil N/K\rceil$

While not (termination criterion)

    Compute the cost $f_i(x_j)$ for each objective i for each solution $x_j\in P$

    For each objective i where $i\in[1,k]$

        $P_i\leftarrow M$ solutions probabilistically selected from P using $f_i(\cdot)$

    Next objective

    $P\leftarrow N$ solutions selected from $\{P_1,\cdots,P_k\}$

    $C\leftarrow N$ children created from recombining solutions from P

    Probabilistically mutate the children in C

    $P\leftarrow C$

Next generation

Figure 11.4. *Outline of VEGA for solving an optimization problem with k objectives and a population size of N*

Figure 11.4 shows that VEGA begins with a population of N candidate solutions that we usually generate randomly. At each generation, we compute the value of all k objective function values for all N solutions. We then use any desired selection scheme to select M solutions, where $M=\lceil N/K\rceil$ is the smallest integer that is greater than or equal to N/K. We perform this selection probabilistically, first using $f_1(\cdot)$ to create population P_1, then using $f_2(\cdot)$ to create population P_2, and so on. After we have created the P_i subpopulations, we combine them to obtain a parent population P. We then recombine the solutions in P to create a set of children C. We can perform recombination using any EA method. We see that the name VEGA is somewhat of an anachronism; depending on the recombination method that we use, we could call it VEBBO if we use BBO migration for recombination.

Initialize a population of candidate solutions $P = \{ x_j \}$ for $j \in [1, N]$
While not (termination criterion)
  Compute the cost $f_i(x_j)$ for each objective i for each solution $x_j \in P$
  For each objective i where $i \in [1,k]$
    $\gamma_{ji} \leftarrow$ rank of x_j with respect to the ith objective function for $j \in [1,N]$
    Immigration rates $\lambda_{ji} \leftarrow \gamma_{ji} / \sum_{q=1}^{N} \gamma_{qi}$ for $j \in [1,N], i \in [1,k]$
    Emigration rates $\mu_{ji} \leftarrow 1 - \lambda_{ji}$ for $j \in [1,N], i \in [1,k]$
    For each solution x_j where $j \in [1,N]$
      For each decision variable $s \in [1, n]$
        $k_i \leftarrow \text{rand}(1, k)$ = uniformly distributed integer between 1 and k
        $\delta \leftarrow \text{rand}(0, 1)$ = uniformly distributed real number between 0 and 1
        If $\delta < \lambda_{j,k_i}$ then
          $k_e \leftarrow \text{rand}(1, k)$ = uniformly distributed integer between 1 and k
          Probabilistically select emigrant x_e, where $\Pr(x_e = x_\beta) = \mu_{\beta,k_e} / \sum_{q=1}^{N} \mu_{q,k_e}$ for $\beta \in [1, N]$
          $x_j(s) \leftarrow x_e(s)$
        End if
      Next decision variable
    Next solution
  Next objective
  Probabilistically mutate the population P as in the standard BBO algorithm
Next generation

Figure 11.5. *Outline of VEBBO for solving an n-dimensional optimization problem with k objectives and a population size of N. At each generation, the best solution x_b with respect to the ith objective value has rank $\gamma_{bi} = 1$, and the worst solution x_w has rank $\gamma_{wi} = N$*

VEBBO first produces a set of subpopulations, one set for each objective function. Then, solutions are selected from the subpopulations to obtain parents, which create children by using the BBO migration method. The outline of VEBBO for a k-objective optimization problem is shown in Figure 11.5, where BBO immigration is based on the k_ith objective function value of each solution, and k_i is a random objective function index at the ith migration trial. Then, emigration is based on the k_eth objective function value of each solution, where k_e is also a random objective function index.

11.2.2. *Non-dominated sorting BBO*

Next, non-dominated sorting BBO (NSBBO) is proposed, which combines BBO with the non-dominated sorting genetic algorithm (NSGA). Recall that NSGA, which was one of the original MOEAs, assigns the cost of each solution based on its dominance level [DEB 02, SRI 94]. First, all solutions are copied to a temporary population T. Then, we find all non-dominated solutions in T; these solutions, which are denoted as the set B, are assigned the lowest cost value. Recall that a solution x is dominated by a solution x^* if x^* performs at least as good as x in all objectives, and performs better than x in at least one objective. A solution is called non-dominated if there are no solutions in the population (T in this case) that dominate it. Next, B is removed from T, and we find all non-dominated solutions in the reduced set T. These solutions are assigned the second-lowest cost value. This process is repeated to obtain a cost for each solution that is based on its level of non-domination. Figure 11.6 gives an outline of NSGA.

Figure 11.6 shows that we begin with a population of N candidate solutions, usually generated randomly. At each generation, we compute the value of all k objective function values for all N solutions. We copy the solutions to a temporary population T. We assign a cost value of 1 to all solutions that are non-dominated. We remove all of those solutions from T, find all the solutions in the reduced set T that are non-dominated, and assign them a cost value of 2. We repeat this process until all solutions have been assigned a cost value based on their level of domination. We then use the cost values $\phi(\cdot)$ in Figure 11.6 to perform selection, and we recombine selected solutions using any desired EA recombination method. Finally, we mutate the child population, replace the parents with the children and continue to the next generation.

```
Initialize a population of candidate solutions P = {x_j} for j ∈ [1, N]
While not (termination criterion)
    Temporary population T ← P
    Non-domination level c ← 1
    While |T| > 0
        B ← non-dominated solutions in T
        Cost φ(x) ← c for all x ∈ B
        Remove B from T
        c ← c + 1
    End while
    C ← N children created from recombining the solutions in P
    Probabilistically mutate the children in C
    P ← C
Next generation
```

Figure 11.6. *Outline of NSGA for solving an optimization problem with k objectives. We use the cost values $\phi(x_j)$ to select parents for recombination*

NSGA-II is a modification of NSGA [DEB 02]. NSGA-II computes the cost of a solution by taking into account not only the solutions that dominate it, but also the solutions that it dominates. For each solution, we also compute a crowding distance by finding the distance to the nearest solutions along each objective function dimension. We use the crowding distance to modify the fitness of each solution. NSGA sets the crowding distance of each solution equal to the average distance to its nearest neighbors along each objective function dimension. For example, suppose that we have N solutions. Further suppose that solution x has the objective function vector:

$$f(x) = [f_1(x), \cdots, f_k(x)] \qquad [11.14]$$

For each objective function dimension, we find the closest larger value and the closest smaller value in the population as follows:

$$\begin{aligned} f_j^-(x) &= \max_{x^*}\left\{f_j\left(x^*\right) \text{ such that } f_j\left(x^*\right) < f_j(x)\right\} \\ f_j^+(x) &= \min_{x^*}\left\{f_j\left(x^*\right) \text{ such that } f_j\left(x^*\right) > f_j(x)\right\} \end{aligned} \qquad [11.15]$$

We then compute the crowding distance of x as:

$$d(x) = \sum_{j=1}^{k}\left(f_j^+(x) - f_j^-(x)\right) \qquad [11.16]$$

Solutions that are in more crowded regions of the objective function space tend to have a smaller crowding distance. Solutions at the extreme values of the objective function space have an infinite crowding distance:

$$d(x) = \infty \text{ for } x \in \left\{\arg\min_{x^*} f_i(x^*) \cup \arg\max_{x} f_i(x^*) \text{ for all } i \in [1, k]\right\} \qquad [11.17]$$

The crowding distance of x corresponds to half of the perimeter of the largest hypercube, called a cuboid [DEB 00a, DEB 00b], whose boundaries do not extend beyond the objective function space coordinates of the nearest neighbors of x in each dimension.

Now that the crowding distance has been computed for each solution in the population, we use crowding distance as a secondary sorting parameter for obtaining the rank of each solution. As in the NSGA algorithm of Figure 11.6, we rank each solution on the basis of its non-domination level, but we also use a more fine-grained ranking metric using crowding distance. That is, x is ranked better than x^* if $\phi(x) < \phi(x^*)$, or if $\phi(x) = \phi(x^*)$ and $d(x) > d(x^*)$. While NSGA uses $\phi(x)$ to select parents for recombination in Figure 11.6, NSGA-II instead uses the ranks described above to select parents for recombination.

Next we combine BBO with NSGA by changing the recombination logic in NSGA to BBO migration operations, which results in the NSBBO algorithm shown in Figure 11.7.

Initialize a population of candidate solutions $P = \{x_j\}$ for $j \in [1, N]$
While not (termination criterion)
    Temporary population $T \leftarrow P$
    Non-domination level $c \leftarrow 1$
    While the temporary population size $|T| > 0$
        $B \leftarrow$ non-dominated solutions in T
        Cost $f(x) \leftarrow c$ for all $x \in B$
        Remove B from T
        $c \leftarrow c + 1$
    End while
    Immigration rates $\lambda_j \leftarrow f(x_j) / \sum_{q=1}^{N} f(x_q)$ for $j \in [1, N]$
    Emigration rates $\mu_j \leftarrow 1 - \lambda_j$ for $j \in [1, N]$
    For each solution x_j where $j \in [1, N]$
        For each decision variable $s \in [1, n]$
            $\delta \leftarrow \text{rand}(0, 1)$ = uniformly distributed real number between 0 and 1
            If $\delta < \lambda_j$ then
                Probabilistically select emigrant x_e, where $\Pr(x_e = x_\beta) = \mu_\beta / \sum_{q=1}^{N} \mu_q$ for $\beta \in [1, N]$
                $x_j(s) \leftarrow x_e(s)$
            End if
        Next decision variable
    Next solution
    Probabilistically mutate the population P as described in the standard BBO algorithm
Next generation

Figure 11.7. *Outline of NSBBO for solving an n-dimensional optimization problem with k objectives and a population size of N*

11.2.3. *Niched Pareto BBO*

Niched Pareto BBO (NPBBO) combines BBO with the niched Pareto genetic algorithm (NPGA). NPGA was one of the original MOEAs and is similar to NSGA in its assignment of cost on the basis of domination [HOR 94]. NPGA is an attempt to reduce the computational effort of NSGA. Two candidate solutions x_1 and x_2 are randomly selected from the population, and then a subset S of the population is also randomly selected, which is typically around 10% of the population. If one of the solutions x_1 and x_2 is dominated by any of the solutions in S, and the other is not, then the non-dominated solution, denoted as x_0, wins the tournament and is selected for recombination. If both solutions x_1 and x_2 are dominated by at least one solution in S, or both solutions are not dominated by any solutions in S, then fitness sharing is used to decide the tournament winner; that is, the solution in the least crowded region of the objective function space wins the tournament. This selection process can be described as follows:

$$\begin{aligned} d_i &= \left| x^* \in S : x^* \succ x_i \right| \quad \text{for } i \in [1, 2] \\ c_i &= \text{Crowding distance of } x_i \text{ for } i \in [1, 2] \\ r &= \begin{cases} x_1 & \text{if} \begin{cases} (d_1 = 0) \text{ and } (d_2 > 0), \text{ or} \\ (d_1 > 0) \text{ and } (d_2 > 0) \text{ and } (c_1 < c_2), \text{ or} \\ (d_1 = 0) \text{ and } (d_2 = 0) \text{ and } (c_1 < c_2) \end{cases} \\ x_2 & \text{otherwise} \end{cases} \end{aligned} \qquad [11.18]$$

where d_i is the number of solutions that dominate x_i, c_i is the crowding distance of x_i and r is the solution (either x_1 or x_2) that we select for recombination. The crowding distance c_i could be computed by equation [11.16]. The crowding distance is smaller for solutions that are in more crowded regions of the search space or the objective function space. Figure 11.8 gives an outline of NPGA.

```
Initialize a population of candidate solutions P = {x_j} for j ∈ [1, N]
While not (termination criterion)
    R ← ∅
    While |R| < N
        Randomly select two solutions x_1 and x_2 from P
        Randomly select a population subset S ⊂ P
        Use equation [11.18] to select r from {x_1, x_2}
      P ← {R, r}
    End while
    Recombine the solutions in R to obtain N children
    Probabilistically mutate the children
   P ← children
Next generation
```

Figure 11.8. *Outline of NPGA for solving an optimization problem with k objectives and a population size of N*

Initialize a population of candidate solutions $P = \{x_j\}$ for $j \in [1, N]$
While not (termination criterion)
  Temporary population $T \leftarrow \phi$
  While the temporary population size $|T| < N$
    Randomly select two solutions x_1 and x_2 from P
    Randomly select a population subset $S \subset P$
    Use equation [11.18] to select x_0 from $\{x_1, x_2\}$
    $T \leftarrow \{T, x_0\}$
  End while
  For each solution $x_j \in T$, where $j \in [1, N]$
    For each decision variable $s \in [1, n]$
      $\delta \leftarrow \text{rand}(0, 1)$ = uniformly distributed real number between 0 and 1
      If $\delta < 1/N$ then
        Probabilistically select emigrant x_e, where $\Pr(x_e = x_\beta) = 1/N$ for $\beta \in [1, N]$
        $x_j(s) \leftarrow x_e(s)$
      End if
    Next decision variable
  Next solution
  Probabilistically mutate the population P as described in the standard BBO algorithm
Next generation

Figure 11.9. *Outline of NPBBO for solving an n-dimensional optimization problem with k objectives and a population size of N*

11.2.4. *Strength Pareto BBO*

Strength Pareto BBO (SPBBO) combines BBO with the strength Pareto evolutionary algorithm (SPEA). SPEA was one of the original MOEAs, and was the first MOEA to explicitly use elitism [ZIT 99, ZIT 04]. Of course, any of the MOEAs can be implemented with elitism, but for some reason most of them did not incorporate elitism when originally introduced. Elitism is usually a common-sense option in both single-objective and multi-objective EAs. Also, elitism is theoretically necessary to guarantee convergence in some MOEAs [RUD 00].

SPEA maintains all non-dominated solutions that are found during the evolutionary process in an archive. Whenever we find a non-dominated solution, we copy it to the archive. We assign a strength value $S(\alpha)$ to each archived solution based on the number of solutions in the population that α dominates:

$$S(\alpha) = \frac{\left| x \in \{P\} \text{ such that } \alpha \succ x \right|}{N+1} \text{ for all } \alpha \in A \qquad [11.19]$$

where P is the set of candidate solutions, N is the size of P and A is the archive set. Note that $S(a) \in [0,1)$. For each solution x in P, we find the set $\alpha(x)$ of all archived solutions that dominate it. We then compute the raw cost of x, denoted as $R(x)$, as the sum of the strengths of the solutions in $\alpha(x)$:

$$R(x) = 1 + \sum_{x^* \in \alpha(x)} s(x^*), \text{ for all } x \in P$$
$$\text{where } \alpha(x) = \{x^* \in A : x^* \succ x\}. \qquad [11.20]$$

Adding one in equation [11.20] ensures that $R(x) \geq 1$, which in turn ensures that $R(x) > S(\alpha)$ for all $x \in P$ and all $\alpha \in A$. Note that if x has a low raw cost, then x is a high-performing solution.

As mentioned above, at each generation, all solutions in $\{P, A\}$ that are non-dominated are added to the archive A. However, this can result in unbounded growth of the archive. SPEA handles this potential problem with a clustering method [ZIT 99]. Supposing that the archive has $|A|$ solutions, we define each solution as a cluster. We then merge the two closest clusters into a single cluster so that the cluster count of A is reduced by one. We repeat this process until the archive contains N_A clusters, which is the desired archive size. Finally, we retain only one point from each cluster, usually the one that is closest to the cluster center. Figure 11.10 gives an outline of SPEA, where N is the population size, N_A is the maximum archive size and usually $N_A < N$.

Initialize a population of candidate solutions $P = \{x_j\}$ for $j \in [1, N]$
Initialize the archive A as the empty set
While not (termination criterion)
    Copy non-dominated solutions from P to A:
    $A \leftarrow \{A \cup \{x \in P : \nexists (x^* \in \{P, A\} : x^* \succ x)\}$
    Remove dominated solutions from A
    While $|A| < N_A$
        Use a clustering method to remove a solution from A
    End while
    Use equation [11.20] to calculate the cost of each solution in P
    Select parents from $\{P, A\}$
    Use a recombination method to create children C from the parents
    Probabilistically mutate the child population C
    Use a replacement method to replace solutions in P with solutions from C
Next generation

Figure 11.10. *Outline of SPEA for solving an optimization problem with k objectives and a population size of N*

We can modify Figure 11.10 for BBO by changing the "Select parents" statement and the "Use a recombination method" statement. We do this by calculating migration rates with the raw cost of equation [11.20]. We can then implement BBO migration using these rates. In this section, we take the SPEA approach in which parents can be selected from both the population P and the archive A. This results in the SPBBO algorithm of Figure 11.11.

EXAMPLE 11.4.–

In this example, we present simulation results for the MOBBO algorithms presented above. We test four MOBBO algorithms (VEBBO, NSBBO, NPBBO and SPBBO) on a set of 10 unconstrained functions and 10 constrained functions from the CEC 2009 benchmark set [ZHA 08]. These functions are summarized in Appendix C, where U01–U10 are unconstrained multi-objective benchmark functions and C01–C10 are constrained multi-objective benchmark functions. U01–U07 and C01–C07 are two-objective problems, and U08–U10 and C08–C10 are three-objective problems. The constrained multi-objective benchmark functions include one or two inequality constraints.

Initialize a population of candidate solutions $P=\{x_j\}$ for $j\in[1, N]$

Initialize the archive A as the empty set

While not (termination criterion)

Copy non-dominated solutions from P to A :

$A \leftarrow \{A \bigcup \{x\in P : \nexists(x^* \in \{P, A\} : x^* \succ x)\}$

Remove dominated solutions from A

While $|A| < N_A$

Use a clustering method to remove a solution from A

End while

Use equation [11.19] to calculate the strength $S(\alpha)$ of each solution $\alpha\in A$

Calculate the cost $R(\alpha)\leftarrow 1-S(\alpha)$ for each $\alpha\in A$

Use equation [11.20] to calculate the cost $R(x)$ of each solution $x\in P$

Immigration rates $\lambda_j \leftarrow R(x_j)/\sum_{q=1}^{|P|} R(x_q)$ for all $x_j\in P$

Emigration rates $\mu_j \leftarrow R(x_j)/\sum_{q=1}^{|P|+|A|} R(x_q)$ for all $x_j\in \{P, A\}$

For each solution $x_j\in P$, where $j\in[1, N]$

For each decision variable $s\in[1, n]$

$r\leftarrow$ rand $(0,1)$

If $r<\lambda_j$ then

Probabilistically select emigrant x_e,

where $\Pr(x_e = x_m)=\mu_m / \sum_{q=1}^{|P|+|A|}\mu_q$ for $x_m\in\{P, A\}$

$x_j(s)\leftarrow x_e(s)$

End if

Next decision variable

Next solution

Probabilistically mutate the population C as described in the standard BBO algorithm

Use a replacement method to replace solutions in P with solutions from C

Next generation

Figure 11.11. *Outline of SPBBO for solving an n-dimensional optimization problem with k objectives and a population size of N*

We use a population size of 150 and a mutation rate of 0.01 per solution decision variable per generation. If mutation occurs, the mutated value of the new independent variable is uniformly distributed in the search space. For constrained multi-objective benchmark functions, we incorporate constraints into MOEAs in the same way that we incorporate them into single-objective EAs (see Chapter 9). We evaluate each algorithm 30 times, with a maximum number of function evaluations equal to 300,000 for each simulation. We use the hypervolume and normalized hypervolume as performance metrics.

Tables 11.2 and 11.3 summarize the performance comparisons of the four MOBBO algorithms. It can be seen from Table 11.2 that for unconstrained benchmark functions, SPBBO performs best on six functions (U01, U03, U04, U05, U07 and U09) and NPBBO performs best on the other four functions (U02, U06, U08 and U10). It can be seen from Table 11.3 that for constrained benchmark functions, SPBBO performs best on six functions (C01, C03, C04, C06, C07 and U09), NPBBO performs best on three functions (C02, C05 and C10) and VEBBO performs best on C08. In summary, we can say that SPBBO performs better than the other three MOBBO algorithms for both the unconstrained and constrained multi-objective benchmark functions.

Functions	Multi-objective BBO			
	VEBBO	NSBBO	NPBBO	SPBBO
U01	(58.59, 0.26)	(73.61, 0.33)	(51.56, 0.23)	**(42.51, 0.19)**
U02	(24.39, 0.26)	(30.58, 0.33)	**(17.37, 0.18)**	(22.35, 0.24)
U03	(288.9, 0.24)	(398.4, 0.33)	(232.9, 0.21)	**(231.6, 0.21)**
U04	(8.549, 0.24)	(11.45, 0.31)	(8.373, 0.23)	**(8.012, 0.22)**
U05	(286.7, 0.35)	(282.9, 0.34)	(130.8, 0.16)	**(121.4, 0.15)**
U06	(697.5, 0.27)	(795.4, 0.31)	**(485.4, 0.19)**	(570.3, 0.22)
U07	(55.72, 0.29)	(57.96, 0.30)	(40.51, 0.21)	**(40.40, 0.21)**
U08	(550.4, 0.22)	(763.2, 0.31)	**(428.8, 0.17)**	(711.3, 0.29)
U09	(2,003.4, 0.26)	(2,447.6, 0.31)	(1,870.1, 0.24)	**(1,518.6, 0.19)**
U10	(2,526.9, 0.24)	(3,505.4, 0.33)	**(1,811.7, 0.17)**	(2,768.2, 0.26)

Table 11.2. *MOBBO results for 10 unconstrained multi-objective benchmark functions. The table shows the hypervolume and normalized hypervolume, averaged over 30 simulations. The best results in each row are shown in bold*

Functions	Multi-objective BBO			
	VEBBO	NSBBO	NPBBO	SPBBO
C01	(6.240, 0.30)	(6.545, 0.31)	(4.254, 0.20)	**(3.915, 0.19)**
C02	(15.53, 0.26)	(20.41, 0.34)	**(8.040, 0.14)**	(15.37, 0.26)
C03	(723.6, 0.29)	(730.5, 0.29)	(529.6, 0.21)	**(502.1, 0.20)**
C04	(9.890, 0.26)	(15.30, 0.40)	(7.199, 0.19)	**(6.223, 0.16)**
C05	(39.29, 0.21)	(73.61, 0.39)	**(30.89, 0.16)**	(44.65, 0.24)
C06	(0.280, 0.21)	(0.561, 0.43)	(0.238, 0.18)	**(0.231, 0.18)**
C07	(54.11, 0.18)	(152.1, 0.51)	(50.28, 0.17)	**(42.15, 0.14)**
C08	**(121.6, 0.21)**	(186.1, 0.32)	(123.7, 0.21)	(166.7, 0.28)
C09	(137.0, 0.23)	(176.9, 0.30)	(155.2, 0.26)	**(124.6, 0.21)**
C10	(1,439.5, 0.21)	(2,631.5, 0.38)	**(1,334.8, 0.19)**	(1,478.1, 0.21)

Table 11.3. *MOBBO results for 10 constrained multi-objective benchmark functions. The table shows the hypervolume and normalized hypervolume, averaged over 30 simulations. The best results in each row are shown in bold*

11.3. Real-world applications

In this section, we formulate a real-world automated warehouse scheduling problem as a constrained MOP. Then, we use the MOBBO algorithms from the previous section to solve the problem.

11.3.1. *Warehouse scheduling model*

Warehousing is an important part of production supply chain management, and serves as the backbone in many manufacturing enterprises. Warehousing keeps stocks of products until they are ready to be delivered to the market. A delay in product delivery may lead to the failure of production supply chains. Efficient warehouse management contributes to the timely delivery of the product [CHO 13, YEU 10, YEU 11, MA 15b]. Modern warehouses are equipped with storage and retrieval (S/R) machines to pick up products from an input/output (I/O) location and store them at specific locations, and then to retrieve outgoing products from other storage locations and deliver them to the I/O location. Although S/R machines enhance warehouse management, scheduling is a challenging and vital task [LER 15, WAN 11a, WAN 11b]. The time and cost of product allocation and delivery are important variables to consider during warehouse scheduling.

Warehouse scheduling is a typical NP-hard problem, which is one of the most challenging types of combinatorial optimization problems [GAG 12]. Since this problem is so important for production supply chain success, more research needs to be conducted to make automated warehouse scheduling more robust and efficient.

The layout of the automated warehouse system is shown in Figure 11.12 [YAN 13], and is called a multi-aisle automated storage and retrieval system (multi-aisle AS/RS) with a curve-going S/R machine. It includes six types of components: S/R machine, picking aisles, cross warehouse aisle, storage racks (SRs), rolling conveyor and I/O location. As shown in the figure, the S/R machine can go in and out at both ends of every picking aisle, pick up products at the I/O location and store them at specific storage units in SRs, and then retrieve outgoing products from other storage units and deliver them to the I/O location. The aim of an automated warehouse scheduling system is to optimize scheduling efficiency.

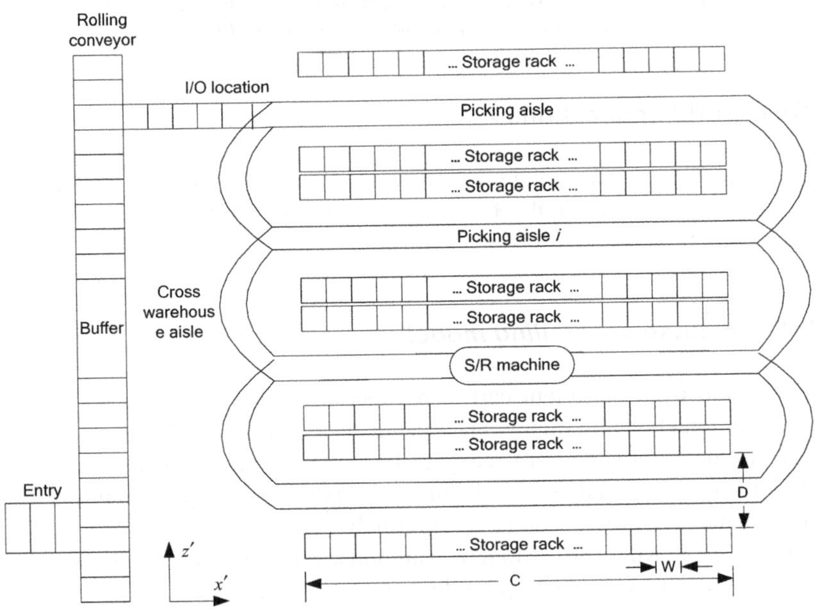

Figure 11.12. *Layout of the warehouse system. Reprinted from [MA 15b] with permission from Elsevier*

In this model, there are many storage units in each SR. There are a storage products, whose storage units are denoted as $(p_{u1}, p_{u2}, p_{u3}, \cdots, p_{ua})$. There are b outgoing products, whose storage units are denoted as $(p_{o1}, p_{o2}, p_{o3}, \cdots, p_{ob})$. In general, the numbers and units of the storage products are not the same as those of the outgoing products; namely, $a \neq b$ and $p_{ui} \neq p_{oi}$.

Suppose that the S/R machine can hold N_0 products at a time. Then, the S/R machine picks up N_0 or fewer products from the I/O location and puts them into storage units. Next, it retrieves N_0 or fewer outgoing products from the other storage units, and delivers them to the I/O location. For a storage products and b outgoing products, there are $a+b$ storage units, namely $(p_1, p_2, \cdots, p_a, p_{a+1} \cdots, p_{a+b})$, and the S/R machine needs to execute $\max\{\lceil a/N_0 \rceil, \lceil b/N_0 \rceil\}$ tasks to store and transport all products, where $\lceil \varphi \rceil$ denotes the smallest integer greater than or equal to φ. When each task corresponds to one route, the automated warehouse scheduling problem is translated into the optimization problem of selecting the optimal $\max\{\lceil a/N_0 \rceil, \lceil b/N_0 \rceil\}$ routes from all possible routes to complete the storage/retrieval tasks while satisfying real-world constraints.

In Figure 11.12, x' and z' are the two directions in the horizontal plane. The velocities in the x' and z' directions are called horizontal velocities, and the corresponding horizontal velocities are v_x and v_z, which are equal to each other. y' is the vertical direction and the corresponding vertical velocity is v_y, which is independent from v_x. The distance between adjacent SRs in the same picking aisle is D, and the number of storage units in each SR is C. The width, length and height of each storage unit are denoted as W, L and H, respectively. The Euclidean coordinates of each storage unit are denoted as $p(x', y', z')$, and the I/O location of the S/R machine is denoted as $p_0(0, 0, 0)$. The following assumptions are made:

– the multi-aisle AS/RS is divided into picking aisles with SRs on both sides, so there are double SRs between the picking aisles and a single SR along each warehouse wall;

– there is one S/R machine;

– the S/R machine is able to move along the cross warehouse aisle by using the curved rails at the end of the picking aisles;

– the S/R machine travels at a constant velocity both in the horizontal and vertical directions;

– the S/R machine simultaneously begins lifting and traveling in the pick aisle;

– the input station dwell-point strategy is used. That is, the S/R machine stays at the input location when it is idle;

– the randomized storage assignment policy is used. That is, any storage location within the S/R is equally likely to be selected for the storage or retrieval request;

– as a first-order approximation, the pickup and set down times, and additional overhead times for manipulating the S/R machine, are ignored.

DEFINITION 11.1.– *If the S/R machine travels the route* $m \in \left[1, max\left(\lceil a/N_0, b/N_0 \rceil\right)\right]$, *then* $l_m = 1$; *otherwise,* $l_m = 0$. *If the storage unit* p_i *belongs to route* m, *then* $g_{im} = 1$; *otherwise,* $g_{im} = 0$.

DEFINITION 11.2.– *If the S/R machine travels from storage unit* $p_i\left(x_i', y_i', z_i'\right)$ *to another storage unit* $p_j\left(x_j', y_j', z_j'\right)$ *for* $i, j \in [1, a+b]$, *then* $e_{ij} = 1$; *otherwise,* $e_{ij} = 0$. *The travel distance* d_{ij} *and time* t_{ij} *of the S/R machine is denoted as*:

$$d_{ij} = \begin{cases} W\times\left|x_i' - x_j'\right| + H\times\left|y_i' - y_j'\right| & \text{if } z_i' = z_j' \\ \min\begin{pmatrix} W\times\left|x_i' - x'\right| + H\times\left|y_i' - y_j'\right| + D\times\left|z_i' - z_j'\right|, \\ W\times\left|\left(C - x_i'\right) - \left(C - x_j'\right)\right| + H\times\left|y_i' - y_j'\right| + D\times\left|z_i' - z_j'\right| \end{pmatrix} & \text{if } z_i' \neq z_j' \end{cases} \quad [11.21]$$

and:

$$t_{ij} = \begin{cases} \max\left(\left(W\times\left|x_i' - x_j'\right|\right)/v_x, \left(H\times\left|y_i' - y_j'\right|\right)/v_y\right) & \text{if } z_i' = z_j' \\ \max\left(\min\begin{pmatrix} \left(W\times\left|x_i' - x_j'\right| + D\times\left|z_i' - z_j'\right|\right)/v_x, \\ \left(W\times\left|\left(C - x_i'\right) - \left(C - x_j'\right)\right| + D\times\left|z_i' - z_j'\right|\right)/v_x \end{pmatrix}, \left(H\times\left|y_i' - y_j'\right|\right)/v_y\right) & \text{if } z_i' \neq z_j' \end{cases} \quad [11.22]$$

The first expressions of each of the above two equations denote that the two storage units are the same SR because $z_i' = z_j'$. The second expressions in each of the equations denote that the two storage units are not the same SR because $z_i' \neq z_j'$, so we need to first compute the minimum distance that the S/R machine drives between the two ends of a picking aisle to compute travel distance and travel time.

DEFINITION 11.3.– *The execution time of each task must be less than or equal to the specified scheduling time* T_m*, in which it will not affect the scheduling quality. If it is larger than* T_m*, it will affect the scheduling quality by an amount equal to the product of the time exceeding* T_m *and the weight coefficient* w_m.

Now the mathematic model of the automated warehouse scheduling problem is formulated as an MOP.

DEFINITION 11.4.– *Suppose the warehouse throughput capacity is* Q *and the number of products in the warehouse is* q*. The automated warehouse scheduling problem has two objectives: the scheduling quality effect should be minimized, and the travel distance should be minimized. The two objectives are defined as follows:*

$$\min f(e) = \min\left(f_1(e), f_2(e)\right) \quad [11.23]$$

$$f_1(e) = \sum_{m=1}^{max\{\lceil a/N_0\rceil, \lceil b/N_0\rceil\}} \left(\max\left\{0, \left(\sum_{i=1}^{a+b}\sum_{j=1}^{a+b} t_{ij} e_{ij} g_{im} g_{jm}\right) - T_m\right\} \cdot w_m \right) \cdot l_m \quad [11.24]$$

$$f_2(e) = \sum_{m=1}^{max\{\lceil a/N_0\rceil, \lceil b/N_0\rceil\}} \left(\sum_{i=1}^{a+b}\sum_{j=1}^{a+b} d_{ij} e_{ij} g_{im} g_{jm} \right) \cdot l_m \quad [11.25]$$

where equation [11.23] denotes that two objectives are to be minimized, equation [11.24] denotes the scheduling quality effect and equation [11.25] denotes the travel distance.

The solution must also satisfy the following constraints:

Total number of routes:

$$\sum_{m=1}^{max\{\lceil a/N_0\rceil, \lceil b/N_0\rceil\}} l_m = \max\left\{\lceil a/N_0\rceil, \lceil b/N_0\rceil\right\} \quad [11.26]$$

Total number of storage and outgoing products:

$$\sum_{m=1}^{max\{\lceil a/N_0\rceil, \lceil b/N_0\rceil\}} \sum_{j=1}^{a+b} g_{jm} l_m = a+b \quad [11.27]$$

Each storage or outgoing product is handled exactly once:

$$\sum_{m=1}^{max\{\lceil a/N_0 \rceil, \lceil b/N_0 \rceil\}} g_{jm} = 1, \quad \text{for } j \in \{1, 2, \cdots, a+b\} \qquad [11.28]$$

S/R machine load:

$$\sum_{j}^{a+b} g_{jm} \leq 2N_0, \text{ for } r \in \{1, \cdots, \max\{\lceil a/N_0 \rceil, \lceil b/N_0 \rceil\}\} \qquad [11.29]$$

Input location:

$$\sum_{m=1}^{max\{\lceil a/N_0 \rceil, \lceil b/N_0 \rceil\}} e_{0_j l_m} = \lceil a/N_0 \rceil, \quad \text{for } j \in \{1, 2, \cdots, a+b\} \qquad [11.30]$$

Output location:

$$\sum_{m=1}^{max\{\lceil a/N_0 \rceil, \lceil b/N_0 \rceil\}} e_{i0 l_m} = \lceil b/N_0 \rceil, \quad \text{for } i \in \{1, 2, \cdots, a+b\} \qquad [11.31]$$

Storage product priorities:

$$p_u \in p_{oj} \quad \text{for } m\left(p_i, p_j, \cdots, p_u\right) \qquad [11.32]$$

Throughput capacity:

$$q + a + b \leq Q \qquad [11.33]$$

Equation [11.26] constrains the number of S/R machine routes to the total number of tasks, equation [11.27] constrains the total number of storage and outgoing products, equation [11.28] constrains each storage and outgoing product to exactly one route, equation [11.29] constrains the total number of storage and outgoing products to no more than twice the S/R machine capacity each route, equations [11.30] and [11.31] constrain the I/O location where the output location is the same as the input location since the S/R machine returns to the input location after each task is completed, equation [11.32] constrains storage products to be handled before outgoing products and equation [11.33] constrains the number of products to be no more than the throughput capacity of the warehouse.

Now that a warehouse model and scheduling problem have been presented, the following section applies MOBBO algorithms to solve the problem.

11.3.2. *Optimization of warehouse scheduling*

For the automated warehouse scheduling model described in the previous section, we set the width $W = 0.3$ m, length $H = 0.4$ m and height $H = 0.4$ m for each storage unit, the distance between adjacent SRs $D = 1.8$ m, the number of storage units in each SR $C = 75$, the warehouse throughput capacity $Q = 600$ and the number of products in the warehouse $q = 250$. We set the S/R machine capacity $N_0 = 4$, its horizontal velocity $v_x = 1$ m/s and its vertical velocity $v_y = 0.5$ m/s. The required time for each task is 120 s. All these parameters are taken from a real-world automated warehouse scheduling problem. We consider five implementation schemes. Scheme 1 (number of storage products $a = 20$ and number of outgoing products $b = 20$) is described as follows:

p_1: (12, 6, 2, 50, 1), p_2: (51, 8, 1, 45, 1), p_3: (40, 3, 5, 46, 1), p_4: (37, 3, 4, 113, 1),
p_5: (14, 7, 1, 40, 1), p_6: (13, 5, 2, 50, 1), p_7: (45, 5, 5, 50, 1), p_8: (18, 7, 3, 33, 1),
p_9: (7, 3, 2, 35, 1), p_{10}: (11, 4, 3, 110, 1), p_{11}: (15, 2, 1, 34, 1), p_{12}: (36, 8, 4, 40, 1),
p_{13}: (8, 6, 5, 47, 1), p_{14}: (56, 2, 3, 34, 1), p_{15}: (42, 8, 3, 30, 1), p_{16}: (23, 4, 1, 50, 1),
p_{17}: (4, 6, 2, 50, 1), p_{18}: (13, 1, 1, 36, 1), p_{19}: (39, 6, 4, 42, 1), p_{20}: (45, 6, 5, 55, 1)
p_{21}: (50, 8, 1, 67, 2), p_{22}: (3, 1, 2, 74, 2), p_{23}: (55, 4, 3, 68, 2), p_{24}: (6, 7, 4, 85, 2),
p_{25}: (17, 4, 5, 74, 2), p_{26}: (60, 7, 3, 80, 2), p_{27}: (35, 6, 2, 67, 2), p_{28}: (15, 2, 1, 70, 2),
p_{29}: (57, 4, 2, 80, 2), p_{30}: (2, 1, 4, 62, 2), p_{31}: (25, 8, 2, 76, 2), p_{32}: (5, 2, 4, 64, 2),
p_{33}: (17, 5, 3, 76, 2), p_{34}: (41, 2, 5, 91, 2), p_{35}: (19, 3, 2, 70, 2), p_{36}: (20, 1, 1, 82, 2),
p_{37}: (42, 6, 2, 88, 2), p_{38}: (32, 6, 3, 75, 2), p_{39}: (9, 7, 1, 82, 2), p_{40}: (58, 5, 4, 69, 2)

Each storage or retrieval operator is denoted by (x', y', z', w, u), where x', y' and z' are the Euclidean coordinates of each storage unit (meters), w is the weighting coefficient described in Definition 11.3 and affects the scheduling quality, $u = 1$ indicates a storage product, and $u = 2$ indicates an outgoing product.

Scheme 2 ($a = 20, b = 16$) includes all the storage products of Scheme 1 but only the first 16 outgoing products of Scheme 1. Scheme 3 ($a = 20, b = 12$) includes all the storage products of Scheme 1 but only the first 12 outgoing products. Scheme 4 ($a = 16, b = 20$) includes the first 16 storage products of Scheme 1 and all of the outgoing products. Scheme 5 ($a = 12, b = 20$) includes the first 12 storage products of Scheme 1 and all of the outgoing products.

The tuning parameters of the MOBBO algorithms are the same as those used in Example 11.4. The optimization results are summarized in Table 11.4. It can be seen from Table 11.4 that SPBBO performs best for all of the schemes except Scheme 4, for which NPBBO is the best because of its shortest travel distance and its lowest scheduling quality effect.

Problem	(a, b)	VEBBO		NSBBO		NPBBO		SPBBO	
		Distance	Effect	Distance	Effect	Distance	Effect	Distance	Effect
Scheme 1	(20, 20)	129.3	2,591.3	122.2	2,504.3	124.3	2,555.3	**121.7**	**2,369.4**
Scheme 2	(20, 16)	92.5	1,531.0	98.3	1,562.7	89.7	1,453.6	**80.5**	**1,392.8**
Scheme 3	(20, 12)	71.0	826.2	63.2	857.7	62.3	798.7	**61.1**	**764.3**
Scheme 4	(16, 20)	109.4	1,823.8	97.6	1,467.9	**88.0**	**1,295.5**	88.2	1,306.0
Scheme 5	(12, 20)	69.8	840.0	60.5	839.4	67.7	908.6	**60.3**	**837.0**

Table 11.4. *MOBBO results for five varieties of the automated warehouse scheduling problem. "Distance" denotes the shortest travel distance, which is measured in meters, and "Effect" denotes the lowest scheduling quality effect. The best results in each row are shown in bold*

A sample SPBBO scheduling route output is shown in Table 11.5 for Scheme 1. It can be seen that the automated warehouse scheduling problem is divided into five routes, and each route includes eight storage units, where the first four storage units are used to store products and the last four storage units are used to retrieve products.

Route	Scheduling orders
1	$p_9 \rightarrow p_1 \rightarrow p_8 \rightarrow p_{15} \rightarrow p_{27} \rightarrow p_{21} \rightarrow p_{36} \rightarrow p_{33}$
2	$p_{13} \rightarrow p_{11} \rightarrow p_{19} \rightarrow p_2 \rightarrow p_{26} \rightarrow p_{37} \rightarrow p_{25} \rightarrow p_{32}$
3	$p_{10} \rightarrow p_6 \rightarrow p_4 \rightarrow p_3 \rightarrow p_{23} \rightarrow p_{40} \rightarrow p_{31} \rightarrow p_{24}$
4	$P_{16} \rightarrow p_{18} \rightarrow p_{20} \rightarrow p_{14} \rightarrow p_{34} \rightarrow p_{38} \rightarrow p_{28} \rightarrow p_{30}$
5	$p_{17} \rightarrow p_5 \rightarrow p_{12} \rightarrow p_7 \rightarrow p_{29} \rightarrow p_{35} \rightarrow p_{39} \rightarrow p_{22}$

Table 11.5. *Scheduling orders for Scheme 1 as optimized by the SPBBO algorithm. "Route" denotes the route number index and "Scheduling orders" denotes the scheduling orders that the S/R machine implements each route*

11.4. Conclusion

In this chapter, we discussed some of the most popular MOEAs and associated ideas, and showed how BBO combines with these MOEA ideas to produce in multi-objective BBO algorithms, including VEBBO, NSBBO, NPBBO and SPBBO. Combinations of BBO with various MOEAs have served as a template for the extension of any other EA to multi-objective optimization. In addition, we have provided a warehouse scheduling problem as a real-world application example to illustrate the optimization performance of the multi-objective BBO algorithms.

This chapter is not intended to provide a complete exposition of the subject of MOEAs, but has shown only how BBO can be modified to solve MOPs. Many other MOEAs have been proposed and new ones are continually appearing in related literature. Coello Coello [COE 06] gives an interesting, high-level, historical view of MOEAs. He maintains an exhaustive and useful Web-based bibliography of papers related to multi-objective evolutionary optimization, and his bibliography included more than 10,000 references as of May 2016 [COE 16a, COE 16b].

In this concluding section, we mention some important topics for future MOEA research, including the following:

Hybridization of MOEAs with local search strategies

The incorporation of local search strategies in MOEAs is an important topic. In particular, MOEAs can be hybridized with derivative-based algorithms or other local search methods to fine-tune the optimization results. Such algorithms are called memetic algorithms because they involve the use of problem-specific information in the hybridized algorithm. Memetic strategies seem to be used a lot in single-objective optimization [ONG 07], but they have not yet been used much in MOPs, although there are a few exceptions [JAS 06].

MOEAs for many objectives (more than three)

The design of MOEAs for many objectives is another important area for future research. Some results have been published in this area, but the more challenging problem is not necessarily the approximation of the Pareto set but rather how to help human decision makers choose a solution from an MOEA's Pareto set approximation. Some research on many-objective problem emphasizes their special challenges [FLE 05], but other research shows that it is actually easier to find a good Pareto set approximation for problems with many objectives [SCH 11].

However, even though a Pareto set approximation may be easier to find with more objectives, the EA will also require more candidate solutions. For example, if we suppose that 10 candidate solutions can give a good Pareto set approximation for a two-objective problem, then we probably need at least 100 individuals in the two-objective MOEA. This means that we might need 10^k individuals for a k-objective MOP, which means that we might need 100,000 individuals for a relatively small five-objective MOP. So the problem with many-objective problems is not the theoretical difficulty of approximating the Pareto set, but the practical difficulties of computational effort and of approximating high-dimensional surfaces with only a few points.

MOEA theory and mathematical models

Theoretical results for MOEAs are sparse and so there is a lot of room for contributions in this area. Rudolph and Agapie [RUD 00] provide a preliminary Markov model for MOEAs, and a few other researchers have studied MOEA theory [ZIT 10], but compared to single-objective EAs, theoretical studies for MOEAs are sparse.

12

Hybrid BBO Algorithms

Hybrid evolutionary algorithms (EAs) are attractive alternatives to standard EAs. The combination of several algorithms in hybrid EAs allows it to exploit the strength of each algorithm. It has been shown that by properly selecting the constituent algorithms and hybridization strategies, hybrid EAs can outperform their constituent algorithms due to their synergy. This characteristic is strong motivation for the study of hybrid EAs. Many hybrid EAs have been proposed to improve performance and to find global optima. Although some of these improvements are significant, the development of new hybrid EAs and strategies is worthy of further investigation.

Current research directions in hybrid EAs involve several major areas. The first area is the determination of how to hybridize a given set of EAs into a single algorithm; that is, how to determine the hybridization strategy. The second area is the determination of which EAs to combine in a hybrid algorithm. The third area is the application of hybrid EAs to special types of optimization problems, such as constrained optimization and multi-objective optimization. The fourth area is the application of hybrid EAs to real-world optimization problems. The goal of this chapter is to address the first and second areas; that is, we emphasize the mechanism of hybridization to improve the optimization performance of EAs.

Overview of the chapter

In this chapter, we propose several hybrid EAs by combining some popular search strategies or EAs with BBO. Section 12.1 outlines how opposition-based learning (OBL) can be incorporated into an EA, and in particular how it can be used to improve the performance of BBO. Section 12.2 describes the incorporation of various types of local search methods into BBO to improve performance. Section 12.3 describes the hybridization of BBO with other EAs, including the use of biogeography-based hybridization strategies at both the iteration level and the algorithm level. The concluding section of this chapter provides references to additional resources and suggests several important topics for future hybrid BBO research.

12.1. Opposition-based BBO

This section first presents some general definitions of opposition as related to mathematics. We then discuss how opposition can be extended to EAs, including BBO.

12.1.1. *Opposition definitions and concepts*

We first discuss definitions and concepts related to the opposite of a scalar or vector. We begin by considering scalars. We begin by assuming that a variable x is defined on the domain $[a,b]$, and that the center of the domain is c :

$$\begin{aligned} &x \in [a, b] \quad \text{where } a < b \\ &c = (a+b)/2 \end{aligned} \qquad [12.1]$$

We can think of several different ways to define the opposite of a scalar x [TIZ 08]. For example, the reflected opposite of x is defined as

$$x_o = a + b - x. \qquad [12.2]$$

This means that x_o is the same distance as x from the center of the domain:

$$c - x = x_o - c \qquad [12.3]$$

Figure 12.1 illustrates the reflected opposite.

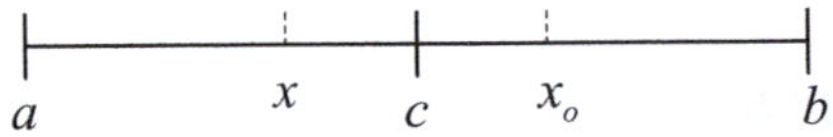

Figure 12.1. *Illustration of the reflected opposite of a scalar x. The scalar x is defined on the domain $[a,b]$, and c is the center of the domain. The reflected opposite x_o is the same distance as x from c*

Next we extend the reflected opposite to vectors in a simple but straightforward way. Suppose that x is a D-dimensional vector defined on a rectangular domain, that is, x_i defined on the domain $[a_i, b_i]$, and the center of the domain of x_i is c_i:

$$\begin{aligned} x &= [x_1 \; \dots \; x_D] \\ \text{where } x_i &\in [a_i, b_i] \text{ and } a_i < b_i \text{ for } i \in [1, D] \\ c_i &= (a_i + b_i)/2 \text{ for } i \in [1, D] \end{aligned} \qquad [12.4]$$

The reflected opposite of x is defined as

$$\begin{aligned} x_o &= [x_{o1} \; \dots \; x_{oD}] \\ \text{where} \quad x_{oi} &= a_i + b_i - x_i \text{ for } i \in [1, D]. \end{aligned} \qquad [12.5]$$

Now we define three other types of opposites: quasi opposite, super opposite and quasi reflected opposite. As before, we consider the scalar $x \in [a, b]$ with c as the center of its domain.

The quasi opposite of x is defined as follows [TIZ 08]:

$$x_{qo} = \text{rand}(c, x_o) \qquad [12.6]$$

where x_o is the standard reflected opposite defined in equation [12.2]. That is, x_{qo} is the realization of a random number that is uniformly distributed on $[c, x_o]$. Note that we define the rand function in such a way that its result is independent of the order of its arguments; that is, the notations rand (c, x_o) and rand (x_o, c) are equivalent.

The super opposite of x is defined as follows [TIZ 08]:

$$x_{so} = \begin{cases} \text{rand}(x_o, b) & \text{if } x < c \\ \text{rand}(a, x_o) & \text{if } x > c \end{cases} \qquad [12.7]$$

That is, x_{so} is the realization of a random number that is uniformly distributed between x_o and the domain boundary that is farthest from x. This definition is not complete because it does not define x_{so} for the case $x = c$, but that special situation can be handled by arbitrarily changing one of the inequalities in equation [12.7] so that it includes both equality and inequality.

The quasi reflected opposite of x is defined as follows [ERG 09, ERG 14, ERG 15]:

$$x_{qr} = \text{rand}(x, c) \quad [12.8]$$

That is, x_{qr} is the realization of a random number that is uniformly distributed between x and c. Note that the use of word "reflected" in the term "quasi reflected" is not related to the word "reflected" in the term "reflected opposite" (see equation [12.2]).

Figure 12.2 illustrates four different methods of opposition. We can extend these definitions to vectors by following the procedures presented in equations [12.4] and [12.5].

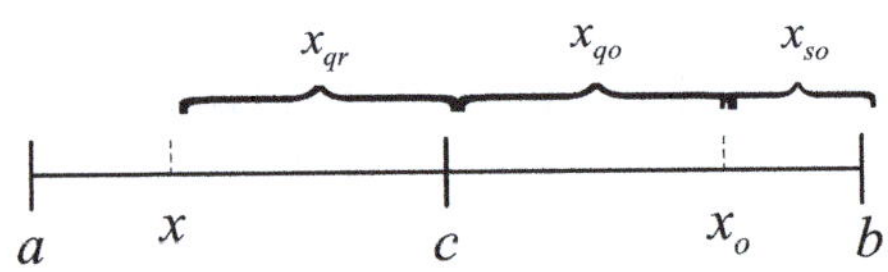

Figure 12.2. *Suppose we have a scalar $x \in [a,b]$. The opposite of x is x_o, which is obtained by reflecting x across the center of domain c. The quasi opposite of x is x_{qo}, which is obtained by generating a random number between c and x_o. The super opposite of x is x_{so}, which is obtained by generating a random number between x_o and the domain boundary that is farthest from x. The quasi reflected opposite of x is x_{qr}, which is obtained by generating a random number between x and c*

12.1.2. Oppositional BBO

Now we show how the oppositional concepts presented above can be used in BBO. We combine the standard BBO algorithm of Figure 3.5 in Chapter 3 with OBL to obtain oppositional BBO (OBBO) [ERG 09]. Figure 12.3 shows an outline of the OBBO algorithm. Note that the algorithm of Figure 12.3 is identical to that of Figure 3.5 except for the pseudo-code between the lines "Comment: Begin Opposition Logic" and "Comment: End Opposition Logic".

```
Initialize a population of candidate solutions { x_k } for k ∈ [1, N]
While not (termination criterion)
    For each x_k , set emigration rate μ_k proportional to the fitness of x_k , where μ_k is
    normalized to [0, 1]
    For each x_k , set immigration rate λ_k = 1 − μ_k
    { z_k } ← { x_k }
    For each solution z_k
        For each decision variable s
            Use λ_k to probabilistically decide whether to immigrate to z_k
            If immigrating then
                Use {μ_i} to probabilistically select the emigrating solution x_j
                z_k(s) ← x_j(s)
            End if
        Next decision variable
        Probabilistically decide whether to mutate z_k
    Next solution
    Comment: Begin Opposition Logic
    r ← U[0, 1]
    If r < J_r then
        Use { z_k } to create opposite population {z̄_k}
        {z_k} ← the best N solutions from {z_k} ∪ {z̄_k}
    End if
    Comment: End Opposition Logic
    { x_k } ← { z_k }
Next generation
```

Figure 12.3. *Outline of OBBO with a population size of N. $\{x_k\}$ is the population of solutions and $\{z_k\}$ is a temporary population of solutions. x_k is the kth candidate solution and $x_k(s)$ is the decision variable s of x_k. J_r is the jumping rate, which is described below*

We see from the OBBO algorithm that OBL can augment any EA. One simple approach to use an OBL with any EA is to perform the following steps:

1) When the N solutions of the EA population are initialized, N opposite solutions are created, each opposite solution corresponding to one of the N original solutions. Given our $2N$ solutions (N original solutions and N opposite solutions), we keep the best N as the initial population of the opposition-based EA.

2) We perform a standard implementation of an EA. As we have seen earlier, this involves a loop of cost function evaluations, migration and mutation. By definition, the loop executes once per generation.

3) Once every few generations, we compute the opposite of each of the N solutions. Of these $2N$ solutions (N standard EA solutions and N opposite solutions), we keep the best N for the next EA generation. At each generation, we perform this step with probability $J_r \in [0,1]$, which is a tuning parameter called the jumping rate.

We have to make some decisions when implementing an opposition-based EA.

1) Which EA should we use? Answering this question means that we must choose all of the tuning parameters of the EA.

2) What type of opposition should we use? We can choose one of at least four OBL methods, including reflected opposite, quasi opposite, super opposite and reflected quasi opposite. We can also create new OBL methods to combine with EAs.

3) What value should we use for the jumping rate J_r? The jumping rate is a tuning parameter. We do not have many guidelines for the value of J_r, but we do not want to make it too high. The reason that we periodically create an opposite population is to explore new areas of the search space. However, we do not want to create an opposite population every generation because then we would just be repeatedly jumping back and forth in the search space, which would waste function evaluations. Results from opposition-based differential evolution (DE) indicate that $J_r = 0.3$ provides a good balance [RAH 08].

12.1.3. *Experimental results*

In this section, we investigate the performance of OBBO on a representative set of 13 benchmark functions. These functions are briefly described in Table 3.2 in Chapter 3. A more detailed description of these functions can be found in Appendix A. All benchmark functions are minimization problems. For the OBBO algorithms, we combine BBO with the four OBL methods described above: reflected opposite, quasi opposite, super opposite and reflected quasi opposite. We call the resulting algorithms reflected OBBO, quasi OBBO, super OBBO and quasi reflected OBBO, respectively.

For these comparisons, the parameters used in the OBBO algorithms and standard BBO are the same: population size 50, maximum immigration rate and maximum emigration rate 1, and maximum mutation rate 0.01 with a mutated value randomly chosen from a uniform distribution in the search domain. We use linear migration curves. The jumping rate sets $J_r = 0.3$ in the OBBO algorithms. We terminated after a maximum of 20,000 fitness function evaluations. We run 25 Monte Carlo simulations on each benchmark to obtain representative performances.

Table 12.1 summarizes the performance of the OBBO algorithms and the standard BBO algorithm on these benchmarks, which shows the mean minimum values found by each algorithm. From Table 12.1, we see that quasi reflected OBBO performs the best on 6 of the 13 benchmark functions (F01, F03, F05, F07, F11 and F12), quasi OBBO performs the best on four functions (F04, F08, F09 and F13) and super OBBO performs the other two functions (F02 and F10). For function F06, all OBBO algorithms obtain the global optimum. The table indicates that OBL accelerates BBO performance, and that OBBO significantly outperforms BBO. It also indicates that quasi reflection is the preferred oppositional method for OBBO.

Function	BBO	Reflected OBBO	Quasi OBBO	Super OBBO	Quasi reflected OBBO
F01	1.04E-04	2.31E-04	6.76E-05	5.37E-04	**1.25E-05**
F02	6.29E-15	6.73E-14	8.90E-15	**1.53E-15**	9.05E-15
F03	3.84E+01	5.89E+00	3.26E+00	6.33E+00	**1.83E+00**
F04	6.35E-15	3.12E-15	**1.25E-15**	4.16E-15	5.79E-15
F05	8.85E-01	6.90E-02	6.25E-03	4.16E-03	**2.08E-03**
F06	**0.00E+00**	**0.00E+00**	**0.00E+00**	**0.00E+00**	**0.00E+00**
F07	8.61E-09	2.01E-10	6.78E-12	2.15E-12	**5.44E-14**
F08	4.97E+00	4.35E-02	**3.47E-02**	6.82E-02	7.80E-02
F09	8.77E-01	1.27E-03	**1.24E-04**	6.92E-04	5.65E-04
F10	4.18E-01	8.86E-02	8.90E-02	**2.31E-02**	5.66E-02
F11	6.53E+00	3.44E+00	5.12E+00	4.32E+00	**1.80E+00**
F12	4.64E-32	6.18E-32	6.79E-32	5.21E-32	**2.44E-32**
F13	8.47E-32	7.29E-32	**1.23E-32**	6.87E-32	6.04E-32

Table 12.1. *Comparison of results for BBO and OBBO algorithms. The table shows the average of the best performance of 25 Monte Carlo simulations. The best value in each row is indicated in bold*

12.2. BBO with local search

This section discusses several local search methods that can be incorporated into BBO to improve its performance [SIM 14]. One drawback of BBO is its poor local search ability, so we discuss local search operators that can be included in BBO, including gradient descent, boundary search, grid search and Latin hypercube search.

12.2.1. *Local search methods*

A. Gradient descent

Gradient descent is a first-order optimization algorithm. To find a local minimum of a function using gradient descent, we take steps that are proportional to the negative of the gradient of the function at the current search point. If we instead take steps that are proportional to the gradient, we approach a local maximum of the function; the procedure is then known as gradient ascent. The outline of local search using gradient descent is given in Figure 12.4.

If $\text{FE} > \alpha\ \text{FE}_{\max}$ or $(f_{\min}(g) - f_{\min}(g+1)) / f_{\min}(g) < \varepsilon_1$ then

  For the best N_g candidate solutions x_k ($k = 1$ to N_g)

    $x_k \leftarrow x_k - \gamma \cdot \nabla f(x_k)$

  Next solution

End if

Figure 12.4. *Outline of gradient descent on each candidate solution of an EA, performed under two conditions which are described below*

Figure 12.4 shows that gradient descent is implemented on the best N_g solutions in the population by subtracting a term $\gamma \cdot \nabla f(x_k)$, where γ is a step size which is allowed to change every generation, and $\nabla f(\cdot)$ is the first-order derivative of the function with respect to its decision variable. The purpose of gradient descent is to move against the gradient and down toward the minimum. The figure shows that gradient descent is implemented under two conditions. The first condition involves FE, which is the current number of function evaluations that have been performed so far, and $\text{FE}_{\max}$, which is the maximum function evaluation limit. $\alpha \in [0,1]$ is a factor that determines when gradient descent is activated. We typically use $\alpha = 1/2$ so that gradient descent is activated when we have used 50% or more of our allotted function evaluations. The second condition involves $f_{\min}(g)$, which is the minimum function value obtained by BBO during the

$(g+1)$st (current) generation. The quantity $(f_{\min}(g)-f_{\min}(g+1))/f_{\min}(g)$ indicates the relative improvement in the best function value (taken over all candidate solutions) found by BBO from the gth generation to the $(g+1)$st generation. ε_1 is a threshold that determines when gradient descent is activated. This condition assumes that the cost value $f(\cdot)$ is positive for all candidate solutions. We typically use $\varepsilon_1 = 0.1$ so that gradient descent is activated whenever the best candidate solution in the population improves by less than 10% from one generation to the next.

B. Boundary search

Many real-world optimization problems have their solution on the boundary of the search space. This is not surprising because we normally expect to obtain an optimization goal by using all of the available energy, or force, or some other resource.

We implement boundary search in BBO as follows. If any of the decision variables of the best solution in the population are within a certain threshold of the search space boundary, then we move the decision variable to the search space boundary and perform local search (gradient descent) on the other decision variables. This idea is shown in Figure 12.5.

If $(f_{\min}(g)-f_{\min}(g+1))/f_{\min}(g)<\varepsilon_2$ then
    For the best N_g candidate solutions x_k ($k=1$ to N_g)
        For each decision variable s
            If $(U_s - x_k(s)) < \alpha_s (U_s - L_s)$ then
                $x_k(s) \leftarrow U_s$
            End if
            If $(x_k(s) - L_s) < \alpha_s (U_s - L_s)$ then
                $x_k(s) \leftarrow L_s$
            End if
        Next decision variable
        Perform gradient descent on unbounded dimensions of x_k
    Next solution
End if

Figure 12.5. *Outline of boundary search combined with gradient descent*

Figure 12.5 shows that boundary search is implemented under similar conditions as gradient descent in Figure 12.4. That is, boundary search is implemented whenever the best solution in the population improves by a factor of less than ε_2 from one generation to the next. We implement boundary search for the best N_g solutions in the population. If any decision variable of the best N_g solutions is within a factor α_s of the upper boundary of the search domain, we set that decision variable equal to the upper boundary. Similarly, if any decision variable of the best N_g solutions is within a factor α_s of the lower boundary of search domain, we set that decision variable equal to the lower boundary. Then, we perform gradient descent on those solutions. However, we perform gradient descent only on decision variables that are not equal to a search space boundary.

C. Grid search

The next type of search that we implement is grid search. This search systematically covers the search space, as shown in Figure 12.6.

If $(f_{\min}(g) - f_{\min}(g+1)) / f_{\min}(g) < \varepsilon_3$ then
    $\Delta = \alpha_0(U_s - L_s)$
    For the best N_g candidate solutions x_k $(k = 1$ to $N_g)$
        $w_k = x_k$
        For each decision variable s
            While $w_k(s) \geq L_s$
                $x_k \leftarrow \text{argmin}\{f(x_k), f(w_k)\}$
                $w_k(s) \leftarrow w_k(s) - \Delta$
            End while
        Next decision variable
        $w_k = x_k$
        For each decision variable s
            While $w_k(s) \leq U_s$
                $x_k \leftarrow \text{argmin}\{f(x_k), f(w_k)\}$
                $w_k(s) \leftarrow w_k(s) + \Delta$
            End while
        Next decision variable
    Next solution
End if

Figure 12.6. *Outline of grid search*

Figure 12.6 shows that grid search is implemented under similar conditions as gradient descent in Figure 12.4 and boundary search in Figure 12.5. Grid search is implemented whenever the best solution in the population improves by a factor of less than ε_3 from one generation to the next. We implement grid search for the best N_g solutions in the population. Figure 12.6 shows that for the best N_g solutions, we increment or decrement each decision variable by a specific fraction α_0 of the search space size. For example, if $\alpha_0 = 0.1$, grid search decreases the value of a given decision variable of x_k by an increment equal to 10% of the search space size, one increment at a time, until it reaches the lower boundary of the search space. Grid search then increases the values of a given decision variable until it reaches the upper boundary of the search space. Grid search performs this process for each decision variable and updates the solution with the best value that it finds.

D. Latin hypercube search

Latin hypercube sampling divides a domain into intervals in each dimension, and then places sample points in such a way that each interval in each dimension contains only one sample point [SIM 13a]. This idea is illustrated in Figure 12.7.

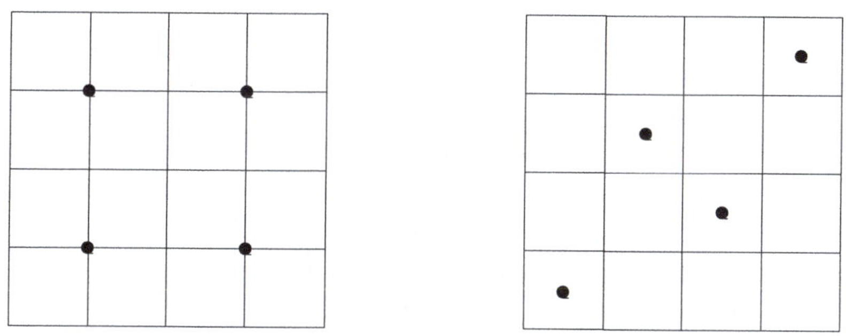

Figure 12.7. *The figure on the left shows uniform sampling of four points in a search domain. The figure on the right shows Latin hypercube sampling. Note that there is only one point in each row and only one point in each column*

Latin hypercube sampling can sometimes capture the unpredictable, unknown nature of a function better than uniform sampling. Also, Latin hypercube sampling is more efficient than uniform sampling. With uniform sampling of a D-dimensional search space where each dimension is divided into n intervals, we need x^D sample points, but with Latin hypercube sampling, we need only n sample points.

We perform Latin hypercube sampling every G_L generations around the current best solution in the population if we seem to be converging to an optimum. We perform this search under the following two conditions, both of which must be satisfied.

The first condition is that the best solution in the population has a cost that is less than βe, where e is the function value required for success and β is a scale factor. Note that this requires that we know *a priori* what function value is required for success. Although this is not always the case, in real-world problems, we often know the target value of our optimization problem ahead of time. We typically set $G_L = 10$ and $\beta = 1000$.

The second condition is that the best solution in the population is not improving sufficiently fast. That is,

$$(f_{\min}(g - G_L) - f_{\min}(g)) / f_{\min}(g - G_L) < \varepsilon_4 \quad [12.9]$$

where $f_{\min}(g)$ is the cost function value of the best solution in the population and ε_4 is a relative tolerance.

We perform a Latin hypercube search within a domain of size $\alpha_L(U - L)$ that is centered at the best solution in the population, where U and L are the upper and lower search space bounds, and α_L defines the relative size of the hypercube within which we search. We divide each dimension of the search space into n evenly spaced points within the search range, and then find n search points within the search range. We then perform gradient descent on the best n_L of those solutions. We combine these solutions with the N-member population to obtain a temporary population size of $N + n_L$. We then save the best n of these solutions as the population of the next generation.

E. BBO with local search

The hybrid BBO algorithm, including all of the new components discussed above, is summarized in Figure 12.8. Each method in Figure 12.8 executes regardless of the success or failure of the previous methods in the algorithm. Note that the performance of BBO might be affected if the methods in Figure 12.8 are implemented in a different order.

For each solution x_k, calculate λ_k and μ_k based on the fitness of x_k

$\{z_k\} \leftarrow \{x_k\}$

Use standard BBO migration to update z_k

Apply gradient descent if needed

Apply boundary search if needed

Apply grid search if needed

Apply Latin hypercube search if needed

$\{x_k\} \leftarrow \{z_k\}$

Figure 12.8. *One iteration of the BBO algorithm with local search*

12.2.2. *Simulation results*

In this section, we investigate the performance of the BBO algorithm with local search for the same benchmark functions as those used above. All the parameter settings of BBO are the same as those described previously. Furthermore, to address the relative importance of each of local search methods, including gradient descent, boundary search, grid search and Latin hypercube search, we conduct the following simple test. First, we run standard BBO, and BBO with all components of local search. Then, we run BBO with all components of local search except the first one; that is, we skip gradient descent in the BBO algorithm with local search. Next, we run BBO with all components of local search except the second one; that is, we skip boundary search in the BBO algorithm with local search. In general, we run BBO with all components of local search except the *n*th one, for $n \in [1, 4]$.

Table 12.2 summarizes the performance of BBO, BBO with all local search components, and BBO with the *n*th component omitted for $n \in [1, 4]$, on the benchmark functions. The table shows the mean minimum errors found by each algorithm over 25 Monte Carlo simulations. The table shows that BBO with all local search components performs the best on 9 of the 13 benchmark functions, and BBO with the nth component of local search removed (for each $n \in [1, 4]$) performs better than standard BBO for the majority of the functions. These results indicate that local search can significantly improve the performance of BBO. From Table 12.2, we see that except for standard BBO, BBO without gradient descent performs the worst on most of the functions, and BBO without grid search performs better than all of the other cases of BBO with the nth component omitted. This means that gradient descent is the most important component of BBO with local search, and grid search is the least important component. We conclude that BBO might not be very effective unless it is augmented with a local search method such as gradient descent.

Function	BBO	BBO with all local search	No gradient	No boundary	No grid	No Latin
F01	1.04E-04	**2.14E-08**	6.70E-01	7.11E-04	4.43E-06	2.16E-04
F02	6.29E-15	**1.65E-18**	1.29E-16	2.39E-16	6.78E-18	4.92E-16
F03	3.84E+01	3.75E+00	5.43E+00	5.33E+00	**2.50E+00**	3.39E+00
F04	6.35E-15	**4.23E-16**	7.55E-15	1.98E-15	1.86E-15	2.43E-15
F05	8.85E-01	**1.19E-05**	3.23E-01	6.86E-02	6.08E-05	3.85E-02
F06	**0.00E+00**	**0.00E+00**	**0.00E+00**	**0.00E+00**	**0.00E+00**	**0.00E+00**
F07	8.61E-09	**2.06E-12**	8.39E-09	2.43E-10	7.23E-12	8.08E-10
F08	4.97E+00	**5.76E-02**	7.45E-01	1.08E-01	9.44E-02	3.75E-01
F09	8.77E-01	**2.33E-04**	1.28E-02	5.84E-03	2.53E-04	1.66E-03
F10	4.18E-01	**7.01E-03**	9.04E-01	3.22E-02	7.08E-03	6.21E-02
F11	6.53E+00	**3.11E-01**	8.77E+00	7.24E+00	5.36E-01	4.21E+00
F12	4.64E-32	**0.00E+00**	2.54E-32	1.98E-32	**0.00E+00**	1.66E-32
F13	8.47E-32	**0.00E+00**	1.25E-32	2.00E-32	**0.00E+00**	2.30E-32

Table 12.2. *Comparison between BBO, BBO with all local search components and BBO with the nth local search component omitted for each $n \in [1, 4]$. The best value in each row is indicated in bold*

12.3. BBO with other EAs

This section discusses the combination of BBO with other popular EAs, which results in hybrid BBO algorithms [MA 14a, MA 14b]. Biogeography-based hybridization is discussed at both the iteration level and the algorithm level, which is based on migration behaviors in biogeography. Popular EAs that could be hybridized with BBO include evolution strategy (ES), genetic algorithm (GA), DE, particle swarm optimization (PSO), and so on. Here we present a general biogeography-based hybridization strategy and demonstrate its performance on a representative set of benchmarks.

12.3.1. *Iteration-level hybridization*

Iteration-level hybridization is a straightforward method in which various EAs are executed in sequence. Iteration-level hybridization divides the search procedure into two stages:

1) In the first stage, one EA with high convergence speed is used to shrink the search region to promising areas;

2) In the second stage, another EA with good exploration ability is used to explore the previously limited area more extensively to find better solutions.

Iteration-level hybridization will perform at least as well as one algorithm alone, and more often will perform better due to the synergy of exploration and exploitation. Previous studies have found that this kind of hybrid EA improves optimization performance [JAI 05]. Another attractive feature of iteration-level hybridization is that its structure is simple and easily programmed.

Here we implement iteration-level hybridization by combining popular EAs with BBO, in which a popular EA is used in the first stage to obtain good candidate solutions, and then BBO is used in the second stage to improve the candidate solutions obtained by the first EA. The goal of this hybridization approach is to balance the exploration and exploitation ability of various EAs. A general flowchart of iteration-level hybridization is shown in Figure 12.9. The main procedure of iteration-level hybridization is shown in Figure 12.10.

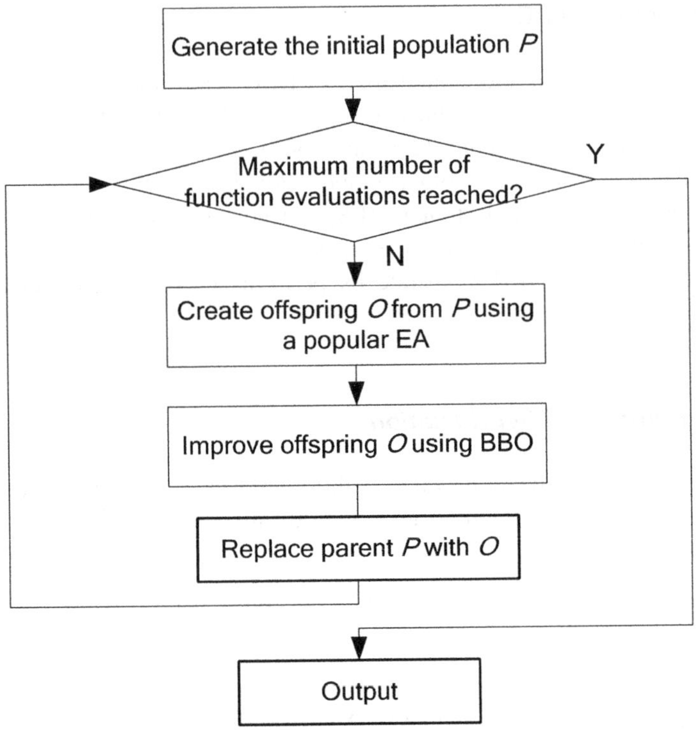

Figure 12.9. *Flowchart of iteration-level hybridization combining a popular EA with BBO, where P is the parent population and O is the offspring population. Reprinted from [MA 14] with permission from Elsevier*

Randomly initialize the parent population P

Evaluate the fitness of all candidate solutions in P

While not (termination criterion)

  Execute a popular EA (for example, DE or PSO) to create offspring population O

  Evaluate the fitness of each solution in offspring population O

  Calculate the immigration rate λ and emigration rate μ of each solution

  Perform one generation of BBO as shown in Figure 3.5 in Chapter 3 to

  improve the solutions in the offspring population O

  Replace the parent population P with the offspring population O

Next generation

Figure 12.10. *Outline of iteration-level hybridization of BBO with other EAs*

EXAMPLE 12.1.–

Figure 12.11 shows one generation of an iteration-level hybridization of DE and BBO, which is a special case of Figure 12.10. Note that Figure 12.10 is provided for purposes of illustration. Any EA could be hybridized at the iteration level with BBO, in which case Figure 12.11 would be modified accordingly.

$\{z_k\} \leftarrow \{x_k\}$

For each candidate solution z_k ($k = 1$ to N)

  For each decision variable s

    Pick three random solutions x_{r1}, x_{r2} and x_{r3} mutually distinct from each other and from z_k

    Pick a random index n between 1 and N

    Use CR (probabilistic) or n (deterministic) to decide on recombination

    If recombination then

      $z_k(s) \leftarrow x_{r1}(s) + F \cdot (x_{r2}(s) - x_{r3}(s))$

    End if

  Next decision variable

  Evaluate the fitness of each candidate solution z_k in the population

  For each z_k define emigration rate μ_k proportional to the fitness of z_k, where $\mu_k \in [0,1]$

  For each candidate solution z_k define immigration rate $\lambda_k = 1 - \mu_k$

  For each decision variable s

    Use λ_k to probabilistically decide whether to immigrate to z_k

    If immigrating then

      Use $\{\mu\}$ to probabilistically select the emigrating solution y_j

      $z_k(s) \leftarrow x_j(s)$

    End if

  Next decision variable

Next solution

$\{x_k\} \leftarrow \{z_k\}$

Figure 12.11. *One generation of an iteration-level hybridization of DE and BBO, where N is the population size. $\{x_k\}$ and $\{z_k\}$ comprise the entire population of candidate solutions, x_k is the kth candidate solution, and $x_k(s)$ is the sth decision variable of x_k. CR and F are the probability of crossover and the scaling factor of DE, respectively [DAS 11a, DAS 11b]*

12.3.2. *Algorithm-level hybridization*

Algorithm-level hybridization is a method that involves several subpopulations running independently and periodically exchanging information with each other [JAI 05]; see Figure 12.12. The information exchange provides the mechanism to enhance a given subpopulation with the improvements achieved in other subpopulations. Therefore, algorithm-level hybridization will perform at least as well as each constituent algorithm, and more often it will perform better due to the exchange of information among the algorithms.

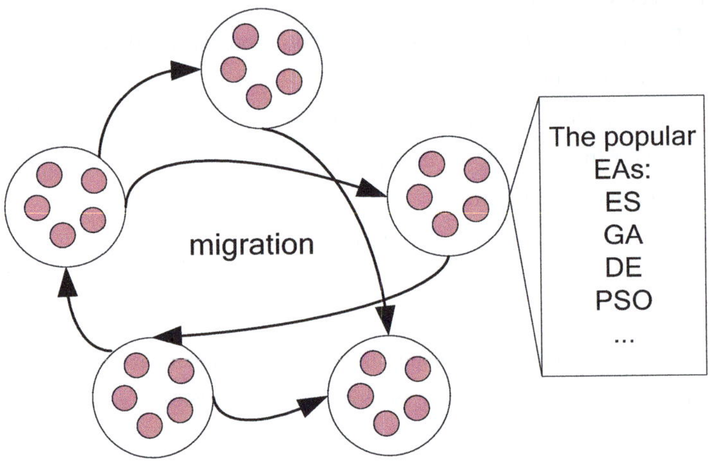

Figure 12.12. *Illustration of algorithm-level hybridization. Reprinted from [MA 14a, MA 14b] with permission from Elsevier. For a color version of this figure, see www.iste.co.uk/ma-simon/evolutionary.zip*

An important aspect of algorithm-level hybridization is the migration strategy, which is configured by various parameters. (Note that "migration" in this context is different from standard BBO migration.)

1) Migration frequency: how often is information shared between algorithms?

2) Migration rate: how much information migrates between algorithms?

3) Information selection: what information is selected to migrate between algorithms?

4) Migration topology: which subpopulations exchange information with each other?

A study of migration parameters has been presented in Jaimes and Coello Coello [JAI 05], but that strategy should be adjusted based on *a priori* knowledge of the problem. It is hard to obtain useful information about the best migration strategy in most real-world problems.

In our approach to algorithm-level hybridization, the fitness of each solution is used by BBO to determine the migration strategy between each algorithm, including migration frequency, migration rate, information selection and migration topology. That is, even though BBO is not one of the constituent EAs, the BBO approach is used to migrate information between EAs. This migration strategy is naturally adaptive because of the information-exchanging mechanism of BBO. The advantage of this method is that BBO determines the migration parameter settings of algorithm-level hybridization, and interaction with a human decision maker does not need to occur during optimization. This hybridization approach, which combines recently developed EAs using biogeography-based strategies, has the common features of algorithm-level hybridization but also has the distinctive migration characteristics of BBO.

A flowchart of algorithm-level hybridization combining popular EAs using biogeography is shown in Figure 12.13, where three subpopulations are used, although fewer or more could also be used, depending on the application. Subpopulations execute independently and generate their own offspring subpopulations. The offspring subpopulations are combined with biogeography-based migration. In this way, algorithm-level hybridization will always keep the solutions that are improved by migration, which will lead to better optimization performance than that achieved by a single constituent algorithm. The main procedure of this approach to algorithm-level hybridization is shown in Figure 12.14.

Note that in iteration-level hybridization, we combine various EAs with BBO, and in algorithm-level hybridization, we combine various EA populations using ideas from biogeography. In algorithm-level hybridization, we do not necessarily combine a particular EA with BBO; instead, we use the BBO migration strategy to combine multiple EAs. In this approach, various EAs are taken as the baseline algorithms, and then we use the migration mechanism of BBO to adaptively improve the solutions. That is, the constituent EAs generate offspring solutions each generation, and then we use the BBO migration operator to exchange information between these solutions.

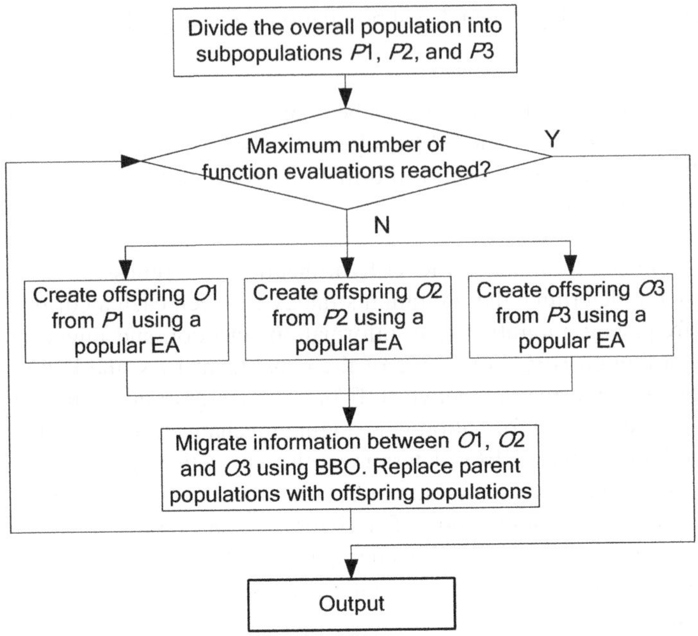

Figure 12.13. *Flowchart of algorithm-level hybridization combining popular EAs with biogeography-based migration, where there are three subpopulations. P1, P2, P3 are parent subpopulations, and O1, O2, O3 are offspring subpopulations. Reprinted from [MA 14a, MA 14b] with permission from Elsevier*

Randomly initialize the overall population P and divide it into subpopulations P_i
($i = 1\ldots n$, where n denotes the number of subpopulations)
Evaluate the fitness of all candidate solutions in P

While not (termination criterion)
    For each subpopulation P_i
        Perform an independent EA to create offspring subpopulation O_i
    Next subpopulation
    Evaluate the fitness of offspring population O, which is composed of all subpopulations O_i
    Calculate the immigration rate λ and emigration rate μ for each offspring in population O
    For each offspring subpopulation O_i
        Immigrate solution information from the overall offspring population O using one generation of BBO as shown in Figure 3.5 in Chapter 3
    Next subpopulation
Next generation

Figure 12.14. *Outline of algorithm-level hybridization of BBO with other EAs*

```
{ z_k } ← { x_k }
Divide { z_k } into subpopulations P_i (i = 1 to n )
For each subpopulation P_i
    For each candidate solution z_ik (k = 1 to K )
        For each decision variable s
            Pick three random solutions x_r1, x_r2 and x_r3 mutually distinct from each other
            and from z_ik
            Pick a random index n between 1 and the population size
            Use CR (probabilistic) or n (deterministic) to decide on recombination
            If recombination then
                z_k(s) ← x_r1(s) + F·(x_r2(s) − x_r3(s))
            End if
        Next decision variable
    Next solution
Next subpopulation
Evaluate the fitness of each candidate solution z_ik in all subpopulations
For each z_ik define emigration rate μ_ik proportional to fitness of z_ik, where μ_ik ∈ [0,1]
For each candidate solution z_ik define immigration rate λ_ik = 1 − μ_ik
For each subpopulation P_i
    For each candidate solution z_ik
        For each decision variable s
            Use λ_ik to probabilistically decide whether to immigrate to z_ik
            If immigrating then
                    Use {μ} to probabilistically select the emigrating solution x_j from
                    the combined population
                z_k(s) ← x_j(s)
            End if
        Next decision variable
    Next solution
Next subpopulation
{ x_k } ← { z_k }
```

Figure 12.15. *One generation of an algorithm-level hybridization of DE and BBO, where Pi is the subpopulation, n is the number of subpopulations and K is the size of each subpopulation. $\{x_k\}$ and $\{z_k\}$ are the entire population of candidate solutions, xk is the kth candidate solution and xk(s) is the sth decision variable of xk. CR and F are the probability of crossover and the scaling factor of DE, respectively*

EXAMPLE 12.2.–

Figure 12.15 shows one generation of an algorithm-level hybridization of DE and BBO, which is a special case of Figure 12.14. Any set of EAs could be hybridized at the algorithm level, in which case Figure 12.15 would be modified accordingly.

12.3.3. *Experimental results*

In this section, we investigate the performance of hybrid EAs for the same benchmark functions as those used above. We evaluate the performance of iteration-level and algorithm-level hybridization combining various EAs with BBO. The EAs that we use include ES, GA, DE and PSO. We choose these algorithms because they are some of the most popular EAs. The four algorithms that we choose form a representative set rather than a complete set. We could hybridize many other algorithms besides these four. However, the goal here is not to be exhaustive, rather to demonstrate a general biogeography-based hybridization strategy on a representative set of constituent algorithms and benchmarks.

In summary, we investigate the following hybrid EAs: GA/BBO-I, GA/BBO-A, ES/BBO-I, ES/BBO-A, DE/BBO-I, DE/BBO-A, PSO/BBO-I and PSO/BBO-A, where I and A denote iteration-level hybridization and algorithm-level hybridization, respectively. For example, GA/BBO-I denotes iteration-level hybridization of GA and BBO, as given in Figures 12.10 and 12.11, where DE in the algorithm is replaced with GA. Similarly, GA/BBO-A denotes algorithm-level hybridization of GA and BBO, as given in Figures 12.14 and 12.15, where DE in the algorithm is replaced with GA. Similar statements can be made for each of the other hybrid BBO algorithms.

The next step is to set the parameters of each constituent algorithm. For BBO, we use the same parameters as those used in the previous experiments. For the other EAs, we use the same parameters as those used in section 3.3 in Chapter 3. For algorithm-level hybridization, we use three subpopulations, and each subpopulation implements the same EA. The population size of each subpopulation in algorithm-level hybridization is 50, so the total population size is 150. For fair comparisons, the population size of iteration-level hybridization is also set to 150. We use a maximum of 20,000 fitness function evaluations.

Tables 12.3 and 12.4 summarize the performance of the algorithms, and show the mean minimum errors found by each algorithm over 25 Monte Carlo simulations. From Table 12.3, we see that for each group of hybrid algorithms (for example, considering GA, GA/BBO-I and GA/BBO-A as one group), iteration-level hybridization and algorithm-level hybridization perform better than the constituent algorithm for all benchmark functions. This indicates that hybrid biogeography-based algorithms improve performance for these benchmark functions.

We also note from Tables 12.3 and 12.4 that algorithm-level hybrid algorithms perform better than iteration-level hybrid algorithms for most of the benchmark functions. This result indicates the superiority of algorithm-level hybridization over iteration-level hybridization. This may be due to the interacting subpopulations that comprise algorithm-level hybridization, which is a structure that other research has also found to be highly efficient for global optimization [DAS 11a, DAS 11b, LAS 10].

Function	GA	GA/BBO-I	GA/BBO-A	ES	ES/BBO-I	ES/BBO-A
F01	1.43E-03	6.79E-04	**3.16E-05**	5.67E+04	6.37E+02	**5.43E+01**
F02	3.80E+02	5.78E-05	**4.43E-07**	4.21E+03	4.65E-02	**3.36E-02**
F03	4.43E+03	2.54E+01	**3.54E+00**	5.80E+02	7.75E+02	**6.28E+01**
F04	4.10E+00	6.65E-12	**4.31E-14**	1.22E+01	**4.12E-10**	8.08E-06
F05	6.73E+01	8.98E-00	**4.67E-00**	2.34E+03	**5.34E-02**	7.80E-02
F06	3.24E+00	2.15E-02	**5.30E-04**	6.78E+03	1.28E+01	**6.85E+00**
F07	5.33E+02	4.34E-08	**8.94E-09**	4.21E+02	**1.75E-02**	6.77E-01
F08	1.05E+02	2.78E+01	**3.26E+00**	4.32E+01	**2.59E+00**	8.93E+00
F09	2.76E+01	1.54E-01	**5.44E-02**	5.38E+02	1.26E+02	**1.25E+02**
F10	3.40E-02	3.22E-04	**3.28E-05**	1.45E+00	4.41E-01	**8.15E-03**
F11	3.76E+02	7.80E+01	**1.69E+01**	7.89E+01	6.74E+01	**4.24E+00**
F12	3.54E-32	**0.00E+00**	**0.00E+00**	1.12E-07	**7.35E-10**	8.95E-10
F13	6.06E-31	**0.00E+00**	**0.00E+00**	7.82E-07	3.35E-09	**1.28E-09**

Table 12.3. *Comparison of results of hybrid algorithms combining BBO with GA and ES based on iteration-level hybridization and algorithm-level hybridization. The best value in each row is indicated in bold*

Function	DE	DE/BBO-I	DE/BBO-A	PSO	PSO/BBO-I	PSO/BBO-A
F01	3.16E+03	5.67E+02	**7.98E+01**	3.68E-01	1.32E-03	**6.65E-05**
F02	5.38E+02	3.12E-11	**3.25E-12**	2.25E+02	9.03E-11	2.33E-11
F03	2.16E+02	**5.67E+00**	5.87E+00	1.76E+02	7.25E+01	**4.31E+00**
F04	2.34E+01	7.89E-10	**6.55E-12**	6.78E-01	5.34E-13	**6.74E-15**
F05	3.41E+01	2.18E-01	**4.32E-02**	6.74E+00	**3.22E-02**	5.87E-02
F06	2.15E+02	4.33E+00	**1.26E-01**	**0.00E+00**	**0.00E+00**	**0.00E+00**
F07	2.38E+01	6.84E+00	**7.86E-04**	2.58E+00	**1.23E-08**	4.90E-06
F08	1.25E+01	1.22E-02	**3.25E-03**	8.81E-01	3.19E-03	**5.70E-05**
F09	1.16E+02	3.90E+00	**5.79E-01**	4.54E+02	**3.32E+01**	5.32E+01
F10	3.87E+00	5.43E+00	**1.25E-02**	7.53E-01	5.66E-02	**7.17E-03**
F11	2.31E+01	6.78E+00	**8.94E-01**	3.30E+00	7.16E+00	**1.81E+00**
F12	2.53E-05	1.26E-08	**2.51E-10**	2.62E-32	**0.00E+00**	**0.00E+00**
F13	2.51E-07	**6.39E-08**	6.89E-08	9.52E-32	**0.00E+00**	**0.00E+00**

Table 12.4. *Comparison of results of hybrid algorithms combining BBO with DE and PSO based on iteration-level hybridization and algorithm-level hybridization. The best value in each row is indicated in bold*

12.4. Conclusion

In this chapter, we discussed hybrid BBO algorithms that combine BBO with various popular search methods, including local search and global search, and including learning strategies and hybrid strategies. First, we showed how OBL could be incorporated into BBO to produce OBBO algorithms. Then, we showed how several local search methods could be incorporated into BBO to produce hybrid BBO algorithms. Finally, we showed the hybridization of BBO with other popular EAs at both the iteration level and the algorithm level based on the information exchange mechanism of biogeography. The simulation results indicate that all of these hybrid BBO algorithms can improve the optimization performance of BBO. In addition, these combinations of BBO with various search methods can serve as a template for the improvement or hybridization of any other EA.

This chapter is not intended to provide a complete exposition of the subject of hybrid BBO algorithms, but has only shown how BBO can be combined with various search methods, and particularly how hybridization can improve the performance of BBO. More hybrid BBO algorithms have been proposed and new ones are continually appearing in the literature [BHA 10, BOU 11a, BOU 11b, CHA 12, WAN 11b]. As of June 2016, INSPEC included more than 400 publications about hybrid BBO algorithms.

In this concluding section, we mention some important topics for future research of hybrid EAs.

New hybridization approaches

Just as ideas from BBO have been used in this chapter to develop new hybridization approaches, ideas from other EAs could also be used to develop new hybridization approaches. For example, information exchange mechanisms based on DE, PSO, GA or any other EA could be used to combine EAs running in parallel [BLU 11, NIK 09]. The use of BBO as a hybridization strategy should motivate the investigation of other EAs as hybridization strategies, and then these hybridization strategies could be compared with one another.

Adaptation in hybrid EAs

Adaptation might be a promising way to pursue improved performance in hybrid EAs [WAN 09, ZHA 11]. Because there are usually many tuning parameters in a hybrid EA, we need to decrease the number of manual tuning parameters. Adaptation could be implemented in different ways. For example, since an OBBO population tends to converge to good solutions as the generation count increases, perhaps OBL should be implemented more often in the early stages of BBO operation and less often in the later stages. This could be done by making the jump rate a decreasing function of the generation number. Other ways of implementing adaptation in hybrid EAs might include changing the types of search methods based on the generation count or based on individual fitness values.

Applications of hybrid EAs

Another important topic is the application of hybrid EAs to real-world problems. Many application results have been published [MAK 10, MON 12, NGU 07, PRO 11], but more challenging problems are found in various fields of society and science. Research on multi-objective problems, constrained problems and large-scale problems emphasizes their special challenges, and we need to find good hybrid EAs to solve these problems.

Theory and mathematical models

Theoretical results for hybrid EAs are sparse and so there is a lot of room for contributions in this area. For example, we could use the Markov models and dynamic system models from the previous chapters to compare various hybrid EAs on an analytical level rather than relying on simulation.

APPENDICES

Appendix A

Unconstrained Benchmark Functions

The problem is to minimize $f(x)$ over all x. We use x^* to represent the optimizing value of x, and $f(x^*)$ is the minimum value of $f(x)$:

$$x^* = \arg\min_x f(x) \quad [A.1]$$

Many of the benchmarks that we present in this section are from Yao *et al.* [YAO 99]. Detailed information about the unconstrained benchmarks and evaluation metrics for EA competitions at the 2005 IEEE Congress on Evolutionary Computation can be found in Suganthan *et al.* [SUG 05]. For the benchmarks presented here, their dimensionality can be varied so that performance can be explored as a function of the number of dimensions n.

F01 Sphere Function

$$\begin{aligned} f(x) &= \sum_{i=1}^{n} x_i^2, \quad |x_i| \le 100 \\ x^* &= 0 \\ f(x^*) &= 0 \end{aligned} \quad [A.2]$$

F02 Schwefel's Problem 2.22

$$\begin{aligned} f(x) &= \sum_{i=1}^{n} |x_i| + \prod_{i=1}^{n} |x_i|, \quad |x_i| \le 10 \\ x^* &= 0 \\ f(x^*) &= 0 \end{aligned} \quad [A.3]$$

F03 Schwefel's Problem 1.2

$$f(x) = \sum_{i=1}^{n} \left(\sum_{j=1}^{i} x_j \right)^2, \quad |x_i| \le 100$$
$$x^* = 0 \qquad [A.4]$$
$$f(x^*) = 0$$

F04 Schwefel's Problem 2.21

$$f(x) = \max_i \left(|x_i| : i \in \{1, \cdots, n\} \right) \quad |x_i| \le 100$$
$$x^* = 0 \qquad [A.5]$$
$$f(x^*) = 0$$

F05 Generalized Rosenbrock's Function

$$f(x) = \sum_{i=1}^{n-1} \left[100(x_{i+1} - x_i^2)^2 + (x_i - 1)^2 \right] \quad |x_i| \le 30$$
$$x^* = [1, \ldots, 1] \qquad [A.6]$$
$$f(x^*) = 0$$

F06 Step Function

$$f(x) = \sum_{i=1}^{n} \left(\lfloor x_i + 0.5 \rfloor \right)^2 \quad |x_i| \le 100$$
$$x^* = 0 \qquad [A.7]$$
$$f(x^*) = 0$$

F07 Quartic Function

$$f(x) = \sum_{i=1}^{n} i x_i^4 \quad |x_i| \le 1.28$$
$$x^* = 0 \qquad [A.8]$$
$$f(x^*) = 0$$

F08 Generalized Schwefel's Problem 2.26

$$
\begin{aligned}
f(x) &= -\sum_{i=1}^{n} x_i \sin\sqrt{|x_i|} \quad |x_i| \le 500 \\
x^* &= [420.9687, \cdots 420.9687] \\
f(x^*) &= -12569.5
\end{aligned} \tag{A.9}
$$

F09 Generalized Rastrigin's Function

$$
\begin{aligned}
f(x) &= \sum_{i=1}^{n}\left[x_i^2 - 10\cos(2\pi x_i) + 10\right] \quad |x_i| \le 5.12 \\
x^* &= 0 \\
f(x^*) &= 0
\end{aligned} \tag{A.10}
$$

F10 Ackley's Function

$$
\begin{aligned}
f(x) &= 20 + e - 20\exp(-0.2\sum_{i=1}^{n} x_i^2 / n) - \exp(\sum_{i=1}^{n}(\cos 2\pi x_i)/n) \quad |x_i| \le 32 \\
x^* &= 0 \\
f(x^*) &= 0
\end{aligned} \tag{A.11}
$$

F11 Generalized Griewank's Function

$$
\begin{aligned}
f(x) &= 1 + \sum_{i=1}^{n} x_i^2 / 4000 - \prod_{i=1}^{n}\cos(x_i / \sqrt{i}) \quad |x_i| \le 600 \\
x^* &= 0 \\
f(x^*) &= 0
\end{aligned} \tag{A.12}
$$

F12 Generalized Penalized Function 1

$$
\begin{aligned}
f(x) &= \frac{\pi}{n}\left\{10\sin^2(\pi y_1) + \sum_{i=1}^{n-1}(y_i-1)^2[1+10\sin^2(\pi y_{i+1})] + (y_n-1)^2\right\} \\
&\quad + \sum_{i=1}^{n} u_i \qquad |x| \leq 50 \\
u_i &= \begin{cases} k(x_i-a)^m & x_i > a \\ 0 & -a \leq x_i \leq a \\ k(-x_i-a)^m & x_i < -a \end{cases} \\
y_i &= 1 + (x_i+1)/4 \\
x^* &= [1,\cdots,1] \\
f(x^*) &= 0
\end{aligned} \qquad [A.13]
$$

Note that in this benchmark, values for k, a and m are not given, but we usually use $k=100$, $a=10$ and $m=4$.

F13 Generalized Penalized Function 2

$$
\begin{aligned}
f(x) &= \sum_{i=1}^{n} u_i + 0.1\{\sin^2(3\pi x_1) + (x_n-1)^2 \\
&\quad + \sum_{i=1}^{n-1}(x_i-1)^2[1+\sin^2(3\pi x_{i+1})]\} \qquad |x_i| \leq 50 \\
u_i &= \begin{cases} k(x_i-a)^m & x_i > a \\ 0 & -a \leq x_i \leq a \\ k(-x_i-a)^m & x_i < a \end{cases} \\
x^* &= [1,\cdots,1] \\
f(x^*) &= 0
\end{aligned} \qquad [A.14]
$$

Note that in this benchmark, like the generalized penalized function 1, values for k, a and m are not given but we usually use $k=100$, $a=5$ and $m=4$.

Appendix B

Constrained Benchmark Functions

A constrained optimization problem involves the minimization of $f(x)$ over all x such that $x \in \Gamma \in R^n$, where Γ is the feasible set and n is the problem dimension. We use x^* to represent the optimizing value of x, and $f(x^*)$ is the constrained minimum value of $f(x)$:

$$\begin{aligned} & x^* = \arg\min_x f(x) \\ \text{such that } & g_i(x) \leq 0 \text{ for } i \in [1,m] \\ \text{and } & h_j(x) = 0 \text{ for } j \in [1,p] \end{aligned} \qquad [B.1]$$

This problem includes $(m + p)$ constraints, m of which are inequality constraints and p of which are equality constraints. In this section, we only show simple constrained benchmarks, while detailed information and evaluation metrics for EA competitions at the 2006 and 2010 *IEEE Congress on Evolutionary Computation* can be found in Liang *et al.* [LIA 06] and Mallipeddi and Suganthan [MAL 10].

The G01 Function

$$\begin{aligned} & f(x) = 5\sum_{i=1}^{4} x_i - 5\sum_{i=1}^{4} x_i^2 - \sum_{i=5}^{13} x_i \\ & g_1(x) = 2x_1 + 2x_2 + x_{10} + x_{11} - 10 \leq 0 \\ & g_2(x) = 2x_1 + 2x_3 + x_{10} + x_{11} - 10 \leq 0 \\ & g_3(x) = 2x_2 + 2x_3 + x_{11} + x_{12} - 10 \leq 0 \end{aligned} \qquad [B.2]$$

$$g_4(x) = -8x_1 + x_{10} \le 0$$
$$g_5(x) = -8x_2 + x_{11} \le 0$$
$$g_6(x) = -8x_3 + x_{12} \le 0$$
$$g_7(x) = -2x_4 - x_5 + x_{10} \le 0$$
$$g_8(x) = -2x_6 - x_7 + x_{11} \le 0$$
$$g_9(x) = -2x_8 - x_9 + x_{12} \le 0$$

where $0 \le x_i \le 1\,(i = 1,...,9), 0 \le x_i \le 100\,(i = 10,11,12)$ and $0 \le x_{13} \le 1$. The optimum solution is $f(x^*) = -15$.

The G02 Function

$$f(x) = -\left| \frac{\sum_{i=1}^{n} \cos^4(x_i) - 2\prod_{i=1}^{n} \cos^2(x_i)}{\sqrt{\sum_{i=1}^{n} ix_i^2}} \right|$$
$$g_1(x) = 0.75 - \prod_{i=1}^{n} x_i \le 0 \qquad [B.3]$$
$$g_2(x) = \sum_{i=1}^{n} x_i - 0.75n \le 0$$

where $n = 20$ and $0 < x_i \le 10\,(i = 1,...,n)$. The optimum solution is $f(x^*) = -0.80361910412559$.

The G03 Function

$$f(x) = -(\sqrt{n})^n \prod_{i=1}^{n} x_i \qquad [B.4]$$
$$h(x) = \sum_{i=1}^{n} x_i^2 - 1 = 0$$

where $n = 10$ and $0 \le x_i \le 1\,(i = 1,...,n)$. The optimum solution is $f(x^*) = -1.00050010001000$.

The G04 Function

$$
\begin{aligned}
f(x) &= 5.3578547x_3^2 + 0.8356891x_1x_5 + 37.293239x_1 - 40792.141 \\
g_1(x) &= 85.334407 + 0.0056858x_2x_5 + 0.0006262x_1x_4 - 0.0022053x_3x_5 - 92 \le 0 \\
g_2(x) &= -85.334407 - 0.0056858x_2x_5 - 0.0006262x_1x_4 + 0.0022053x_3x_5 \le 0 \\
g_3(x) &= 80.51249 + 0.0071317x_2x_5 + 0.0029955x_1x_2 + 0.0021813x_3^2 - 110 \le 0 \\
g_4(x) &= -80.51249 - 0.0071317x_2x_5 - 0.0029955x_1x_2 - 0.0021813x_3^2 + 90 \le 0 \\
g_5(x) &= 9.300961 + 0.0047026x_3x_5 + 0.0012547x_1x_3 + 0.0019085x_3x_4 - 25 \le 0 \\
g_6(x) &= -9.300961 - 0.0047026x_3x_5 - 0.0012547x_1x_3 - 0.0019085x_3x_4 + 20 \le 0
\end{aligned}
\qquad [B.5]
$$

where $78 \le x_1 \le 102, 33 \le x_2 \le 45$ and $27 \le x_i \le 45\ (i = 3,4,5)$. The optimum solution is $f(x^*) = -3.066553867178332e+004$.

The G05 Function

$$
\begin{aligned}
f(x) &= 3x_1 + 0.000001x_1^3 + 2x_2 + (0.000002/3)x_2^3 \\
g_1(x) &= -x_4 + x_3 - 0.55 \le 0 \\
g_2(x) &= -x_3 + x_4 - 0.55 \le 0 \\
h_3(x) &= 1000\sin(-x_3 - 0.25) + 1000\sin(-x_4 - 0.25) + 894.8 - x_1 = 0 \\
h_4(x) &= 1000\sin(x_3 - 0.25) + 1000\sin(x_3 - x_4 - 0.25) + 894.8 - x_2 = 0 \\
h_5(x) &= 1000\sin(x_4 - 0.25) + 1000\sin(x_4 - x_3 - 0.25) + 1294.8 = 0
\end{aligned}
\qquad [B.6]
$$

where $0 \le x_1 \le 1200, 0 \le x_2 \le 1200, -0.55 \le x_3 \le 0.55$ and $-0.55 \le x_4 \le 0.55$. The optimum solution is $f(x^*) = 5126.4967140071$.

The G06 Function

$$
\begin{aligned}
f(x) &= (x_1 - 10)^3 + (x_2 - 20)^3 \\
g_1(x) &= -(x_1 - 5)^2 - (x_2 - 5)^2 + 100 \le 0 \\
g_2(x) &= (x_1 - 6)^2 + (x_2 - 5)^2 - 82.81 \le 0
\end{aligned}
\qquad [B.7]
$$

where $13 \le x_1 \le 100$ and $0 \le x_2 \le 100$. The optimum solution is $f(x^*) = -6961.81387558015$.

The G07 Function

$$
\begin{aligned}
f(x) &= x_1^2 + x_2^2 + x_1x_2 - 14x_1 - 16x_2 + (x_3 - 10)^2 \\
&\quad + 4(x_4 - 5)^2 + (x_5 - 3)^2 + 2(x_6 - 1)^2 + 5x_7^2 \\
&\quad + 7(x_8 - 11)^2 + 2(x_9 - 10)^2 + (x_{10} - 7)^2 + 45 \\
g_1(x) &= -105 + 4x_1 + 5x_2 - 3x_7 + 9x_8 \le 0 \\
g_2(x) &= 10x_1 - 8x_2 - 17x_7 + 2x_8 \le 0 \\
g_3(x) &= -8x_1 + 2x_2 + 5x_9 - 2x_{10} - 12 \le 0 \\
g_4(x) &= 3(x_1 - 2)^2 + 4(x_2 - 3)^2 + 2x_3^2 - 7x_4 - 120 \le 0 \\
g_5(x) &= 5x_1^2 + 8x_2 + (x_3 - 6)^2 - 2x_4 - 40 \le 0 \\
g_6(x) &= x_1^2 + 2(x_2 - 2)^2 - 2x_1x_2 + 14x_5 - 6x_6 \le 0 \\
g_7(x) &= 0.5(x_1 - 8)^2 + 2(x_2 - 4)^2 + 3x_5^2 - x_6 - 30 \le 0 \\
g_8(x) &= -3x_1 + 6x_2 + 12(x_9 - 8)^2 - 7x_{10} \le 0
\end{aligned}
\qquad \text{[B.8]}
$$

where $-10 \le x_i \le 10 \ (i = 1,...,10)$. The optimum solution is $f(x^*) = 24.30620906818$.

The G08 Function

$$
\begin{aligned}
f(x) &= -\frac{\sin^3(2\pi x_1)\sin(2\pi x_2)}{x_1^3(x_1 + x_2)} \\
g_1(x) &= x_1^2 - x_2 + 1 \le 0 \\
g_2(x) &= 1 - x_1 + (x_2 - 4)^2 \le 0
\end{aligned}
\qquad \text{[B.9]}
$$

where $0 \le x_1 \le 10$ and $0 \le x_2 \le 10$. The optimum solution is $f(x^*) = -0.0958250414180359$.

The G09 Function

$$
\begin{aligned}
f(x) &= (x_1-10)^2+5(x_2-12)^2+x_3^{\,4}+3(x_4-11)^2 \\
&\quad +10x_5^{\,6}+7x_6^{\,2}+x_7^{\,4}-4x_6x_7-10x_6-8x_7 \\
g_1(x) &= -127+2x_1^{\,2}+3x_2^{\,4}+x_3+4x_4^{\,2}+5x_5 \le 0 \\
g_2(x) &= -282+7x_1+3x_2+10x_3^{\,2}+x_4-x_5 \le 0 \\
g_3(x) &= -196+23x_1+x_2^{\,2}+6x_6^{\,2}-8x_7 \le 0 \\
g_4(x) &= 4x_1^{\,2}+x_2^{\,2}-3x_1x_2+2x_3^{\,2}+5x_6-11x_7 \le 0
\end{aligned}
\qquad [B.10]
$$

where $-10 \le x_i \le 10$ for $i = 1,\cdots,7$. The optimum solution is $f(\vec{x}^*) = 680.630057374402$.

The G10 Function

$$
\begin{aligned}
f(x) &= x_1+x_2+x_3 \\
g_1(x) &= -1+0.0025(x_4+x_6) \le 0 \\
g_2(x) &= -1+0.0025(x_5+x_7-x_4) \le 0 \\
g_3(x) &= -1+0.01(x_8-x_5) \le 0 \\
g_4(x) &= -x_1x_6+833.33252x_4+100x_1-83333.333 \le 0 \\
g_5(x) &= -x_2x_7+1250x_5+x_2x_4-1250x_4 \le 0 \\
g_6(x) &= -x_3x_8+1250000+x_3x_5-2500x_5 \le 0
\end{aligned}
\qquad [B.11]
$$

where $100 \le x_1 \le 10000, 1000 \le x_i \le 10000\ (i = 2,3)$ and $10 \le x_i \le 1000\ (i = 4,...,8)$. The optimum solution is $f(x^*) = 7049.24802052867$.

The G11 Function

$$
\begin{aligned}
f(x) &= x_1^{\,2}+(x_2-1)^2 \\
h(x) &= x_2-x_1^{\,2} = 0
\end{aligned}
\qquad [B.12]
$$

where $-1 \le x_1 \le 1$ and $-1 \le x_2 \le 1$. The optimum solution is $f(x^*) = 0.7499$.

The G12 Function

$$\begin{aligned} f(x) &= -(100-(x_1-5)^2-(x_2-5)^2-(x_3-5)^2)/100 \\ g(x) &= (x_1-p)^2+(x_2-q)^2+(x_3-r)^2-0.0625 \le 0 \end{aligned} \qquad [B.13]$$

where $0 \le x_i \le 10\ (i=1,2,3)$ and $p,q,r=1,2,\cdots,9$. The optimum solution is $f(x^*)=-1$.

The G13 Function

$$\begin{aligned} f(x) &= e^{x_1x_2x_3x_4x_5} \\ h_1(x) &= x_1^2+x_2^2+x_3^2+x_4^2+x_5^2-10=0 \\ h_2(x) &= x_2x_3-5x_4x_5=0 \\ h_3(x) &= x_1^3+x_2^3+1=0 \end{aligned} \qquad [B.14]$$

where $-2.3 \le x_i \le 2.3\ (i=1,2)$ and $-3.2 \le x_i \le 3.2\ (i=3,4,5)$. The optimum solution is $f(x^*)=0.053941514041898$.

The G14 Function

$$\begin{aligned} f(x) &= \sum_{i=1}^{10} x_i \left(c_i + \ln \frac{x_i}{\sum_{j=1}^{10} x_j} \right) \\ h_1(x) &= x_1+2x_2+2x_3+x_6+x_{10}-2=0 \\ h_2(x) &= x_4+2x_5+x_6+x_7-1=0 \\ h_3(x) &= x_3+x_7+x_8+2x_9+x_{10}-1=0 \end{aligned} \qquad [B.15]$$

where $0 < x_i \le 10\ (i=1,\ldots,10)$ and $c_1=-6.089, c_2=-17.164, c_3=-34.054, c_4=-5.914$, $c_5=-24.721$, $c_6=-14.986$, $c_7=-24.1$, $c_8=-10.708$, $c_9=-26.662$, $c_{10}=-22.179$. The optimum solution is $f(x^*)=-47.7648884594915$.

The G15 Function

$$\begin{aligned}
f(x) &= 1000 - x_1^2 - 2x_2^2 - x_3^2 - x_1x_2 - x_1x_3 \\
h_1(x) &= x_1^2 + x_2^2 + x_3^2 - 25 = 0 \\
h_2(x) &= 8x_1 + 14x_2 + 7x_3 - 56 = 0
\end{aligned} \qquad \text{[B.16]}$$

where $0 \le x_i \le 10\ (i = 1,2,3)$. The optimum solution is $f(x^*) = 961.715022289961$.

The G16 Function

$$\begin{aligned}
f(x) &= 0.000117y_{14} + 0.1365 + 0.00002358y_{13} \\
&\quad + 0.000001502y_{16} + 0.0321y_{12} + 0.004324y_5 \\
&\quad + 0.0001\frac{c_{15}}{c_{16}} + 37.48\frac{y_2}{c_{12}} - 0.0000005843y_{17} \\
g_1(x) &= \frac{0.28}{0.72}y_5 - y_4 \le 0 \\
g_2(x) &= x_3 - 1.5x_2 \le 0 \\
g_3(x) &= 3496\frac{y_2}{c_{12}} - 21 \le 0 \\
g_4(x) &= 110.6 + y_1 - \frac{62212}{c_{17}} \le 0 \\
g_5(x) &= 213.1 - y_1 \le 0 \\
g_6(x) &= y_1 - 405.23 \le 0 \\
g_7(x) &= 17.505 - y_2 \le 0 \\
g_8(x) &= y_2 - 1053.6667 \le 0 \\
g_9(x) &= 11.275 - y_3 \le 0 \\
g_{10}(x) &= y_3 - 35.03 \le 0 \\
g_{11}(x) &= 214.228 - y_4 \le 0 \\
g_{12}(x) &= y_4 - 665.585 \le 0 \\
g_{13}(x) &= 7.458 - y_5 \le 0 \\
g_{14}(x) &= y_5 - 584.463 \le 0 \\
g_{15}(x) &= 0.961 - y_6 \le 0
\end{aligned} \qquad \text{[B.17]}$$

$$g_{16}(x) = y_6 - 265.916 \le 0$$
$$g_{17}(x) = 1.612 - y_7 \le 0$$
$$g_{18}(x) = y_7 - 7.046 \le 0$$
$$g_{19}(x) = 0.146 - y_8 \le 0$$
$$g_{20}(x) = y_8 - 0.222 \le 0$$
$$g_{21}(x) = 107.99 - y_9 \le 0$$
$$g_{22}(x) = y_9 - 273.366 \le 0$$
$$g_{23}(x) = 922.693 - y_{10} \le 0$$
$$g_{24}(x) = y_{10} - 1286.105 \le 0$$
$$g_{25}(x) = 926.832 - y_{11} \le 0$$
$$g_{26}(x) = y_{11} - 1444.046 \le 0$$
$$g_{27}(x) = 18.766 - y_{12} \le 0$$
$$g_{28}(x) = y_{12} - 537.141 \le 0$$
$$g_{29}(x) = 1072.163 - y_{13} \le 0$$
$$g_{30}(x) = y_{13} - 3247.039 \le 0$$
$$g_{31}(x) = 8961.448 - y_{14} \le 0$$
$$g_{32}(x) = y_{14} - 26844.086 \le 0$$
$$g_{33}(x) = 0.063 - y_{15} \le 0$$
$$g_{34}(x) = y_{15} - 0.386 \le 0$$
$$g_{35}(x) = 71084.33 - y_{16} \le 0$$
$$g_{36}(x) = y_{16} - 140000 \le 0$$
$$g_{37}(x) = 2802713 - y_{17} \le 0$$
$$g_{38}(x) = y_{17} - 12146108 \le 0$$

where

$$y_1 = x_2 + x_3 + 41.6$$
$$c_1 = 0.024x_4 - 4.62$$
$$y_2 = \frac{12.5}{c_1} + 12$$

$$c_2 = 0.0003535x_1^2 + 0.5311x_1 + 0.08705y_2x_1$$
$$c_3 = 0.052x_1 + 78 + 0.002377y_2x_1$$
$$y_3 = \frac{c_2}{c_3}$$
$$y_4 = 19y_3$$
$$c_4 = 0.04782(x_1 - y_3) + \frac{0.1956(x_1 - y_3)^2}{x_2} + 0.6376y_4 + 1.594y_3$$
$$c_5 = 100x_2$$
$$c_6 = x_1 - y_3 - y_4$$
$$c_7 = 0.950 - \frac{c_4}{c_5}$$
$$y_5 = c_6c_7$$
$$y_6 = x_1 - y_5 - y_4 - y_3$$
$$c_8 = 0.995(y_5 + y_4)$$
$$y_7 = \frac{c_8}{y_1}$$
$$y_8 = \frac{c_8}{3798}$$
$$c_9 = y_7 - \frac{0.0663y_7}{y_8} - 0.3153$$
$$y_9 = \frac{96.82}{c_9} + 0.321y_1$$
$$y_{10} = 1.29y_5 + 1.258y_4 + 2.29y_3 + 1.71y_6$$
$$y_{11} = 1.71x_1 - 0.452y_4 + 0.580y_3$$
$$c_{10} = \frac{12.3}{752.3}$$
$$c_{11} = (1.75y_2)(0.995x_1)$$
$$c_{12} = 0.995y_{10} + 1998$$
$$y_{12} = c_{10}x_1 + \frac{c_{11}}{c_{12}}$$
$$y_{13} = c_{12} - 1.75y_2$$
$$y_{14} = 3623 + 64.4x_2 + 58.4x_3 + \frac{146312}{y_9 + x_5}$$

$$c_{13} = 0.995y_{10} + 60.8x_2 + 48x_4 - 0.1121y_{14} - 5095$$
$$y_{15} = \frac{y_{13}}{c_{13}}$$
$$y_{16} = 148000 - 331000y_{15} + 40y_{13} - 61y_{15}y_{13}$$
$$c_{14} = 2324y_{10} - 28740000y_2$$
$$y_{17} = 14130000 - 1328y_{10} - 531y_{11} + \frac{c_{14}}{c_{12}}$$
$$c_{15} = \frac{y_{13}}{y_{15}} - \frac{y_{13}}{0.52}$$
$$c_{16} = 1.104 - 0.72y_{15}$$
$$c_{17} = y_9 + x_5$$

and where $704.4148 \le x_1 \le 906.3855$, $68.6 \le x_2 \le 288.88$, $0 \le x_3 \le 134.75$, $193 \le x_4 \le 287.0966$ and $25 \le x_5 \le 84.1988$. The optimum solution is $f(x^*) = -1.90515525853479$.

The G17 Function

$$f(x) = f_1(x_1) + f_2(x_2)$$
$$f_1(x_1) = \begin{cases} 30x_1 & 0 \le x_1 < 300 \\ 31x_1 & 300 \le x_1 \le 400 \end{cases} \qquad \text{[B.18]}$$
$$f_2(x_2) = \begin{cases} 28x_2 & 0 \le x_2 < 100 \\ 29x_2 & 100 \le x_2 < 200 \\ 30x_2 & 200 \le x_2 < 1000 \end{cases}$$

$$h_1(x) = -x_1 + 300 - \frac{x_3x_4}{131.078}\cos(1.48477 - x_6) + \frac{0.90798x_3^2}{131.078}\cos(1.47588)$$
$$h_2(x) = -x_2 - \frac{x_3x_4}{131.078}\cos(1.48477 + x_6) + \frac{0.90798x_4^2}{131.078}\cos(1.47588)$$
$$h_3(x) = -x_5 - \frac{x_3x_4}{131.078}\sin(1.48477 + x_6) + \frac{0.90798x_4^2}{131.078}\sin(1.47588)$$
$$h_4(x) = 200 - \frac{x_3x_4}{131.078}\sin(1.48477 - x_6) + \frac{0.90798x_3^2}{131.078}\sin(1.47588)$$

where $0 \le x_1 \le 400$, $0 \le x_2 \le 1000$, $340 \le x_3 \le 420$, $340 \le x_4 \le 420$, $-1000 \le x_5 \le 1000$ and $0 \le x_6 \le 0.5236$. The optimum solution is $f\left(x^*\right) = 8853.53967480648$.

The G18 Function

$$
\begin{aligned}
& f(x) = -0.5(x_1x_4 - x_2x_3 + x_3x_9 - x_5x_9 + x_5x_8 - x_6x_7) \\
& g_1(x) = x_3^{\,2} + x_4^{\,2} - 1 \le 0 \\
& g_2(x) = x_9^{\,2} - 1 \le 0 \\
& g_3(x) = x_5^{\,2} + x_6^{\,2} - 1 \le 0 \\
& g_4(x) = x_1^{\,2} + (x_2 - x_9)^2 - 1 \le 0 \\
& g_5(x) = (x_1 - x_5)^2 + (x_2 - x_6)^2 - 1 \le 0 \\
& g_6(x) = (x_1 - x_7)^2 + (x_2 - x_8)^2 - 1 \le 0 \\
& g_7(x) = (x_3 - x_5)^2 + (x_4 - x_6)^2 - 1 \le 0 \\
& g_8(x) = (x_3 - x_7)^2 + (x_4 - x_8)^2 - 1 \le 0 \\
& g_9(x) = x_7^{\,2} + (x_8 - x_9)^2 - 1 \le 0 \\
& g_{10}(x) = x_2x_3 - x_1x_4 \le 0 \\
& g_{11}(x) = -x_3x_9 \le 0 \\
& g_{12}(x) = x_5x_9 \le 0 \\
& g_{13}(x) = x_6x_7 - x_5x_8 \le 0
\end{aligned}
\qquad [B.19]
$$

where $-10 \le x_i \le 10\ (i = 1,\ldots,8)$ and $0 \le x_9 \le 20$. The optimum solution is $f\left(x^*\right) = -0.866025403784439$.

The G19 Function

$$
\begin{aligned}
& f(x) = \sum_{j=1}^{5}\sum_{i=1}^{5} c_{ij}x_{(10+i)}x_{(10+j)} + 2\sum_{j=1}^{5} d_j x_{(10+j)}^3 - \sum_{i=1}^{10} b_i x_i \\
& g_j(x) = -2\sum_{i=1}^{5} c_{ij}x_{(10+i)} - 3d_j x_{(10+j)}^2 - e_j + \sum_{i=1}^{10} a_{ij}x_i \le 0 \quad j = 1,\cdots,5
\end{aligned}
\qquad [B.20]
$$

where $b = [-40, -2, -0.25, -4, -4, -1, -40, -60, 5.1]$ and the remaining data is given in Table B.1. The bounds are $0 \le x_i \le 10\ (i = 1,...,15)$. The optimum solution is $f(x^*) = 32.6555929502463$.

j	1	2	3	4	5
e_j	−15	−27	−36	−18	−12
c_{1j}	30	−20	−10	32	−10
c_{2j}	−20	39	−6	−31	32
c_{3j}	−10	−6	10	−6	−10
c_{4j}	32	−31	−6	39	−20
c_{5j}	−10	32	−10	−20	30
d_j	4	8	10	6	2
a_{1j}	−16	2	0	1	0
a_{2j}	0	−2	0	0.4	2
a_{3j}	−3.5	0	2	0	0
a_{4j}	0	−2	0	−4	−1
a_{5j}	0	−9	−2	1	−2.8
a_{6j}	2	0	−4	0	0
a_{7j}	−1	−1	−1	−1	−1
a_{8j}	−1	−2	−3	−2	−1
a_{9j}	1	2	3	4	5
a_{10j}	1	1	1	1	1

Table B.1. *Data set for benchmark function G19*

The G20 Function

$$
\begin{aligned}
f(x) &= \sum_{i=1}^{24} a_i x_i \\
g_i(x) &= \frac{(x_i + x_{(i+12)})}{\sum_{j=1}^{24} x_j + e_i} \le 0 \quad i = 1, 2, 3 \\
g_i(x) &= \frac{(x_{(i+3)} + x_{(i+15)})}{\sum_{j=1}^{24} x_j + e_i} \le 0 \quad i = 4, 5, 6 \\
h_i(x) &= \frac{x_{(i+12)}}{b_{(i+12)} \sum_{j=13}^{24} \frac{x_j}{b_j}} - \frac{c_i x_i}{40 b_i \sum_{j=1}^{12} \frac{x_j}{b_j}} = 0 \quad i = 1, \cdots, 12 \\
h_{13}(x) &= \sum_{i=1}^{24} x_i - 1 = 0 \\
h_{14}(x) &= \sum_{i=1}^{12} \frac{x_i}{d_i} + k \sum_{i=13}^{24} \frac{x_i}{b_i} - 1.671 = 0
\end{aligned} \qquad [B.21]
$$

where $k = (0.7302)(530)(14.7/40)$ and the data set is detailed in Table B.2. The bounds are $0 \le x_i \le 10\ (i = 1, \ldots, 24)$. This solution is a little infeasible and no feasible solution is found so far.

i	a_i	b_i	c_i	d_i	e_i
1	0.0693	44.094	123.7	31.244	0.1
2	0.0577	58.12	31.7	36.12	0.3
3	0.05	58.12	45.7	34.784	0.4
3	0.2	137.4	14.7	92.7	0.3
4	0.26	120.9	84.7	82.7	0.6
5	0.55	170.9	27.7	91.6	0.3
7	0.06	62.501	49.7	56.708	–
8	0.1	84.94	7.1	82.7	–
9	0.12	133.425	2.1	80.8	–
10	0.18	82.507	17.7	64.517	–
11	0.1	46.07	0.85	49.4	–

12	0.09	60.097	0.64	49.1	–
13	0.0693	44.094	–	–	–
14	0.0577	58.12	–	–	–
15	0.05	58.12	–	–	–
16	0.2	137.4	–	–	–
17	0.26	120.9	–	–	–
18	0.55	170.9	–	–	–
19	0.06	62.501	–	–	–
20	0.1	84.94	–	–	–
21	0.12	133.425	–	–	–
22	0.18	82.507	–	–	–
23	0.1	46.07	–	–	–
24	0.09	60.097	–	–	–

Table B.2. *Data set for benchmark function G20*

The G21 Function

$$\begin{aligned}
&f(x) = x_1 \\
&g_1(x) = -x_1 + 35x_2^{0.6} + 35x_3^{0.6} \le 0 \\
&h_1(x) = -300x_3 + 7500x_5 - 7500x_6 - 25x_4x_5 + 25x_4x_6 + x_3x_4 = 0 \\
&h_2(x) = 100x_2 + 155.365x_4 + 2500x_7 - x_2x_4 - 25x_4x_7 - 15536.5 = 0 \\
&h_3(x) = -x_5 + \ln(-x_4 + 900) = 0 \\
&h_4(x) = -x_6 + \ln(x_4 + 300) = 0 \\
&h_5(x) = -x_7 + \ln(-2x_4 + 700) = 0
\end{aligned} \quad [B.22]$$

where $0 \le x_1 \le 1000$, $0 \le x_2, x_3 \le 40$, $100 \le x_4 \le 300$, $6.3 \le x_5 \le 6.7, 5.9 \le x_6 \le 6.4$, and $4.5 \le x_7 \le 6.25$. The optimum solution is $f(x^*) = 193.724510070035$.

The G22 Function

$$
\begin{aligned}
& f(x)=x_1 \\
& g_1(x)=-x_1+x_2^{0.6}+x_3^{0.6}+x_4^{0.6}\leq 0 \\
& h_1(x)=x_5-100000x_8+1\times10^7=0 \\
& h_2(x)=x_6+100000x_8-100000x_9=0 \\
& h_3(x)=x_7+100000x_9-5\times10^7=0 \\
& h_4(x)=x_5+100000x_{10}-3.3\times10^7=0 \\
& h_5(x)=x_6+100000x_{11}-4.4\times10^7=0 \\
& h_6(x)=x_7+100000x_{12}-6.6\times10^7=0 \\
& h_7(x)=x_5-120x_2x_{13}=0 \\
& h_8(x)=x_6-80x_3x_{14}=0 \\
& h_9(x)=x_7-40x_4x_{15}=0 \\
& h_{10}(x)=x_8-x_{11}+x_{16}=0 \\
& h_{11}(x)=x_9-x_{12}+x_{17}=0 \\
& h_{12}(x)=-x_{18}+\ln(x_{10}-100)=0 \\
& h_{13}(x)=-x_{19}+\ln(-x_8+300)=0 \\
& h_{14}(x)=-x_{20}+\ln(x_{16})=0 \\
& h_{15}(x)=-x_{21}+\ln(-x_9+400)=0 \\
& h_{16}(x)=-x_{22}+\ln(x_{17})=0 \\
& h_{17}(x)=-x_8-x_{10}+x_{13}x_{18}-x_{13}x_{19}+400=0 \\
& h_{18}(x)=x_8-x_9-x_{11}+x_{14}x_{20}-x_{14}x_{21}+400=0 \\
& h_{19}(x)=x_9-x_{12}-4.60517x_{15}+x_{15}x_{22}+100=0
\end{aligned}
\quad [B.23]
$$

where $0\leq x_1\leq 20000$, $0\leq x_2,x_3,x_4\leq 1\times10^6$, $0\leq x_5,x_6,x_7\leq 4\times10^7$, $100\leq x_8\leq 290.99$, $100\leq x_9\leq 399.99$, $100.01\leq x_{10}\leq 300$, $100\leq x_{11}\leq 400$, $100\leq x_{12}\leq 600$, $0\leq x_{13},x_{14},x_{15}\leq 500$, $0.01\leq x_{16}\leq 300$, $0.01\leq x_{17}\leq 400$ and $-4.7\leq x_{18},x_{19},x_{20},x_{21},x_{22}\leq 6.25$. The optimum solution is $f\left(x^*\right)=236.430975504001$.

The G23 Function

$$\begin{aligned}
&f(x) = -9x_5 - 15x_8 + 6x_1 + 16x_2 + 10(x_6 + x_7)\\
&g_1(x) = x_9x_3 + 0.02x_6 - 0.025x_5 \le 0\\
&g_2(x) = x_9x_4 + 0.02x_7 - 0.015x_8 \le 0\\
&h_1(x) = x_1 + x_2 - x_3 - x_4 = 0\\
&h_2(x) = 0.03x_1 + 0.01x_2 - x_9(x_3 + x_4) = 0\\
&h_3(x) = x_3 + x_6 - x_5 = 0\\
&h_4(x) = x_4 + x_7 - x_8 = 0
\end{aligned} \qquad [B.24]$$

where $0 \le x_1, x_2, x_6 \le 300$, $0 \le x_3, x_5, x_7 \le 100$, $0 \le x_4, x_8 \le 200$ and $0.01 \le x_9 \le 0.03$. The optimum solution is $f(x^*) = -400.055099999999584$.

The G24 Function

$$\begin{aligned}
&f(x) = -x_1 - x_2\\
&g_1(x) = -2x_1^4 + 8x_1^3 - 8x_1^2 + x_2 - 2 \le 0\\
&g_2(x) = -4x_1^4 + 32x_1^3 - 88x_1^2 + 96x_1 + x_2 - 36 \le 0
\end{aligned} \qquad [B.25]$$

where $0 \le x_1 \le 3$ and $0 \le x_2 \le 4$. The optimum solution is $f(x^*) =$ -5.50801327159536.

The C01 Function

$$\begin{aligned}
&f(x) = -\left| \frac{\sum_{i=1}^{n} \cos^4(z_i) - 2\prod_{i=1}^{n} \cos^2(z_i)}{\sqrt{\sum_{i=1}^{n} iz^2}} \right|\\
&g_1(x) = 0.75 - \prod_{i=1}^{n} z_i \le 0\\
&g_2(x) = \sum_{i=1}^{n} z_i - 0.75D \le 0\\
&x_i \in [0,10]
\end{aligned} \qquad [B.26]$$

where $z_i = x_i - o_i$ for $i \in [1, n]$. In this function and the functions below, we use o_i to refer to a random offset and M to refer to a random rotation matrix.

The C02 Function

$$\begin{aligned} f(x) &= \max_i(z_i) \\ g_1(x) &= 10 - \frac{1}{n}\sum_{i=1}^{n}[z_i^2 - 10\cos(2\pi z_i) + 10] \le 0 \\ g_2(x) &= \frac{1}{n}\sum_{i=1}^{n}[z_i^2 - 10\cos(2\pi z_i) + 10] - 15 \le 0 \\ h(x) &= \frac{1}{n}\sum_{i=1}^{n}[y_i^2 - 10\cos(2\pi y_i) + 10] - 20 = 0 \\ x_i &\in [-5.12, 5.12] \end{aligned} \qquad [B.27]$$

where $z_i = x_i - o_i$ and $y_i = z_i - 0.5$ for $i \in [1, n]$.

The C03 Function

$$\begin{aligned} f(x) &= \sum_{i=1}^{n-1}\left[100(z_i^2 - z_{i+1})^2 + (z_i^2 - 1)^2\right] \\ h(x) &= \sum_{i=1}^{n-1}(z_i - z_{i+1})^2 = 0 \\ x_i &\in [-1000, 1000] \end{aligned} \qquad [B.28]$$

where $z_i = x_i - o_i$ for $i \in [1, n]$.

The C04 Function

$$\begin{aligned} f(x) &= \max_i(z_i) \\ h_1(x) &= \frac{1}{n}\sum_{i=1}^{n}(z_i \cos(\sqrt{|z_i|}) = 0 \\ h_2(x) &= \sum_{i=1}^{n/2-1}(z_i - z_{i+1})^2 = 0 \\ h_3(x) &= \sum_{i=n/2+1}^{n}(z_i^2 - z_{i+1})^2 = 0 \\ h_4(x) &= \sum_{i=1}^{n} z_i = 0 \\ x_i &\in [-50, 50] \end{aligned} \qquad [B.29]$$

where $z_i = x_i - o_i$ for $i \in [1, n]$.

The C05 Function

$$\begin{aligned} & f(x) = \max_i (z_i) \\ & h_1(x) = \frac{1}{n}\sum_{i=1}^{n}(-z_i \sin(\sqrt{|z_i|}) = 0 \\ & h_2(x) = \frac{1}{n}\sum_{i=1}^{n}(-z_i \cos(\sqrt{|z_i|}) = 0 \\ & x_i \in [-600, 600] \end{aligned} \tag{B.30}$$

where $z_i = x_i - o_i$ for $i \in [1, n]$.

The C06 Function

$$\begin{aligned} & f(x) = \max_i (z_i) \\ & y_i = (z_i + 483.6106156535)M - 483.6106156535 \\ & h_1(x) = \frac{1}{n}\sum_{i=1}^{n}(-y_i \sin(\sqrt{|y_i|}) = 0 \\ & h_2(x) = \frac{1}{n}\sum_{i=1}^{n}(-y_i \cos(0.5\sqrt{|y_i|}) = 0 \\ & x_i \in [-600, 600] \end{aligned} \tag{B.31}$$

where $z_i = x_i - o_i$ for $i \in [1, n]$.

The C07 Function

$$\begin{aligned} & f(x) = \sum_{i=1}^{n-1}\left[100(z_i^2 - z_{i+1})^2 + (z_i - 1)^2\right] \\ & g(x) = 0.5 - \exp(-0.1\sqrt{\frac{1}{n}\sum_{i=1}^{n} y_i^2}) \\ & \qquad - 3\exp(\frac{1}{n}\sum_{i=1}^{n}\cos(0.1 y_i) + \exp(1) \le 0 \\ & x_i \in [-140, 140] \end{aligned} \tag{B.32}$$

where $y_i = x_i - o_i$ and $z_i = x_i + 1 - o_i$ for $i \in [1, n]$.

The C08 Function

$$
\begin{aligned}
f(x) &= \sum_{i=1}^{n-1}\left[100(z_i^2 - z_{i+1})^2 + (z_i - 1)^2\right] \\
g(x) &= 0.5 - \exp(-0.1\sqrt{\frac{1}{n}\sum_{i=1}^{n} y_i^2}) \\
&\quad - 3\exp(\frac{1}{n}\sum_{i=1}^{n}\cos(0.1 y_i) + \exp(1) \le 0 \\
x_i &\in [-140, 140]
\end{aligned}
\qquad [B.33]
$$

where $y_i = (x_i - o_i)M$ and $z_i = x_i + 1 - o_i$ for $i \in [1, n]$.

The C09 Function

$$
\begin{aligned}
f(x) &= \sum_{i=1}^{n-1}\left[100(z_i^2 - z_{i+1})^2 + (z_i - 1)^2\right] \\
h(x) &= \sum_{i=1}^{n}(y \sin\sqrt{|y_i|}) = 0 \\
x_i &\in [-500, 500]
\end{aligned}
\qquad [B.34]
$$

where $y_i = x_i - o_i$ and $z_i = x_i + 1 - o_i$ for $i \in [1, n]$.

The C10 Function

$$
\begin{aligned}
f(x) &= \sum_{i=1}^{n-1}\left[100(z_i^2 - z_{i+1})^2 + (z_i - 1)^2\right] \\
h(x) &= \sum_{i=1}^{n}(y_i \sin\sqrt{|y_i|}) = 0 \\
x_i &\in [-500, 500]
\end{aligned}
\qquad [B.35]
$$

where $y_i = (x_i - o_i)M$ and $z_i = x_i + 1 - o_i$ for $i \in [1, n]$.

The C11 Function

$$f(x)=\frac{1}{n}\sum_{i=1}^{n}\left[-z_i\cos\left(2\sqrt{|z_i|}\right)\right]$$
$$h(x)=\sum_{i}^{n-1}(100(y_i-y_{i+1})^2+(y_i-1)^2)=0 \qquad [B.36]$$
$$x_i\in[-100,100]$$

where $y_i=(x_i-o_i)M$ and $z_i=x_i+1-o_i$ for $i\in[1,n]$.

The C12 Function

$$f(x)=\sum_{i=1}^{n}z_i\sin\sqrt{|z_i|}$$
$$h(x)=\sum_{i=1}^{n}(z_i^2-z_{i+1})^2=0 \qquad [B.37]$$
$$g(x)=\sum_{i=1}^{n}(z_i-100\cos(0.1z_i)+10)\leq 0$$
$$x_i\in[-1000,1000]$$

where $z_i=x_i-o_i$ for $i\in[1,n]$.

The C13 Function

$$f(x)=\frac{1}{n}\sum_{i=1}^{n}\left[-z_i\sin\sqrt{|z_i|}\right] \qquad [B.38]$$
$$g_1(x)=-50+\frac{1}{100n}\sum_{i}^{n}z_i^2\leq 0$$

$$g_2(x) = \frac{50}{n}\sum_{i=1}^{n}\sin(\frac{1}{50}\pi z_i) \le 0$$
$$g_3(x) = 75 - 50\left[\sum_{i=1}^{n}\frac{z_i^2}{4000} - \prod_{i=1}^{n}\cos(\frac{z_i}{\sqrt{i}}) + 1\right] \le 0$$
$$x_i \in [-500, 500]$$

where $z_i = x_i - o_i$ for $i \in [1, n]$.

The C14 Function

$$\begin{aligned}
f(x) &= \sum_{i=1}^{n-1}\left[100(z_i^2 - z_{i+1})^2 + (z_i - 1)^2\right] \\
g_1(x) &= \sum_{i=1}^{n}(-y_i\cos\sqrt{|y_i|}) - n \le 0 \\
g_2(x) &= \sum_{i=1}^{n}(y_i\cos\sqrt{|y_i|}) - n \le 0 \\
g_3(x) &= \sum_{i=1}^{n}(y_i\sin\sqrt{|y_i|}) - 10n \le 0 \\
&x_i \in [-1000, 1000]
\end{aligned} \tag{B.39}$$

where $y_i = x_i - o_i$ and $z_i = x_i + 1 - o_i$ for $i \in [1, n]$.

The C15 Function

$$\begin{aligned}
f(x) &= \sum_{i=1}^{n-1}\left[100(z_i^2 - z_{i+1})^2 + (z_i - 1)^2\right] \\
g_1(x) &= \sum_{i=1}^{n}(-y_i\cos\sqrt{|y_i|}) - n \le 0 \\
g_2(x) &= \sum_{i=1}^{n}(y_i\cos\sqrt{|y_i|}) - n \le 0 \\
g_3(x) &= \sum_{i=1}^{n}(y_i\sin\sqrt{|y_i|}) - 10n \le 0 \\
&x_i \in [-1000, 1000]
\end{aligned} \tag{B.40}$$

where $y_i = (x_i - o_i)M$ and $z_i = x_i + 1 - o_i$ for $i \in [1, n]$.

The C16 Function

$$
\begin{aligned}
f(x) &= \sum_{i=1}^{n} \frac{z_i^2}{4000} - \prod_{i=1}^{n} \cos\left(\frac{z_i}{\sqrt{i}}\right) + 1 \\
g_1(x) &= \sum_{i=1}^{n} [z_i^2 - 100\cos(\pi z_i) + 10] \le 0 \\
g_2(x) &= \prod_{i=1}^{n} z_i \le 0 \\
h_1(x) &= \sum_{i=1}^{n} (z_i \sin\sqrt{|z_i|}) = 0 \\
h_2(x) &= \sum_{i=1}^{n} (-z_i \sin\sqrt{|z_i|}) = 0 \\
x_i &\in [-10, 10]
\end{aligned}
\quad [B.41]
$$

where $z_i = x_i - o_i$ for $i \in [1, n]$.

The C17 Function

$$
\begin{aligned}
f(x) &= \sum_{i=1}^{n-1} (z_i - z_{i+1})^2 \\
g_1(x) &= \prod_{i=1}^{n} z_i \le 0 \\
g_2(x) &= \sum_{i=1}^{n} z_i \le 0 \\
h(x) &= \sum_{i=1}^{n} z_i \sin(4\sqrt{|z_i|}) = 0 \\
x_i &\in [-10, 10]
\end{aligned}
\quad [B.42]
$$

where $z_i = x_i - o_i$ for $i \in [1, n]$.

The C18 Function

$$
\begin{aligned}
f(x) &= \sum_{i=1}^{n-1} (z_i - z_{i+1})^2 \\
g(x) &= \frac{1}{n} \sum_{i=1}^{n} (-z_i \sin\sqrt{|z_i|}) \le 0
\end{aligned}
\quad [B.43]
$$

$$h(x) = \frac{1}{n}\sum_{i=1}^{n}(z_i \sin\sqrt{|z_i|}) = 0$$
$$x_i \in [-50, 50]$$

where $z_i = x_i - o_i$ for $i \in [1, n]$.

Appendix C

Multi-objective Benchmark Functions

A multi-objective optimization problem (MOP) involves the minimization of $f(x)$ over all x, where $f(x)$ is a vector, and x is the n-dimensional decision vector. Vector minimization is undefined in the normal sense of the word, and so we define the Pareto set P_s and the Pareto front P_f in Chapter 11. We can then pose an MOP as the problem of finding the "best" possible P_s and P_f.

Detailed information about the multi-objective benchmarks and evaluation metrics for EA competition at the 2009 IEEE Congress on Evolutionary Computation can be found in Zhang *et al.* [ZHA 08]. The dimension of the independent variable in the benchmarks given below is variable, but the CEC 2009 competition used $n = 30$.

U01 Unconstrained Problem 1

The two objectives to be minimized:

$$
\begin{aligned}
f_1 &= x_1 + \frac{2}{|J_1|}\sum_{j \in J_1}[x_j - \sin(6\pi x_1 + \frac{j\pi}{n})]^2 \\
f_2 &= 1 - \sqrt{x_1} + \frac{2}{|J_2|}\sum_{j \in J_2}[x_j - \sin(6\pi x_1 + \frac{j\pi}{n})]^2
\end{aligned}
\qquad [C.1]
$$

where $J_1 = \{j \mid j \text{ is odd and } 2 \le j \le n\}$ and $J_2 = \{j \mid j \text{ is even and } 2 \le j \le n\}$.

The search space is $[0,1]\times[-1,1]^{n-1}$.

Its Pareto front is

$$0 \le f_1 \le 1,\ f_2 = 1-\sqrt{f_1} \quad \text{[C.2]}$$

Its Pareto set is

$$0 \le x_1 \le 1,\ x_j = \sin(6\pi x_1 + \frac{j\pi}{n}), j = 2,....n \quad \text{[C.3]}$$

U02 Unconstrained Problem 2

The two objectives to be minimized:

$$\begin{aligned} f_1 &= x_1 + \frac{2}{|J_1|}\sum_{j\in J_1} y_j^{\,2} \\ f_2 &= 1-\sqrt{x_1} + \frac{2}{|J_2|}\sum_{j\in J_2} y_j^{\,2} \end{aligned} \quad \text{[C.4]}$$

where

$$J_1 = \{j \mid j \text{ is odd and } 2 \le j \le n\} \text{ and } J_2 = \{j \mid j \text{ is even and } 2 \le j \le n\},$$

and

$$y_j = \begin{cases} x_j - [0.3x_1^{\,2}\cos(24\pi x_1 + \frac{4j\pi}{n}) + 0.6x_1]\cos(6\pi x_1 + \frac{j\pi}{n}) & j \in J_1 \\ x_j - [0.3x_1^{\,2}\cos(24\pi x_1 + \frac{4j\pi}{n}) + 0.6x_1]\sin(6\pi x_1 + \frac{j\pi}{n}) & j \in J_2 \end{cases} \quad \text{[C.5]}$$

Its search space is $[0,1]\times[-1,1]^{n-1}$.

Its Pareto front is

$$0 \le f_1 \le 1, f_2 = 1-\sqrt{f_1} \quad \text{[C.6]}$$

Its Pareto set is

$$x_j = \begin{cases} 0 \le x_1 \le 1 \\ \{0.3x_1^2 \cos(24\pi x_1 + \frac{4j\pi}{n}) + 0.6x_1\} \cos(6\pi x_1 + \frac{j\pi}{n}) \ j \in J_1 \\ \{0.3x_1^2 \cos(24\pi x_1 + \frac{4j\pi}{n}) + 0.6x_1\} \sin(6\pi x_1 + \frac{j\pi}{n}) \ j \in J_2 \end{cases} \quad [C.7]$$

U03 Unconstrained Problem 3

The two objectives to be minimized:

$$\begin{aligned} f_1 &= x_1 + \frac{2}{|J_1|}(4\sum_{j \in J_1} y_j^2 - 2\prod_{j \in J_1} \cos(\frac{20 y_j \pi}{\sqrt{j}}) + 2) \\ f_2 &= 1 - \sqrt{x_1} + \frac{2}{|J_2|}(4\sum_{j \in J_2} y_j^2 - 2\prod_{j \in J_2} \cos(\frac{20 y_j \pi}{\sqrt{j}}) + 2) \end{aligned} \quad [C.8]$$

where J_1 and J_2 are the same as those of U01, and

$$y_j = x_j - x_1^{0.5(1.0 + \frac{3(j-2)}{n-2})}, j = 2, \ldots n \quad [C.9]$$

Its search space is $[0, 1]^n$.

Its Pareto front is

$$f_2 = 1 - \sqrt{f_1}, 0 \le f_1 \le 1 \quad [C.10]$$

Its Pareto set is

$$0 \le x_1 \le 1, \ x_i = x_1^{0.5(1.0 + \frac{3(j-2)}{n-2})}, j = 2, \ldots n \quad [C.11]$$

U04 Unconstrained Problem 4

The two objectives to be minimized:

$$\begin{aligned} f_1 &= x_1 + \frac{2}{|J_1|}\sum_{j \in J_1} h(y_j) \\ f_2 &= 1 - x_1^2 + \frac{2}{|J_2|}\sum_{j \in J_2} h(y_j) \end{aligned} \quad [C.12]$$

where J_1 and J_2 are the same as those of U01, and

$$y_j = x_j - \sin(6\pi x_1 + \frac{j\pi}{n}), j = 2,\ldots n \qquad [C.13]$$

and

$$h(t) = \frac{|t|}{1+e^{2|t|}} \qquad [C.14]$$

Its search space is $[0,1]\times[-2,2]^{n-1}$.

Its Pareto front is

$$0 \le f_1 \le 1,\ f_2 = 1 - f_2^2 \qquad [C.15]$$

Its Pareto set is

$$0 \le x_1 \le 1,\ x_j = \sin(6\pi x_1 + \frac{j\pi}{n}), j = 2,\ldots n \qquad [C.16]$$

U05 Unconstrained Problem 5

The two objectives to be minimized:

$$\begin{aligned} f_1 &= x_1 + (\frac{1}{2N} + \varepsilon)\left|\sin(2N\pi x_1)\right| + \frac{2}{|J_1|}\sum_{j\in J_1} h(y_j) \\ f_2 &= 1 - x_1 + (\frac{1}{2N} + \varepsilon)\left|\sin(2N\pi x_1)\right| + \frac{2}{|J_2|}\sum_{j\in J_2} h(y_j) \end{aligned} \qquad [C.17]$$

where J_1 and J_2 are the same as those of U01, N is an integer ($N = 10$ in the CEC 2009 competition), $\varepsilon > 0$ ($\varepsilon = 0.5$ in the CEC 2009 competition), and

$$y_j = x_j - \sin(6\pi x_1 + \frac{j\pi}{n}), j = 2,\ldots n \qquad [C.18]$$

and

$$h(t) = 2t^2 - \cos(4\pi t) + 1 \qquad \text{[C.19]}$$

The search space is $[0,1]\times[-1,1]^{n-1}$.

Its Pareto front has $(2N+1)$ discrete points:

$$(\frac{i}{2N}, 1-\frac{i}{2N}) \qquad \text{[C.20]}$$

for $i = 0,1,....2N$.

Its Pareto set also contains $(2N+1)$ discrete points, but they cannot be expressed analytically, so we do not show them here.

U06 Unconstrained Problem 6

The two objectives to be minimized:

$$\begin{aligned} f_1 &= x_1 + \max\{0, 2(\frac{1}{2N}+\varepsilon)\sin(2N\pi x_1)\} \\ &+ \frac{2}{|J_1|}(4\sum_{j\in J_1} y_j^2 - 2\prod_{j\in J_1}\cos(\frac{20y_j\pi}{\sqrt{j}})+2) \\ f_2 &= 1 - x_1 + \max\{0, 2(\frac{1}{2N}+\varepsilon)\sin(2N\pi x_1)\} \\ &+ \frac{2}{|J_2|}(4\sum_{j\in J_2} y_j^2 - 2\prod_{j\in J_2}\cos(\frac{20y_j\pi}{\sqrt{j}})+2) \end{aligned} \qquad \text{[C.21]}$$

where N is an integer ($N = 2$ in the CEC 2009 competition), $\varepsilon > 0$ ($\varepsilon = 0.1$ in the CEC 2009 competition) and

$$z_i = \frac{2}{|J_i|}\left(4\sum_{j\in J_1} y_j^2 - 2\prod_{j\in J_1}\cos(\frac{20y_j\pi}{\sqrt{j}})+2\right), i\in[1, 2] \qquad \text{[C.22]}$$

where J_1 and J_2 are the same as those of U01, and

$$y_j = x_j - \sin(6\pi x_1 + \frac{j\pi}{n}), j = 2,....n \quad [C.23]$$

The search space is $[0,1]\times[-1,1]^{n-1}$.

Its Pareto front consists of one isolated point (0, 1), and the following N disconnected parts:

$$f_1 \in \bigcup_{i=1}^{N} [\frac{2i-1}{2N}, \frac{2i}{2N}), \; f_2 = 1 - f_1 \quad [C.24]$$

Its Pareto set consists of discrete points, but they cannot be expressed analytically, so we do not show them here.

U07 Unconstrained Problem 7

The two objectives to be minimized:

$$\begin{aligned} f_1 &= \sqrt[5]{x_1} + \frac{2}{|J_1|}\sum_{j\in J_1} y_j^{\,2} \\ f_2 &= 1 - \sqrt[5]{x_1} + \frac{2}{|J_2|}\sum_{j\in J_2} y_j^{\,2} \end{aligned} \quad [C.25]$$

where J_1 and J_2 are the same as those of U01, and

$$y_j = x_j - \sin(6\pi x_1 + \frac{j\pi}{n}), j = 2,....n \quad [C.26]$$

The search space is $[0,1]\times[-1,1]^{n-1}$.

Its Pareto front is

$$0 \le f_1 \le 1, \; f_2 = 1 - \sqrt{f_1} \quad [C.27]$$

Its Pareto set is

$$0 \le x_1 \le 1,\ x_j = \sin\left(6\pi x_1 + \frac{j\pi}{n}\right), j = 2,\ldots n \qquad [C.28]$$

U08 Unconstrained Problem 8

The three objectives to be minimized:

$$\begin{aligned} f_1 &= \cos(0.5x_1\pi)\cos(0.5x_2\pi) + \frac{2}{|J_1|}\sum_{j\in J_1}\left(x_j - 2x_2\sin(2\pi x_1 + \frac{j\pi}{n})\right)^2 \\ f_2 &= \cos(0.5x_1\pi)\cos(0.5x_2\pi) + \frac{2}{|J_2|}\sum_{j\in J_2}\left(x_j - 2x_2\sin(2\pi x_1 + \frac{j\pi}{n})\right)^2 \\ f_3 &= \sin(0.5x_1\pi) + \frac{2}{|J_3|}\sum_{j\in J_3}\left(x_j - 2x_2\sin(2\pi x_1 + \frac{j\pi}{n})\right)^2 \end{aligned} \qquad [C.29]$$

where

$$\begin{aligned} J_1 &= \{j \mid 3 \le j \le n, \text{and } j-1 \text{ is a multiple of } 3\} \\ J_2 &= \{j \mid 3 \le j \le n, \text{and } j-2 \text{ is a multiple of } 3\} \\ J_3 &= \{j \mid 3 \le j \le n, \text{and } j \text{ is a multiple of } 3\} \end{aligned} \qquad [C.30]$$

The search space is $[0,1]^2 \times [-2,2]^{n-2}$.

Its Pareto front is

$$0 \le f_1, f_2, f_3 \le 1,\ f_1^2 + f_2^2 + f_3^3 = 1 \qquad [C.31]$$

Its Pareto set is

$$0 \le x_1 \le 1,\ 0 \le x_2 \le 1,\ x_j = 2x_2 \sin\left(2\pi x_1 + \frac{j\pi}{n}\right), j = 3,\ldots n \qquad [C.32]$$

U09 Unconstrained Problem 9

The three objectives to be minimized:

$$\begin{aligned} f_1 &= 0.5[\max\{0,(1+\varepsilon)(1-4(2x_1-1)^2)\}+2x_1]x_2 \\ &\quad +\frac{2}{|J_1|}\sum_{j\in J_1}(x_j-2x_2\sin(2\pi x_1+\frac{j\pi}{n}))^2 \\ f_2 &= 0.5[\max\{0,(1+\varepsilon)(1-4(2x_1-1)^2)\}+2x_1]x_2 \\ &\quad +\frac{2}{|J_2|}\sum_{j\in J_2}(x_j-2x_2\sin(2\pi x_1+\frac{j\pi}{n}))^2 \\ f_3 &= 1-x_2+\frac{2}{|J_3|}\sum_{j\in J_3}(x_j-2x_2\sin(2\pi x_1+\frac{j\pi}{n}))^2 \end{aligned} \quad [C.33]$$

where

$$\begin{aligned} J_1 &= \{j \mid 3\le j\le n, \text{and } j-1 \text{ is a multiple of 3}\} \\ J_2 &= \{j \mid 3\le j\le n, \text{and } j-2 \text{ is a multiple of 3}\} \\ J_3 &= \{j \mid 3\le j\le n, \text{and } j \text{ is a multiple of 3}\} \end{aligned} \quad [C.34]$$

and $\varepsilon = 0.1$, which can take any other positive values.

The search space is $[0,1]\times[-2,2]^{n-2}$.

The Pareto front has two parts. The first part is

$$\begin{aligned} & 0\le f_3\le 1 \\ & 0\le f_1\le \frac{1}{4}(1-f_3) \\ & f_2 = 1-f_1-f_3 \end{aligned} \quad [C.35]$$

and the second one is

$$\begin{aligned} & 0\le f_3\le 1 \\ & \frac{3}{4}(1-f_3)\le f_1\le 1 \\ & f_2 = 1-f_1-f_3 \end{aligned} \quad [C.36]$$

The Pareto set is

$$
\begin{aligned}
&x_1 \in [0, 0.25] \cup [0.75, 1], \quad 0 \le x_2 \le 1, \\
&x_j = 2x_2 \sin\,(2\pi x_1 + \frac{j\pi}{n}), j = 3, \ldots n
\end{aligned}
\qquad [C.37]
$$

U10 Unconstrained Problem 10

The three objectives to be minimized:

$$
\begin{aligned}
f_1 &= \cos(0.5x_1\pi)\cos(0.5x_2\pi) + \frac{2}{|J_1|}\sum_{j\in J_1}[4y_j^{\,2} - \cos(8\pi y_j) + 1] \\
f_2 &= \cos(0.5x_1\pi)\sin(0.5x_2\pi) + \frac{2}{|J_2|}\sum_{j\in J_2}[4y_j^{\,2} - \cos(8\pi y_j) + 1] \\
f_3 &= \sin(0.5x_1\pi) + \frac{2}{|J_3|}\sum_{j\in J_3}[4y_j^{\,2} - \cos(8\pi y_j) + 1]
\end{aligned}
\qquad [C.38]
$$

where

$$
\begin{aligned}
J_1 &= \{j | 3 \le j \le n, \text{and } j-1 \text{ is a multiple of 3}\} \\
J_2 &= \{j | 3 \le j \le n, \text{and } j-2 \text{ is a multiple of 3}\} \\
J_3 &= \{j | 3 \le j \le n, \text{and } j \text{ is a multiple of 3}\}
\end{aligned}
\qquad [C.39]
$$

and

$$
y_j = x_j - 2x_2 \sin\,(2\pi x_1 + \frac{j\pi}{n}), j = 3, \ldots n \qquad [C.40]
$$

The search space is $[0,1]^2 \times [-2,2]^{n-2}$.

Its Pareto front is

$$
0 \le f_1, f_2, f_3 \le 1, \;\; f_1^2 + f_2^2 + f_3^3 = 1 \qquad [C.41]
$$

Its Pareto set is

$$
0 \le x_1 \le 1, \;\; 0 \le x_2 \le 1, \quad x_j = 2x_2 \sin\,(2\pi x_1 + \frac{j\pi}{n}), j = 3, \ldots n \qquad [C.42]
$$

C01 Constrained Problem 1

The two objectives to be minimized:

$$f_1 = x_1 + + \frac{2}{|J_1|} \sum_{j \in J_1} (x_j - x_1^{0.5(1.0+\frac{3(j-2)}{n-2})})^2$$
$$f_2 = 1 - x_1 + \frac{2}{|J_2|} \sum_{j \in J_2} (x_j - x_1^{0.5(1.0+\frac{3(j-2)}{n-2})})^2 \qquad [C.43]$$

where $J_1 = \{j | j \text{ is odd and } 2 \le j \le n\}$ and $J_2 = \{j | j \text{ is even and } 2 \le j \le n\}$.

The constraint is

$$f_1 + f_2 - a\left|\sin[N\pi(f_1 - f_2 + 1)]\right| - 1 \ge 0 \qquad [C.44]$$

where N is an integer and $a \ge \frac{1}{2N}$.

The search space is $[0, 1]^n$.

The Pareto front in the objective space consists of $2N+1$ points:

$$(i/2N, 1-i/2N), i = 0,1,....,2N \qquad [C.45]$$

$N = 10, a = 1$ and $n = 10$ in the CEC 2009 competition.

C02 Constrained Problem 2

The two objectives to be minimized:

$$f_1 = x_1 + \frac{2}{|J_1|} \sum_{j \in J_1} (x_j - \sin(6\pi x_1 + \frac{j\pi}{n}))^2$$
$$f_2 = 1 - \sqrt{x_1} + \frac{2}{|J_2|} \sum_{j \in J_2} (x_j - \cos(6\pi x_1 + \frac{j\pi}{n}))^2 \qquad [C.46]$$

where

$J_1 = \{j | j \text{ is odd and } 2 \le j \le n\}$ and $J_2 = \{j | j \text{ is even and } 2 \le j \le n\}$.

The search space is $[0,1]\times[-1,1]^{n-1}$.

The constraint is

$$\frac{t}{1+e^{4|t|}} \geq 0 \qquad [C.47]$$

where

$$t = f_2 + \sqrt{f_1} - a\sin[N\pi(\sqrt{f_1} - f_2 + 1)] - 1 \qquad [C.48]$$

Its Pareto front consists of an isolated Pareto optimal solution (0, 1), and N disconnected parts, the ith part is

$$f_2 = 1 - \sqrt{f_1}, (\frac{2i-1}{2N})^2 \leq f_1 \leq (\frac{2i}{2N})^2, i = 1,....,N \qquad [C.49]$$

$N = 2, a = 1$ and $n = 10$ in the CEC 2009 competition.

C03 Constrained Problem 3

The two objectives to be minimized:

$$\begin{aligned} f_1 &= x_1 + \frac{2}{|J_1|}(4\sum_{j\in J_1} y_j^2 - 2\prod_{j\in J_1}\cos(\frac{20y_j\pi}{\sqrt{j}}) + 2) \\ f_2 &= 1 - x_1^2 + \frac{2}{|J_2|}(4\sum_{j\in J_2} y_j^2 - 2\prod_{j\in J_2}\cos(\frac{20y_j\pi}{\sqrt{j}}) + 2) \end{aligned} \qquad [C.50]$$

where

$J_1 = \{j | j$ is odd and $2 \leq j \leq n\}$ and $J_2 = \{j | j$ is even and $2 \leq j \leq n\}$,

and

$$y_j = x_j - \sin(6\pi x_1 + \frac{j\pi}{n}), j = 2,....n \qquad [C.51]$$

The constraint is

$$f_2 + f_1^2 - a\sin[N\pi(f_1^2 - f_2 + 1)] - 1 \geq 0 \quad [C.52]$$

The search space is $[0,1]\times[-2,2]^{n-1}$.

Its Pareto front consists of an isolated Pareto optimal solution (0, 1), and N disconnected parts, the ith part is

$$f_2 = 1 - f_1^2, \sqrt{\frac{2i-1}{2N}} \leq f_1 \leq \sqrt{\frac{2i}{2N}}, i = 1,\ldots, N \quad [C.53]$$

$N = 2, a = 1$ and $n = 10$ in the CEC 2009 competition.

C04 Constrained Problem 4

The two objectives to be minimized:

$$\begin{aligned} f_1 &= x_1 + \sum_{j\in J_1} h_j(y_j) \\ f_2 &= 1 - x_1 + \sum_{j\in J_2} h_j(y_j) \end{aligned} \quad [C.54]$$

where

$$J_1 = \{j | j \text{ is odd and } 2 \leq j \leq n\} \text{ and } J_2 = \{j | j \text{ is even and } 2 \leq j \leq n\},$$

and

$$y_j = x_j - \sin(6\pi x_1 + \frac{j\pi}{n}), j = 2,\ldots n \quad [C.55]$$

The search space is $[0,1]\times[-2,2]^{n-1}$.

$$h_2(t) = \begin{cases} |t| & \text{if } t < \frac{3}{2}(1 - \frac{\sqrt{2}}{2}) \\ 0.125 + (t-1)^2 & \text{otherwise} \end{cases} \quad [C.56]$$

and

$$h_j(t) = t^2 \quad \text{[C.57]}$$

For $j = 3, 4, ..., n$.

The constraint is

$$\frac{t}{1+e^{4|t|}} \geq 0 \quad \text{[C.58]}$$

where

$$t = x_2 - \sin\left(6\pi x_1 + \frac{2\pi}{n}\right) - 0.5x_1 + 0.25. \quad \text{[C.59]}$$

The Pareto front is

$$f_2 = \begin{cases} 1 - f_1 & \text{if } 0 \leq f_1 \leq 0.5 \\ -0.5f_1 + \dfrac{3}{4} & \text{if } 0.5 \leq f_1 \leq 0.75 \\ 1 - f_1 + 0.125 & \text{if } 0.75 \leq f_1 \leq 1 \end{cases} \quad \text{[C.60]}$$

$n = 10$ in the CEC 2009 competition.

C05 Constrained Problem 5

The two objectives to be minimized:

$$\begin{aligned} f_1 &= x_1 + \sum_{j \in J_1} h_j(y_j) \\ f_2 &= 1 - x_1 + \sum_{j \in J_2} h_j(y_j) \end{aligned} \quad \text{[C.61]}$$

where

$J_1 = \{j | j \text{ is odd and } 2 \leq j \leq n\}$ and $J_2 = \{j | j \text{ is even and } 2 \leq j \leq n\}$,

and

$$y_j = \begin{cases} x_j - 0.8x_1 \cos(6\pi x_1 + \frac{j\pi}{n}), & \text{if } j \in J_1 \\ x_j - 0.8x_1 \sin(6\pi x_1 + \frac{j\pi}{n}), & \text{if } j \in J_2 \end{cases} \qquad [C.62]$$

$$h_2(t) = \begin{cases} |t| & \text{if } t < \frac{3}{2}(1 - \frac{\sqrt{2}}{2}) \\ 0.125 + (t-1)^2 & \text{otherwise} \end{cases} \qquad [C.63]$$

and

$$h_j(t) = 2t^2 - \cos(4\pi t) + 1 \qquad [C.64]$$

for $j = 3, 4, ..., n$.

The search space is $[0,1] \times [-2,2]^{n-1}$

The constraint is

$$x_2 - 0.8x_1 \sin(6\pi x_1 + \frac{2\pi}{n}) - 0.5x_1 + 0.25 \geq 0 \qquad [C.65]$$

The Pareto front is

$$f_2 = \begin{cases} 1 - f_1 & \text{if } 0 \leq f_1 \leq 0.5 \\ -0.5f_1 + \frac{3}{4} & \text{if } 0.5 \leq f_1 \leq 0.75 \\ 1 - f_1 + 0.125 & \text{if } 0.75 \leq f_1 \leq 1 \end{cases} \qquad [C.66]$$

$n = 10$ in the CEC 2009 competition.

C06 Constrained Problem 6

The two objectives to be minimized:

$$\begin{aligned} f_1 &= x_1 + \sum_{j \in J_1} {y_j}^2 \\ f_2 &= (1 - x_1)^2 + \sum_{j \in J_2} {y_j}^2 \end{aligned} \qquad [C.67]$$

where

$$J_1 = \{j | j \text{ is odd and } 2 \le j \le n\} \text{ and } J_2 = \{j | j \text{ is even and } 2 \le j \le n\},$$

and

$$y_j = \begin{cases} x_j - 0.8x_1 \cos(6\pi x_1 + \frac{j\pi}{n}), & \text{if } j \in J_1 \\ x_j - 0.8x_1 \sin(6\pi x_1 + \frac{j\pi}{n}), & \text{if } j \in J_2 \end{cases} \quad \text{[C.68]}$$

The search space is $[0,1]\times[-2,2]^{n-1}$.

The constraint is

$$\begin{aligned} &x_2 - 0.8x_1 \sin(6\pi x_1 + \frac{2\pi}{n}) - sign(0.5(1-x_1) \\ &\quad -(1-x_1)^2)\sqrt{|0.5(1-x_1)-(1-x_1)^2|} \ge 0 \end{aligned} \quad \text{[C.69]}$$

and

$$\begin{aligned} &x_4 - 0.8x_1 \sin(6\pi x_1 + \frac{2\pi}{n}) - sign(0.25(1-x_1) \\ &\quad -0.5(1-x_1))\sqrt{|0.25(1-x_1)-0.5(1-x_1)|} \ge 0 \end{aligned} \quad \text{[C.70]}$$

The Pareto front is

$$f_2 = \begin{cases} (1-f_1)^2 & \text{if } 0 \le f_1 \le 0.5 \\ 0.5(1-f_1) & \text{if } 0.5 \le f_1 \le 0.75 \\ 0.25\sqrt{(1-f_1)} & \text{if } 0.75 \le f_1 \le 1 \end{cases} \quad \text{[C.71]}$$

$n = 10$ in the CEC 2009 competition.

C07 Constrained Problem 7

The two objectives to be minimized:

$$
\begin{aligned}
f_1 &= x_1 + \sum_{j \in J_1} h_j(y_j) \\
f_2 &= (1-x_1)^2 + \sum_{j \in J_2} h_j(y_j)
\end{aligned}
\qquad [C.72]
$$

where

$$J_1 = \{j \mid j \text{ is odd and } 2 \le j \le n\} \text{ and } J_2 = \{j \mid j \text{ is even and } 2 \le j \le n\},$$

and

$$
y_j = \begin{cases} x_j - \cos\,(6\pi x_1 + \dfrac{j\pi}{n}), & \text{if } j \in J_1 \\ x_j - \sin\,(6\pi x_1 + \dfrac{j\pi}{n}), & \text{if } j \in J_2 \end{cases} \qquad [C.73]
$$

$$h_2(t) = h_4(t) = t^2$$

and

$$h_j(t) = 2t^2 - \cos(4\pi t) + 1 \qquad [C.74]$$

for $j = 3, 4, \ldots., n$.

The search space is $[0,1] \times [-2,2]^{n-1}$.

The constraint is

$$
\begin{aligned}
&x_2 - \sin\,(6\pi x_1 + \frac{2\pi}{n}) - sign(0.5(1-x_1) \\
&\quad -(1-x_1)^2)\sqrt{\left|0.5(1-x_1)-(1-x_1)^2\right|} \ge 0
\end{aligned}
\qquad [C.75]
$$

and

$$
\begin{aligned}
&x_4 - \sin\,(6\pi x_1 + \frac{4\pi}{n}) - sign(0.25\sqrt{1-x_1} \\
&\quad -0.5(1-x_1))\sqrt{\left|0.25\sqrt{1-x_1}-0.5(1-x_1)\right|} \ge 0
\end{aligned}
\qquad [C.76]
$$

The Pareto front is

$$f_2 = \begin{cases} (1-f_1)^2 & \text{if } 0 \le f_1 \le 0.5 \\ 0.5(1-f_1) & \text{if } 0.5 \le f_1 \le 0.75 \\ 0.25\sqrt{(1-f_1)} & \text{if } 0.75 \le f_1 \le 1 \end{cases} \quad [C.77]$$

$n = 10$ in the CEC 2009 competition.

C08 Constrained Problem 8

The three objectives to be minimized:

$$\begin{aligned} f_1 &= \cos(0.5x_1\pi)\cos(0.5x_2\pi) + \frac{2}{|J_1|}\sum_{j\in J_1}(x_j - 2x_2\sin(2\pi x_1 + \frac{j\pi}{n}))^2 \\ f_2 &= \cos(0.5x_1\pi)\sin(0.5x_2\pi) + \frac{2}{|J_2|}\sum_{j\in J_2}(x_j - 2x_2\sin(2\pi x_1 + \frac{j\pi}{n}))^2 \\ f_3 &= \sin(0.5x_1\pi) + \frac{2}{|J_3|}\sum_{j\in J_3}(x_j - 2x_2\sin(2\pi x_1 + \frac{j\pi}{n}))^2 \end{aligned} \quad [C.78]$$

where

$$\begin{aligned} J_1 &= \{j \mid 3 \le j \le n, \text{and } j-1 \text{ is a multiple of } 3\} \\ J_2 &= \{j \mid 3 \le j \le n, \text{and } j-2 \text{ is a multiple of } 3\} \\ J_3 &= \{j \mid 3 \le j \le n, \text{and } j \text{ is a multiple of } 3\} \end{aligned} \quad [C.79]$$

The search space is $[0,1]^2 \times [-4,4]^{n-2}$.

The constraint is

$$\frac{f_1^2 + f_2^2}{1-f_3^2} - a\left|\sin[N\pi(\frac{f_1^2 - f_2^2}{1-f_3^2} + 1)]\right| - 1 \ge 0 \quad [C.80]$$

Its Pareto front will have $2N+1$ disconnected parts:

$$\begin{aligned} f_1 &= [\frac{i}{2N}(1-f_3^2)]^{\frac{1}{2}} \\ f_2 &= [1 - f_1^2 - f_3^2]^{\frac{1}{2}} \\ &0 \le f_3 \le 1 \end{aligned} \quad [C.81]$$

for $j = 0,1,....,2N$.

$N = 2, a = 4$ and $n = 10$ in the CEC 2009 competition.

C09 Constrained Problem 9

The three objectives to be minimized:

$$\begin{aligned} f_1 &= \cos(0.5x_1\pi)\cos(0.5x_2\pi) + \frac{2}{|J_1|}\sum_{j\in J_1}(x_j - 2x_2\sin(2\pi x_1 + \frac{j\pi}{n}))^2 \\ f_2 &= \cos(0.5x_1\pi)\sin(0.5x_2\pi) + \frac{2}{|J_2|}\sum_{j\in J_2}(x_j - 2x_2\sin(2\pi x_1 + \frac{j\pi}{n}))^2 \\ f_3 &= \sin(0.5x_1\pi) + \frac{2}{|J_3|}\sum_{j\in J_3}(x_j - 2x_2\sin(2\pi x_1 + \frac{j\pi}{n}))^2 \end{aligned} \quad [C.82]$$

where

$$\begin{aligned} J_1 &= \{j | 3 \le j \le n, \text{and } j-1 \text{ is a multiple of } 3\} \\ J_2 &= \{j | 3 \le j \le n, \text{and } j-2 \text{ is a multiple of } 3\} \\ J_3 &= \{j | 3 \le j \le n, \text{and } j \text{ is a multiple of } 3\} \end{aligned} \quad [C.83]$$

The search space is $[0,1]^2 \times [-2,2]^{n-2}$.

The constraint is

$$\frac{f_1^2 + f_2^2}{1 - f_3^2} - a\sin[N\pi(\frac{f_1^2 - f_2^2}{1 - f_3^2} + 1)] - 1 \ge 0 \quad [C.84]$$

Its Pareto front consists of a curve:

$$\begin{aligned} f_1 &= 0 \\ 0 \le f_2 &\le 1 \\ f_3 &= (1 - f_2^2)^{\frac{1}{2}} \end{aligned} \quad [C.85]$$

and N disconnected nonlinear surfaces, the ith one is

$$
\begin{aligned}
&0 \le f_3 \le 1 \\
&\{\frac{2i-1}{2N}(1-f_3^2)\}^{\frac{1}{2}} \le f_1 \le \{\frac{2i}{2N}(1-f_3^2)\}^{\frac{1}{2}} \\
&f_2 = [1-f_1^2-f_3^2]^{\frac{1}{2}}
\end{aligned}
\qquad [C.86]
$$

$N = 2, a = 3$ and $n = 10$ in the CEC 2009 competition.

C10 Constrained Problem 10

The three objectives to be minimized:

$$
\begin{aligned}
f_1 &= \cos(0.5x_1\pi)\cos(0.5x_2\pi) + \frac{2}{|J_1|}\sum_{j\in J_1}[4y_j^2 - \cos(8\pi y_j) + 1] \\
f_2 &= \cos(0.5x_1\pi)\sin(0.5x_2\pi) + \frac{2}{|J_2|}\sum_{j\in J_2}[4y_j^2 - \cos(8\pi y_j) + 1] \\
f_3 &= \sin(0.5x_1\pi) + \frac{2}{|J_3|}\sum_{j\in J_3}[4y_j^2 - \cos(8\pi y_j) + 1]
\end{aligned}
\qquad [C.87]
$$

where

$$
\begin{aligned}
J_1 &= \{j \mid 3 \le j \le n, \text{and } j-1 \text{ is a multiple of 3}\} \\
J_2 &= \{j \mid 3 \le j \le n, \text{and } j-2 \text{ is a multiple of 3}\} \\
J_3 &= \{j \mid 3 \le j \le n, \text{and } j \text{ is a multiple of 3}\}
\end{aligned}
\qquad [C.88]
$$

and

$$
y_j = x_j - \sin(6\pi x_1 + \frac{j\pi}{n}), j = 2, \dots. n \qquad [C.89]
$$

for $j = 3, \dots., n$.

The search space is $[0,1]^2 \times [-2,2]^{n-2}$.

The constraint is

$$\frac{f_1^2+f_2^2}{1-f_3^2}-a\sin[N\pi(\frac{f_1^2-f_2^2}{1-f_3^2}+1)]-1\geq 0 \quad [C.90]$$

Its Pareto front consists of a curve:

$$\begin{aligned} f_1 &= 0 \\ 0 \leq f_2 &\leq 1 \\ f_3 &= (1-f_2^2)^{\frac{1}{2}} \end{aligned} \quad [C.91]$$

and N disconnected nonlinear surfaces, the ith one is

$$\begin{aligned} 0 \leq f_3 \leq 1 \\ \{\frac{2i-1}{2N}(1-f_3^2)\}^{\frac{1}{2}} \leq f_1 \leq \{\frac{2i}{2N}(1-f_3^2)\}^{\frac{1}{2}} \\ f_2 = [(1-f_1^2-f_3^2)]^{\frac{1}{2}} \end{aligned} \quad [C.92]$$

$N=2, a=1$ and $n=10$ in the CEC 2009 competition.

Bibliography

[ADL 94] ADLER F., NUERNBERGER B., "Persistence in patchy irregular landscapes", *Theoretical Population Biology*, vol. 45, no. 1, pp. 1989–2017, 1994.

[ALE 96] ALEXANDER R., *Optima for Animals*, Princeton University Press, 1996.

[ARN 03] ARNOLD D.-V., BEYER H.-G., "On the effects of outliers on evolutionary optimization", *Lecture Notes in Computer Science*, vol. 2690, no. 2690, pp. 151–160, 2003.

[ASA 11] ASAFUDDOULA M., RAY T., SARKER, R., "An adaptive differential evolution algorithm and its performance on real world optimization problems", *IEEE Congress on Evolutionary Computation*, New Orleans, pp. 1057–1062, 2011.

[BAC 96] BACK T., *Evolutionary Algorithms in Theory and Practice*, Oxford University Press, 1996.

[BAN 90] BANZHAF W., "The 'molecular' traveling salesman", *Biological Cybernetics*, vol. 64, no. 1, pp. 7–14, 1990.

[BEA 99] BEAULIEU N., "On the generalized multinomial distribution, optimal multinomial detectors, and generalized weighted partial decision detectors", *IEEE Transactions on Communications*, vol. 39, no. 2, pp. 193–194, 1999.

[BEL 61] BELLMAN R., *Adaptive Control Processes: A Guided Tour*, Princeton University Press, 1961.

[BEY 02] BEYER H.G., SCHWEFEL H.P., "Evolution strategies: a comprehensive introduction", *Natural Computing*, vol. 1, no. 1, pp. 3–52, 2002.

[BEY 07] BEYER H.G., SENDHOFF B., "Robust optimization – a comprehensive survey", *Computer Methods in Applied Mechanics & Engineering*, vol. 196, no. 33, pp. 3190–3218, 2007.

[BHA 10] BHATTACHARYA A., CHATTOPADHYAY P., "Hybrid differential evolution with biogeography-based optimization for solution of ecomomic load dispath", *IEEE Transactions on Power Systems*, vol. 25, no. 4, pp. 1955–1964, 2010.

[BLU 11] BLUM C., PUCHINER J., RAIDL G. *et al.*, "Hybrid metaheuristics in combinatorial optimization: a survey", *Applied Soft Computing*, vol. 11, no. 6, pp. 4135–4151, 2011.

[BOU 11a] BOUSSAÏD I., CHATTERJEE A., SIARRY P. *et al.*, "Two-stage update biogeography based optimization using differential evolution algorithm (DBBO)", *Computers & Operations Research*, vol. 38, no. 8, pp. 1188–1198, 2011.

[BOU 11b] BOUSSAÏD I., CHATTERJEE A., SIARRY P. *et al.*, "Hybridizing biogeography-based optimization with differential evolution for optimal power allocation in wireless sensor networks", *IEEE Transactions on Vehicle Technology*, vol. 60, no. 5, pp. 2347–2353, 2011.

[BOU 13] BOUSSAÏD I., LEPAGNOT J., SIARRY P., "A survey on optimization metaheuristics", *Information Sciences*, vol. 237, pp. 82–117, 2013.

[BRA 07] BRATTON D., KENNEDY J., "Defining a standard for particle swarm optimization", *IEEE Swarm Intelligence Symposium*, Honolulu, Hawaii, pp. 120–127, 2007.

[BRE 09] BREST J., "Constrained real-parameter optimization with ϵ-self-adaptive differential evolution", in MEZURA-MONTES E. (ed.), *Constraint-Handling in Evolutionary Computation*, Springer, 2009.

[CAR 98] CARLSON S., SHONKWILER R., "Annealing a genetic algorithm over constraints", *IEEE International Conference on Systems, Man, and Cybernetics*, San Diego, pp. 3931–3936, 1998.

[CAS 89] CASWELL H., *Matrix Population Models*, Sinauer Associates, 1989.

[CHA 12] CHATTERJEE A., SIARRY P., NAKIB A. *et al.*, "An improved biogeography-based optimization approach for segmentation of human head CT-scan images employing fuzzy entropy", *Engineering Applications of Artificial Intelligence*, vol. 25, no. 8, pp. 1698–1709, 2012.

[CHO 13] CHOI T.M., YEUNG W.K., CHENG T.C.E., "Scheduling and co-ordination of multi-suppliers single-warehouse-operator single-manufacturer supply chains with variable production rates and storage costs", *International Journal of Production Research*, vol. 51, no. 9, pp. 2593–2601, 2013.

[CHU 92] CHUAN-CHONG C., KHEE-MENG K., *Principles and Techniques in Combinatorics*, World Scientific, Hackensack, 1992.

[COE 02a] COELLO COELLO C., "Theoretical and numerical constraint-handling techniques used with evolutionary algorithms: a survey of the state of the art", *Computer Methods in Applied Mechanics and Engineering*, vol. 191, nos 11–12, pp. 1245–1287, 2002.

[COE 02b] COELLO COELLO C., "Use of a self-adaptive penalty approach for engineering optimization technique", *Computers in Industry*, vol. 41, no. 2, pp. 113–127, 2002.

[COE 06] COELLO COELLO C., "Evolutionary multiobjective optimization: a historical view of the field", *IEEE Computational Intelligence Magazine*, vol. 1, no. 1, pp. 28–36, 2006.

[COE 09] COELLO COELLO C., "Evolutionary multi-objective optimization: some current research trends and toptics that remain to be explored", *Frontiers of Computer Science in China*, vol. 3, no. 1, pp. 18–30, 2009.

[COE 11] COELLO COELLO C., MEZURA-MONTES E., "Constraint-handling in nature-inspired numerical optimization: past, present and future", *Swarm and Evolutionary Computation*, vol. 1, no. 4, pp. 173–194, 2011.

[COE 16a] COELLO COELLO C., "List of references on evolutionary multiobjective optimization", available at: http://delta.cs.cinvestav.mx/~ccoello/EMOO/EMOObib.html, 2016.

[COE 16b] COELLO COELLO C., "List of references on constraint-handling techniques used with evolutionary algorithms", available at: https://www.cs.cinvestav.mx/~constraint/, 2016.

[COU 43] COURANT R., "Variational methods for the solution of problems of equilibrium and vibrations", *Bulletin of the American Mathematical Society*, vol. 49, no. 1, pp. 1–23, 1943.

[COV 67] COVER T., HART P., "Nearest neighbor pattern classification", *IEEE Transactions on Information Theory*, vol. 13, no. 1, pp. 21–27, 1967.

[DAS 10] DAS S., SUGANTHAN P.N., Problem definitions and evaluation criteria for CEC 2011 competition on testing evolutionary algorithms on real world optimization problems, Technical report, Jadavpur University, Nanyang Technological University, 2010.

[DAS 11a] DAS S., SUGANTHAN P.N., "Differential evolution – a survey of the state-of-the-art", *IEEE Transactions on Evolutionary Computation*, vol. 15, no. 1, pp. 4–31, 2011.

[DAS 11b] DAS S., MAITY S., QU B.-Y. *et al.*, "Real-parameter evolutionary multimodal optimization – a survey of the state-of-the-art", *Swarm and Evolutionary Computation*, vol. 1, no. 2, pp. 71–88, 2011.

[DAV 85] DAVIS L., "Job shop scheduling with genetic algorithms", *International Conference on Genetic Algorithms and Their Application*, Pittsburgh, Pennsylvania, pp. 136–140, 1985.

[DAV 93] DAVIS T., PRINCIPE J., "A Markov chain framework for the simple genetic algorithm", *Evolutionary Computation*, vol. 1, no. 3, pp. 269–288, 1993.

[DEB 99] DEB K., AGRAWAL S., "A niched-penalty approach for constraint handling in genetic algorithms", *International Conference on Artificial Neural Nets and Genetic Algorithms*, Portoroz, Slovenia, pp. 235–242, 1999.

[DEB 00a] DEB K., "An efficient constraint handling method for genetic algorithms", *Computer Methods in Applied Mechanics and Engineering*, vol. 186, nos 2–4, pp. 311–338, 2000.

[DEB 00b] DEB K., AGRAWAL S., PRATAP A. *et al.*, "A fast elitist non-dominated sorting genetic algorithm for multi-objective optimization: NSGA-II", in SCHOENAUER M., DEB K., RUDOLPH G. *et al.* (eds), *Parallel Problem Solving from Nature – PPSN VI*, Springer, 2000.

[DEB 02] DEB K., PRATAP A., AGRAWAL S. *et al.*, "A fast and elitist multi-objective optimization genetic algorithm: NSGA-II", *IEEE Transactions on Evolutionary Computation*, vol. 6, no. 2, pp. 182–197, 2002.

[DU 09] DU D., Biogeography-based optimization: synergies with evolutionary strategies, immigration refusal, and Kalman filters, Masters Thesis, Cleveland State University, 2009.

[DU 13] DU D., SIMON D., "Biogeography-based optimization for large scale combinatorial problem", in IGELNIC B., ZURADA J. (eds), *Efficiency and Scalability Methods for Computational Intellect*, IGI Global, 2013.

[EHR 05] EHRGOTT M., *Multicriteria Optimization*, Springer, 2005.

[EIB 01] EIBEN A., "Evolutionary algorithms and constraint satisfaction: definitions, survey, methodology, and research directions", in KALLEL L., NAUDTS B., ROGERS A. (eds), *Theoretical Aspects of Evolutionary Commutating*, Springer, 2001.

[ELT 58] ELTON C., *Ecology of Invasions by Animals and Plants*, Chapman & Hall, London, 1958.

[ERG 09] ERGEZER M., SIMON D., DU D., "Oppositional biogeography-based optimization", *IEEE Conference on Systems, Man, and Cybernetics*, San Antonio, Texas, pp. 1035–1040, 2009.

[ERG 14] ERGEZER M., SIMON D., "Mathematical and experimental analyses of oppositional algorithms", *IEEE Transactions on Cybernetics*, vol. 44, no. 11, pp. 2178–2189, 2014.

[ERG 15] ERGEZER M., SIMON D., "Probabilistic properies of fitness-based quasi-reflection in evolutionary algorithms", *Computer & Operations Research*, vol. 63, pp. 114–124, 2015.

[EUS 03] EUSUFF M., LANSEY K., "Optimization of water distribution network design using the shuffled frog leaping algorithm (SFLA)", *Journal of Water Resources Planning and Management*, vol. 129, no. 3, pp. 210–225, 2003.

[FIT 88] FITZPATRICK J.M., GREFENSTETTE J.J., "Genetic algorithms in noisy environments", *Machine Learning*, vol. 3, no. 2, pp. 101–120, 1988.

[FLE 05] FLEMING P., PURSHOUSE R., LYGOE R., "Many-objective optimization: an engineering design perspective", in COELLO COELLO C., HERNÁNDEZ AGUIRRE A., ZITZLER E. (eds), *Evolutionary Multi-Criterion Optimization*, Springer, 2005.

[FOG 88] FOGEL D., "An evolutionary approach to the traveling salesman problem", *Biological Cybernetics*, vol. 60, no. 2, pp. 139–144, 1988.

[FOG 90] FOGEL D., "A parallel processing approach to a multiple traveling salesman problem using evolutionary programming", *Fourth Annual Parallel Processing Symposium*, Fullerton, California, pp. 318–326, 1990.

[FOX 91] FOX B., MCMAHON M., "Genetic operators for sequencing problems", in RAWLINS G. (ed.), *Foundations of Genetic Algorithms*, Springer, 1991.

[GAG 12] GAGLIARDI J.P., RENAUD J., RUIZ A., "Models for automated storage and retrieval systems: a literature review", *International Journal of Production Research*, vol. 50, no. 24, pp. 7110–7125, 2012.

[GHO 11] GHOSH A., DAS S., CHOWDHURY A. *et al.*, "An improved differential evolution algorithm with fitness-based adaptation of the control parameters", *Information Sciences*, vol. 181, no. 18, pp. 3749–3765, 2011.

[GOL 89] GOLDBERG D., *Genetic Algorithms in Search, Optimization and Machine Learning*, Addison Wesley, 1989.

[GON 11] GONG W., FIALHO Á., CAI Z. *et al.*, "Adaptive strategy selection in differential evolution for numerical optimization: an empirical study", *Information Sciences*, vol. 181, no. 24, pp. 5364–5386, 2011.

[GOT 08] GOTELLI N., *A Primer of Ecology*, Sinauer Associates, 2008.

[GRI 97] GRINSTEAD C., SNELL J., *Introduction to Probability*, American Mathematical Society, Providence, 1997.

[GRO 05] GROOM M., MEFFE G., CARROLL C., *Principles of Conservation Biology*, Sinauer Associates, 2005.

[GUO 01] GUO G., YU S., "The unified method analyzing convergence of genetic algorithms", *Control Theory and Applications*, vol. 18, no. 3, pp. 43–446, 2001.

[HAD 93] HADJ-ALOUANE A., BEAN J., A genetic algorithm for the multiple choice integer program, Technical report, Department of Industrial and Operations Engineering, University of Michigan, 1993.

[HAN 97] HANSKE I., GILPIN M., *Metapopulation Biology*, Academic Press, 1997.

[HAN 99] HANSKI I., "Habitat connectivity, habitat continuity, and metapopulations in dynamic handscapes", *Oikos*, vol. 87, no. 2, pp. 209–219, 1999.

[HAN 09] HANSEN N., NIEDERBERGER A., GUZZELLA L. *et al.*, "A method for handling uncertainty in evolutionary optimization with an application to feedback control of combustion", *IEEE Transactions on Evolutionary Computation*, vol. 13, no. 1, pp. 180–197, 2009.

[HAR 06] HARDING S., *Animate Earth*, Chelsea Green Publishing Company, 2006.

[HE 99] HE J., KANG L., "On the convergence rates of genetic algorithms", *Theoretical Computer Science*, vol. 229, no. 1, pp. 23–39, 1999.

[HOL 75] HOLLAND J., *Adaptation in Natural and Artificial Systems*, University of Michigan Press, 1975.

[HOM 94] HOMAIFAR A., QI C., LAI S., "Constrained optimization via genetic algorithms", *Simulation*, vol. 62, no. 4, pp. 242–253, 1994.

[HOR 94] HORN J., NAFPLIOTIS N., GOLDBERG D., "A niched Pareto genetic algorithm for multiobjective optimization", *IEEE Congress on Evolutionary Computation*, Orlando, Florida, pp. 82–87, 1994.

[IOS 80] IOSIFESCU M., *Finite Markov Processes and Their Applications*, John Wiley & Sons, 1980.

[JAI 05] JAIMES A.L., COELLO COELLO C., "MRMOGA: parallel evolutionary multiobjective optimization using multiple resolutions", *IEEE Congress on Evolutionary Computation*, pp. 2294–2301, 2005.

[JAS 06] JASZKIEWICZ A., ZIELNIEWICZ S., "Pareto memetic algorithm with path relinking for bi-objective traveling salesperson problem", *European Journal of Operational Research*, vol. 193, no. 3, pp. 885–890, 2006.

[JIN 05] JIN Y., BRANKE J., "Evolutionary optimization in uncertain environments – a survey", *IEEE Transactions on Evolutionary Computation*, vol. 9, no. 3, pp. 303–317, 2005.

[JOH 97] JOHNSON D., MCGEOCH L., "The traveling salesman problem: a case study in local optimization", in AARTS E., LENSTRA J. (eds), *Local Search in Combinatorial Optimization*, Princeton University Press, 1997.

[JOI 94] JOINES J., HOUCK C., "On the use of non-stationary penalty functions to solve nonlinear constrained optimization problems with GA's", *IEEE World Congress on Computational Intelligence*, Orlando, Florida, 1994.

[KAL 60] KALMAN R.E., "A new approach to linear filtering a prediction problem", *Journal of Basic Engineering*, vol. 82, pp. 35–45, 1960.

[KAR 07] KARABOGA D., BASTURK B., "A powerful and efficient algorithm for numerical function optimization: artificial bee colony (ABC) algorithm", *Journal Global Optimization*, vol. 39, pp. 459–471, 2007.

[KEE 97] KEEL L., BHATTACHARYYA S., "Robust, fragile, or optimal?", *IEEE Transactions on Automatic Control*, vol. 42, no. 8, pp. 1098–1105, 1997.

[KEN 74] KENNEDY J., SNELL J., THOMPSON G., *Introduction to Finite Mathematics*, Prentice-Hall, 1974.

[KEN 95] KENNEDY J., EBERHART R., "Particle swarm optimization", *IEEE International Conference on Neural Networks*, Perth, WA, pp. 1942–1948, 1995.

[KEY 01] KEYNES R. (ed.), *Charles Darwin's Beagle Diary*, Cambridge University Press, 2001.

[KHA 98] KHATIB W., FLEMING P., "The stud GA: a mini revolution?", in EIBEN A., BACK T., SCHOENAUER M. *et al.* (eds), *Parallel Problem Solving from Nature*, Springer, 1998.

[KHA 03] KHARE V., YAO X., DEB K., "Performance scaling of multi-objective evolutionary algorithms", in FONSECA C., FLEMING P., ZITZLER E. *et al.* (eds), *Evolutionary Multi-Criterion Optimization: Second International Conference*, 2003.

[KLE 04] KLEIDON A., "Amazonian biogeography as a test for Gaia", in SCHNEIDER S., MILLER J., CRIST E. *et al.* (eds), *Scientists Debate Gaia*, MIT Press, 2004.

[KON 06] KONDOH M., "Does foraging adaptation create the positive complexity–stability relationship in realistic food-web structure?", *Journal of Theoretical Biology*, vol. 238, no. 3, pp. 646–651, 2006.

[LAS 10] LASSIG J., SUDHOLT D., "The benefit of migration in parallel evolutionary algorithms", *Genetic and Evolutionary Computation Conference*, pp. 1105–1112, 2010.

[LEG 09] LEGUIZAMÓN G., COELLO COELLO C., "Boundary search for constrained numerical optimization problems", in MEZURA-MONTES E. (ed.), *Constraint-Handling in Evolutionary Computation*, MIT Press, 2009.

[LEI 79] LEIGHTON F., "A graph coloring algorithm for large scheduling problems", *Journal of Research of the National Bureau of Standards*, vol. 84, pp. 489–505, 1979.

[LEN 98] LENTON T., "Gaia and natural selection", *Nature*, vol. 394, no. 6692, pp. 439–447, 1998.

[LER 15] LERHER T., EKREN B.Y., SARI Z. *et al.*, "Simulation analysis of shuttle based storage and retrieval systems", *International Journal of Simulation Modelling*, vol. 14, no. 1, pp. 11–23, 2015.

[LI 03] LI X., SHAO Z., QIAN J., "An optimizing method based on autonomous animals: fish-swarm algorithm", *Systems Engineering – Theory and Practice*, vol. 22, no. 11, pp. 32–38, 2003.

[LIA 06] LIANG J.J., RUNARSSON T.P., MEZURA-MONTES E. *et al.*, Problem definitions and evaluation criteria for the CEC 2006 special session on constrained real-parameter optimization, Technical report, Nanyang Technological University, Singapore, 2006.

[LOM 00a] LOMOLINO M., "A call for a new paradigm of island biogeography", *Global Ecology and Biogeography*, vol. 9, no. 1, pp. 1–6, 2000.

[LOM 00b] LOMOLINO M., "A species-based theory of insular zoogeography", *Global Ecology and Biogeography*, vol. 9, no. 1, pp. 39–58, 2000.

[LOV 90] LOVELOCK J., "Hands up for the Gaia hypothesis", *Nature*, vol. 344, no. 6262, pp. 100–102, 1990.

[LOV 95] LOVELOCK J. (ed.) *Gaia*, Oxford University Press, 1995.

[LUN 86] LUNDY M., MEES A., "Convergence of an annealing algorithm", *Mathematical Programming*, vol. 34, no. 1, pp. 111–124, 1986.

[MA 09] MA H., NI S., SUN M., "Equilibrium species counts and migration model tradeoffs for biogeography-based optimization", *IEEE Conference on Decision and Control and 28th Chinese Control Conference*, Shanghai, China, pp. 3306–3310, 2009.

[MA 10a] MA H., "An analysis of the equilibrium of migration models for biogeography-based optimization", *Information Sciences*, vol. 180, no. 18, pp. 3444–3464, 2010.

[MA 10b] MA H., SIMON D., "Biogeography-based optimization with blended migration for constrained optimization problems", *Genetic and Evolutionary Computation Conference (GECCO)*, Portland, Oregon, USA, pp. 417–418, 2010.

[MA 11a] MA H., SIMON D., "Blended biogeography-based optimization for constrained optimization", *Engineering Applications of Artificial Intelligence*, vol. 24, no. 3, pp. 517–525, 2011.

[MA 11b] MA H., SIMON D., "Analysis of migration models of biogeography-based optimization using Markov theory", *Engineering Applications of Artificial Intelligence*, vol. 24, no. 6, pp. 1052–1060, 2011.

[MA 13a] MA H., SIMON D., FEI M. *et al.*, "On the equivalences and differences of evolutionary algorithms", *Engineering Applications of Artificial Intelligence*, vol. 26, no. 10, pp. 2397–2407, 2013.

[MA 13b] MA H., SIMON D., FEI M. *et al.*, "Variations of biogeography-based optimization and Markov analysis", *Information Sciences*, vol. 220, no. 1, pp. 492–506, 2013.

[MA 14a] MA H., SIMON D., FEI M., "On the convergence of biogeography-based optimization for binary problems", *Mathematical Problems in Engineering*, vol. 2014, 11 pp., 2014.

[MA 14b] MA H., SIMON D., FEI M. *et al.*, "Hybrid biogeography-based evolutionary algorithms", *Engineering Applications of Artificial Intelligence*, vol. 30, pp. 213–224, 2014.

[MA 15a] MA H., FEI M., SIMON D. *et al.*, "Biogeography-based optimization in noisy environments", *Transactions of the Institute of Measurement and Control*, vol. 37, no. 2, pp. 190–204, 2015.

[MA 15b] MA H., SU S., SIMON D. *et al.*, "Ensemble multi-objective biogeography-based optimization with application to automated warehouse scheduling", *Engineering Applications of Artificial Intelligence*, vol. 44, pp. 79–90, 2015.

[MA 16a] MA H., SIMON D., FEI M., "Statistical mechanics approximation of biogeography-based optimization", *Evolutionary Computation*, vol. 24, pp. 3427–3458, 2016.

[MA 16b] MA H., SIMON D., FEI M. *et al.*, "Interactive Markov models of optimization search strategies", *IEEE Transactions on Systems, Man and Cybernetics: Systems*, In press, 2016.

[MA 16c] MA H., YE S., SIMON D. *et al.*, "Conceptual and numerical comparisons of swarm intelligence optimization algorithms", *Soft Computing*, In press, 2016.

[MAC 55] MACARTHUR R., "Fluctuations of animal populations and a measure of community stability", *Ecology*, vol. 36, no. 3, pp. 533–536, 1955.

[MAC 67] MACARTHUR R., WILSON E., *The Theory of Island Biogeography*, Princeton University Press, 1967.

[MAK 10] MAKEYEV O., SAZONOV E., MOKLYACHUK M. *et al.*, "Hybrid evolutionary algorithm for microscrew thread parameter estimation", *Engineering Applications of Artificial Intelligence*, vol. 23, no. 4, pp. 446–452, 2010.

[MAL 10] MALLIPEDDI R., SUGANTHAN P.N., Problem definitions and evaluation criteria for the CEC 2010 competition on constrained real-parameter optimization, Technical report, Nanyang Technological University, Singapore, 2010.

[MAR 96] MARGULIS L., "Gaia is a tough bitch", in BROCKMAN J. (ed.), *The Third Culture: Beyond the Scientific Revolution*, Touchstone, 1996.

[MAY 73] MAY R., *Stability and Complexity in Model Ecosystems*, Princeton University Press, 1973.

[MCC 00] MCCANN K., "The diversity–stability debate", *Nature*, vol. 405, no. 6783, pp. 228–233, 2000.

[MCT 08] MCTAVISH T., RESTREPO D., "Evolving solutions: the genetic algorithm and evolution strategies for finding optimal parameters", in SMOLINSKI T., MILANOVA M., HASSANIEN A. (eds), *Applications of Computational Intelligence in Biology*, Springer, 2008.

[MIC 96a] MICHALEWICZ Z., *Genetic Algorithms + Data Structures = Evolution Programs*, Springer, 1996.

[MIC 96b] MICHALEWICZ Z., SCHOENAUER M., "Evolutionary algorithms for constrained parameter optimization problems", *Evolutionary Computation*, vol. 4, no. 1, pp. 1–32, 1996.

[MIT 05] MITZENMACHER M., UPFAL E., *Probability and Computing: Randomized Algorithms and Probabilistic Analysis*, Cambridge University Press, 2005.

[MON 12] MONGUS D., REPNIK B., MERNIK M. *et al.*, "A hybrid evolutionary algorithm for tuning a cloth-simulation model", *Applied Soft Computing*, vol. 12, no. 1, pp. 266–273, 2012.

[MOR 98] MORALES A., QUEZADA C., "A universal eclectic genetic algorithm for constrained optimization", *Sixth European Congress on Intelligent Techniques and Soft Computing*, Aachen, Germany, pp. 518–522, 1998.

[MÜH 93] MÜHLENBEIN H., SCHLIERKAMP-VOOSEN D., "Predictive models for the breeder genetic algorithm I. Continuous parameter optimization", *Evolutionary Computation*, vol. 1, no. 1, pp. 25–49, 1993.

[NEK 99] NEKOLA C., WHITE S., "The distance decay of similarity in biogeography and ecology", *Journal of Biogeography*, vol. 26, pp. 867–878, 1999.

[NER 08] NERI F., GARCIA X.T., CASCELLA G.L. *et al.*, "Surrogate assisted local search in PMSM drive design", *Compel International Journal of Computations & Mathematics in Electrical*, vol. 27, no. 3, pp. 573–592, 2008.

[NGU 07] NGUYEN H.D., YOSHIHARA I., YAMAMORI K. *et al.*, "Implementation of an effective hybrid GA for large-scale traveling salesman problems", *IEEE Transactions on Systems, Man, and Cybernetics: Systems*, vol. 37, no. 1, pp. 92–99, 2007.

[NIK 09] NIKNAM T., "An efficient hybrid evolutionary algorithm based on PSO and HBMO algorithms for multi-objective distribution feeder reconfiguration", *Energy Conversion and Management*, vol. 50, no. 8, pp. 2074–2082, 2009.

[NIX 92] NIX A., VOSE M., "Modeling genetic algorithms with Markov chains", *Annals of Mathematics and Artificial Intelligence*, vol. 5, no. 1, pp. 79–88, 1992.

[OKU 01] OKUBO A., LEVIN S., *Diffusion and Ecological Problems*, Springer, 2001.

[OLI 87] OLIVER I., SMITH D., HOLLAND J., "A study of permutation crossover operators on the traveling salesman problem", *International Conference on Genetic Algorithms*, Cambridge, Massachusetts, pp. 224–230, 1987.

[ONG 07] ONG Y.-S., KRASNOGOR N., ISHIBUCHI H., "Special issue on memetic algorithms", *IEEE Transactions on System, Man, Cybernetics – Part B: Cybernetics*, vol. 37, no. 1, pp. 2–5, 2007.

[PAS 02] PASSINO K., "Biomimicry of bacterial foraging", *IEEE Control Systems Magazine*, vol. 22, no. 3, pp. 52–67, 2002.

[PIE 04] PIETRO A., WHILE L., BARONE L., "Applying evolutionary algorithms to problems with noisy, time-consuming fitness functions", *2004 Congress on Evolutionary Computation*, Portland, Oregon, USA, pp. 1254–1261, 2004.

[POW 93] POWELL D., SKOLNICK M., "Using genetic algorithms in engineering design optimization with non-linear constraints", *International Conference on Genetic Algorithms*, Urbana, Champaign, Illinois, pp. 424–432, 1993.

[PRO 11] PRODHON C., "A hybrid evolutionary algorithm for the periodic location-routing problem", *European Journal of Operational Research*, vol. 210, no. 2, pp. 204–212, 2011.

[PRÜ 94] PRÜGEL-BENNETT A., SHAPIRO J.L., "An analysis of genetic algorithms using statistic mechanics", *Physics Review Letters*, vol. 72, no. 9, pp. 1305–1324, 1994.

[QU 12] QU B., LIANG J., SUGANTHAN P.N., "Niching particle swarm optimization with local search for multi-modal optimization", *Information Sciences*, vol. 197, no. 15, pp. 131–143, 2012.

[QUA 96] QUAMMEN D., *The Song of the Dodo: Island Biogeography in an Age of Extinctions*, Scribner Press, 1996.

[RAH 08] RAHNAMAYAN S., TIZHOOSH H., SALAMA M., "Opposition-based differential evolution", *IEEE Transactions on Evolutionary Computation*, vol. 12, no. 1, pp. 64–79, 2008.

[RAT 95] RATTRAY M., "The dynamics of a genetic algorithm under stabilizing selection", *Complex Systems*, vol. 9, no. 3, pp. 213–234, 1995.

[RAT 96a] RATTRAY M., SHAPIRO J.L., "The dynamics of a genetic algorithm for a simple learning problem", *Journal of Physics A*, vol. 29, no. 23, pp. 7451–7473, 1996.

[RAT 96b] RATTRAY M., Modeling the dynamics of genetic algorithms using statistical mechanics, PhD Thesis, University of Manchester, 1996.

[RAY 09] RAY T., SINGH H., ISAACS A. *et al.*, "Infeasibility driven evolutionary algorithm for constrained optimization", in MEZURA-MONTES E. (ed.), *Constraint-Handling in Evolutionary Computation*, MIT Press, 2009.

[REC 68] RECHENBERG I., "Cybernetic solution path of an experimental problem", in FOGEL D. (ed.), *Evolutionary Computation: The Fossil Record*, Wiley-IEEE Press, 1968.

[REE 03] REEVES C., ROWE J., *Genetic Algorithms: Principles and Perspectives*, Kluwer Academic Publishers, Norwell, 2003.

[REI 08] REINELT G., "TSPLIB", available at: http://www.iwr.uni-heidelberg.de/groups/comopt- /software/TSPLIB95/, 2008.

[ROS 67] ROSENBERG R., Simulation of genetic populations with biochemical properties, PhD Thesis, University of Michigan, 1967.

[RUD 94] RUDOLPH G., "Convergence analysis of canonical genetic algorithm", *IEEE Transactions on Neural Networks*, vol. 5, no. 1, pp. 96–101, 1994.

[RUD 00] RUDOLPH G., AGAPIE A., "Convergence properties of some multi-objective evolutionary algorithm", *IEEE Congress on Evolutionary Computation*, San Diego, California, pp. 1010–1016, 2000.

[RUN 00] RUNARSSON T., YAO X., "Stochastic ranking for constrained evolutionary optimization", *IEEE Transactions on Evolutionary Computation*, vol. 4, no. 3, pp. 284–294, 2000.

[SAV 08] SAVICKY P., ROBNIK-SIKONJA M., "Learning random numbers: a Matlab anomaly", *Applied Artificial Intelligence*, vol. 22, no. 3, pp. 254–265, 2008.

[SCH 85] SCHAFFER J., "Multiple objective optimization with vector evaluated genetic algorithms", *International Conference on Genetic Algorithms and Their Application*, Pittsburgh, Pennsylvania, pp. 93–100, 1985.

[SCH 93] SCHWEFEL H.P., *Evolution and Optimum Seeking*, John Wiley & Sons, 1993.

[SCH 11] SCHÜTZE O., LARA A., COELLO COELLO C., "On the influence of the number of objectives on the hardness of a multiobjective optimization problem", *IEEE Transactions on Evolutionary Computation*, vol. 15, no. 4, pp. 444–454, 2011.

[SHA 94] SHAPIRO J.L., PRÜGEL-BENNETT A., RATTRAY M., "A statistical mechanical formulation of the dynamics of genetic algorithms", *Lecture Notes in Computer Science*, vol. 865, pp. 17–27, 1994.

[SIM 06] SIMON D., *Optimal State Estimation*, John Wiley & Sons, 2006.

[SIM 08] SIMON D., "Biogeography-based optimization", *IEEE Transactions on Evolutionary Computation*, vol. 12, no. 6, pp. 702–713, 2008.

[SIM 11a] SIMON D., "A dynamic system model of biogeography-based optimization", *Applied Soft Computing*, vol. 11, pp. 5652–5661, 2011.

[SIM 11b] SIMON D., ERGEZER M., DU D. *et al.*, "Markov models for biogeography-based optimization", *IEEE Transactions on Systems, Man, and Cybernetics – Part B: Cybernetics*, vol. 41, no. 1, pp. 299–304, 2011.

[SIM 11c] SIMON D., "A probabilistic analysis of a simplified biogeography-based optimization algorithm", *Evolutionary Computation*, vol. 19, no. 2, pp. 167–188, 2011.

[SIM 11d] SIMON D., RARICK R., ERGEZER M. *et al.*, "Analytical and numerical comparisons of biogeography-based optimization and genetic algorithms", *Information Sciences*, vol. 181, no. 7, pp. 1224–1248, 2011.

[SIM 13a] SIMON D., *Evolutionary Optimization Algorithms*, John Wiley & Sons, Hoboken, 2013.

[SIM 13b] SIMON D., SHAH A., SCHEIDEGGER C., "Distributed learning with biogeography-based optimization: Markov modeling and robot control", *Swarm and Evolutionary Computation*, vol. 10, pp. 12–24, 2013.

[SIM 14] SIMON D., OMRAN M., CLERC M., "Linearized biogeography-based optimization with re-initialization and local search", *Information Sciences*, vol. 267, pp. 140–157, 2014.

[SPE 97] SPEARS W., DE JONG K., "Analyzing GAs using Markov models with semantically ordered and lumped states", in BELEW R., VOSE M. (eds), *Foundations of Genetic Algorithms*, Springer, vol. 4, 1997.

[SRI 94] SRINIVAS N., DEB K., "Multiobjective optimization using nondominated sorting in genetic algorithm", *Evolutionary Computation*, vol. 2, no. 3, pp. 221–248, 1994.

[STR 01] STROUD P., "Kalman-extended genetic algorithm for search in nonstationary environments with noisy fitness evaluations", *IEEE Transactions on Evolutionary Computation*, vol. 5, no. 1, pp. 66–77, 2001.

[SUG 05] SUGANTHAN P., HANSEN N., LIANG J. *et al.*, Problem definitions and evaluation criteria for the CEC 2005 special session on real-parameter optimization, Technical report, Nanyang Technological University, Singapore, 2005.

[SUZ 95] SUZUKI J., "A Markov chain analysis on simple genetic algorithms", *IEEE Transactions on Systems, Man, and Cybernetics, Part B: Cybernetics*, vol. 25, no. 4, pp. 655–659, 1995.

[SUZ 98] SUZUKI J., "A further result on the Markov chain model of genetic algorithms and its application to a simulated annealing-like strategy", *IEEE Transactions on Systems, Man, and Cybernetics, Part B: Cybernetics,* vol. 28, no. 1, pp. 95–102, 1998.

[TAK 09] TAKAHAMA T., SAKAI S., "Solving difficult constrained optimization problems by the ϵ-constrained differential evolution with gradient-based mutation", in MEZURA-MONTES E. (ed.), *Constraint-Handling in Evolutionary Computation*, MIT Press, 2009.

[TAO 98] TAO G., MICHALEWICZ Z., "Inver-over operator for the TSP", in EIBEN A., BACK T., SCHOENAUER M. *et al.* (eds), *Parallel Problem Solving from Nature*, Springer, 1998.

[TIL 94] TILMAN D., MAY R., LEHMAN C. *et al.*, "Habitat destruction and the extinction debt", *Nature*, vol. 371, no. 3, pp. 65–66, 1994.

[TIN 13] TINOCO J., COELLO COELLO C., "hypDE: a hyper-heuristic based on differential evolution for solving constrained optimization problems", in SCHÜTZE O., COELLO COELLO C., TANTAR A.-A. *et al.* (eds), *EVOLVE – A Bridge between Probability, Set Oriented Numerics, and Evolutionary Computation II*, Springer, 2013.

[TIZ 08] TIZHOOSH H., VENTRESCA M., RAHNAMAYAN S., "Opposition-based computing", in TIZHOOSH H., VENTRESCA M. (eds), *Oppositional Concepts in Computational Intelligence*, Springer, pp. 11–28, 2008.

[TRI 95] TRICK M., "Graph coloring instances", available at: http://mat.gsia.cmu.edu/COLOR/instances.html, 1995.

[VAN 87] VAN LAARHOVEN P.J., AARTS E.H., *Simulated Annealing: Theory and Applications*, Springer, 1987.

[VAV 96] VAVAK F., FOGARTY T., "Comparison of steady state and generational genetic algorithms for use in nonstationary environments", *IEEE conference on Evolutionary Computation*, Nagoya, Japan, pp. 192–195, 1996.

[VOL 97] VOLK T., *Gaia's Body: Toward a Physiology of Earth*, Springer, 1997.

[VOS 90] VOSE M., "Formalizing genetic algorithms", *IEEE Workshop on Genetic Algorithms, Neural Networks, and Simulated Annealing Applied to Signal and Image Processing*, Glasgow, Scotland, 1990.

[VOS 91] VOSE M., LIEPINS G., "Punctuated equilibria in genetic search", *Complex Systems*, vol. 5, no. 1, pp. 31–44, 1991.

[VOS 99] VOSE M., "Random heuristic search", *Theoretical Computer Science*, vol. 229, pp. 1224–1248, 1999.

[WAL 06] WALLACE A., *The Geographical Distribution of Animals (Volumes 1 and 2)*, Adamant Media Corporation, 2006.

[WAN 09] WANG Y., CAI Z., ZHOU Y. *et al.*, "Constrained optimization based on hybrid evolutionary algorithm and adaptive constraint-handling technique", *Structural and Multidisciplinary Optimization*, vol. 37, no. 4, pp. 395–413, 2009.

[WAN 11a] WANG L., FANG C., "An effective shuffled frog-leaping algorithm for multi-mode resource-constrained project scheduling problem", *Information Sciences*, vol. 181, no. 20, pp. 4804–4822, 2011.

[WAN 11b] WANG L., XU Y., "An effective hybrid biogeography-based optimization algorithm for parameter estimation of chaotic systems", *Expert Systems with Applications*, vol. 38, no. 12, pp. 15103–15109, 2011.

[WES 87] WESCHE T., GOERTLER G., HUBERT W., "Modified habitat suitability index model for brown trout in southeastern Wyoming", *North American Journal of Fisheries Management*, vol. 7, no. 2, pp. 232–237, 1987.

[WHI 93] WHITTAKER R., BUSH M., "Dispersal and establishment of tropical forest assemblages, Krakatoa, Indonesia", in MILES J., WALTON D. (eds), *Primary Succession on Land*, Blackwell Science, 1993.

[WHI 98] WHITTAKER R., *Island Biogeography*, Oxford University Press, 1998.

[WIE 92] WIENKE D., LUCASIUS C., KATEMAN G., "Multicriteria target vector optimization of analytical procedures using a genetic algorithm: Part I. Theory, numerical simulations and application to atomic emission spectroscopy", *Analytica Chimica Act*, vol. 265, no. 2, pp. 211–225, 1992.

[WIL 93] WILSON P., MACLEOD M., "Low implementation cost IIR digital filter design using genetic algorithms", *IEEE Workshop on Natural Algorithms in Signal Processing*, Chelmsford, England, pp. 4–8, 1993.

[WIN 08] WINCHESTER S., *The Day the World Exploded*, Collins, Melbourne, 2008.

[WU 95] WU J., VANKAT J., "Island biogeography theory and applications", in NIERENBERGY W. (ed.), *Encyclopedia of Environmental Biology*, Academic Press, 1995.

[YAN 08] YANG X.S. (ed.), *Nature-Inspired Metaheuristic Algorithms*, Luniver Press, 2008.

[YAN 13] YANG W., DENG L., NIU Q. *et al.*, "Improved shuffled frog leaping algorithm for solving multi-aisle automated warehouse scheduling optimization", *Communications in Computer and Information Science*, vol. 402, pp. 82–92, 2013.

[YAO 99] YAO X., LIU Y., LIN G., "Evolutionary programming made faster", *IEEE Transactions on Evolutionary Computation*, vol. 3, no. 1, pp. 82–102, 1999.

[YEU 10] YEUNG W.K., CHOI T.M., CHENG T.C.E., "Optimal scheduling in a single-supplier single-manufacturer supply chain with common due windows", *IEEE Transactions on Automatic Control*, vol. 55, pp. 2767–2777, 2010.

[YEU 11] YEUNG W.K., CHOI T.M., CHENG T.C.E., "Supply chain scheduling and coordination with dual delivery modes and inventory storage cost", *International Journal of Production Economics*, vol. 132, pp. 223–229, 2011.

[YU 08] YU X., TANG K., CHEN T. *et al.*, "Empirical analysis of evolutionary algorithms with immigrants schemes for dynamic optimization", *Memetic Computing*, vol. 1, no. 1, pp. 3–24, 2008.

[ZAV 09] ZAVALA A., AGUIRRE A., DIHARCE E., "Continuous constrained optimization with dynamic tolerance using the COPSO algorithm", in MEZURA-MONTES E. (ed.), *Constraint-Handling in Evolutionary Computation*, MIT Press, 2009.

[ZHA 08] ZHANG Q., ZHOU A., ZHAO S. *et al.*, Multi-objective optimization test instances for the CEC 2009 special session and competition, Technical report, University of Essex and Nanyang Technological University, 2008.

[ZHA 11] ZHAO S.Z., SUGANTHAN P.N., DAS S., "Self-adaptive differential evolution with multi-trajectory search for large-scale optimization", *Soft Computing*, vol. 15, no. 11, pp. 2175–2185, 2011.

[ZIT 90] ZITZLER E., THIELE L., "On set-based multiobjective optimization", *IEEE Transactions on Evolutionary Computation*, vol. 14, no. 1, pp. 58-79, 1990.

[ZIT 99] ZITZLER E., THIELE L., "Multiobjective evolutionary algorithms: a comparative case study and the strength Pareto approach", *IEEE Transactions on Evolutionary Computation*, vol. 3, no. 4, pp. 257–271, 1999.

[ZIT 00] ZITZLER E., DEB K., THIELE L., "Comparison of multiobjective evolutionary algorithms: empirical results", *Evolutionary Computation*, vol. 8, no. 2, pp. 173–195, 2000.

[ZIT 03] ZITZLER E., THIELE L., LAUMANNS M. *et al.*, "Performance assessment of multiobjective optimizers: an analysis and review", *IEEE Transactions on Evolutionary Computation*, vol. 7, no. 2, pp. 117–132, 2003.

[ZIT 04] ZITZLER E., LAUMANNS M., BLEULER S., "A tutorial on evolutionary multiobjective optimization", *Lecture Notes in Economics and Mathematical Systems*, vol. 535, no. 3, pp. 3–37, 2004.

Index

J, K, L

M, N, O

P, Q, R

S, T

U, V, W

Other titles from

iSTE

in

Computer Engineering

2016

BLUM Christian, FESTA Paola
Metaheuristics for String Problems in Bio-informatics
(Metaheuristics Set – Volume 6)

DEROUSSI Laurent
Metaheuristics for Logistics
(Metaheuristics Set – Volume 4)

DHAENENS Clarisse and JOURDAN Laetitia
Metaheuristics for Big Data (Metaheuristics Set – Volume 5)

LABADIE Nacima, PRINS Christian, PRODHON Caroline
Metaheuristics for Vehicle Routing Problems
(Metaheuristics Set – Volume 3)

LEROY Laure
Eyestrain Reduction in Stereoscopy

LUTTON Evelyne, PERROT Nathalie, TONDA Albert
Evolutionary Algorithms for Food Science and Technology
(Metaheuristics Set – Volume 7)

MAGOULÈS Frédéric, ZHAO Hai-Xiang
Data Mining and Machine Learning in Building Energy Analysis

RIGO Michel
Advanced Graph Theory and Combinatorics

2015

BARBIER Franck, RECOUSSINE Jean-Luc
COBOL Software Modernization: From Principles to Implementation with the BLU AGE® Method

CHEN Ken
Performance Evaluation by Simulation and Analysis with Applications to Computer Networks

CLERC Maurice
Guided Randomness in Optimization (Metaheuristics Set – Volume 1)

DURAND Nicolas, GIANAZZA David, GOTTELAND Jean-Baptiste, ALLIOT Jean-Marc
Metaheuristics for Air Traffic Management (Metaheuristics Set – Volume 2)

MAGOULÈS Frédéric, ROUX François-Xavier, HOUZEAUX Guillaume
Parallel Scientific Computing

MUNEESAWANG Paisarn, YAMMEN Suchart
Visual Inspection Technology in the Hard Disk Drive Industry

2014

BOULANGER Jean-Louis
Formal Methods Applied to Industrial Complex Systems

BOULANGER Jean-Louis
Formal Methods Applied to Complex Systems: Implementation of the B Method

GARDI Frédéric, BENOIST Thierry, DARLAY Julien, ESTELLON Bertrand, MEGEL Romain
Mathematical Programming Solver based on Local Search

KRICHEN Saoussen, CHAOUACHI Jouhaina
Graph-related Optimization and Decision Support Systems

LARRIEU Nicolas, VARET Antoine
Rapid Prototyping of Software for Avionics Systems: Model-oriented Approaches for Complex Systems Certification

OUSSALAH Mourad Chabane
Software Architecture 1
Software Architecture 2

PASCHOS Vangelis Th
Combinatorial Optimization – 3-volume series, 2nd Edition
Concepts of Combinatorial Optimization – Volume 1, 2nd Edition
Problems and New Approaches – Volume 2, 2nd Edition
Applications of Combinatorial Optimization – Volume 3, 2nd Edition

QUESNEL Flavien
Scheduling of Large-scale Virtualized Infrastructures: Toward Cooperative Management

RIGO Michel
Formal Languages, Automata and Numeration Systems 1: Introduction to Combinatorics on Words
Formal Languages, Automata and Numeration Systems 2: Applications to Recognizability and Decidability

SAINT-DIZIER Patrick
Musical Rhetoric: Foundations and Annotation Schemes

TOUATI Sid, DE DINECHIN Benoit
Advanced Backend Optimization

2013

ANDRÉ Etienne, SOULAT Romain
The Inverse Method: Parametric Verification of Real-time Embedded Systems

BOULANGER Jean-Louis
Safety Management for Software-based Equipment

DELAHAYE Daniel, PUECHMOREL Stéphane
Modeling and Optimization of Air Traffic

FRANCOPOULO Gil
LMF — Lexical Markup Framework

GHÉDIRA Khaled
Constraint Satisfaction Problems

ROCHANGE Christine, UHRIG Sascha, SAINRAT Pascal
Time-Predictable Architectures

WAHBI Mohamed
Algorithms and Ordering Heuristics for Distributed Constraint Satisfaction Problems

ZELM Martin *et al.*
Enterprise Interoperability

2012

ARBOLEDA Hugo, ROYER Jean-Claude
Model-Driven and Software Product Line Engineering

BLANCHET Gérard, DUPOUY Bertrand
Computer Architecture

BOULANGER Jean-Louis
Industrial Use of Formal Methods: Formal Verification

BOULANGER Jean-Louis
Formal Method: Industrial Use from Model to the Code

CALVARY Gaëlle, DELOT Thierry, SÈDES Florence, TIGLI Jean-Yves
Computer Science and Ambient Intelligence

MAHOUT Vincent
Assembly Language Programming: ARM Cortex-M3 2.0: Organization, Innovation and Territory

MARLET Renaud
Program Specialization

SOTO Maria, SEVAUX Marc, ROSSI André, LAURENT Johann
Memory Allocation Problems in Embedded Systems: Optimization Methods

2011

BICHOT Charles-Edmond, SIARRY Patrick
Graph Partitioning

BOULANGER Jean-Louis
Static Analysis of Software: The Abstract Interpretation

CAFERRA Ricardo
Logic for Computer Science and Artificial Intelligence

HOMES Bernard
Fundamentals of Software Testing

KORDON Fabrice, HADDAD Serge, PAUTET Laurent, PETRUCCI Laure
Distributed Systems: Design and Algorithms

KORDON Fabrice, HADDAD Serge, PAUTET Laurent, PETRUCCI Laure
Models and Analysis in Distributed Systems

LORCA Xavier
Tree-based Graph Partitioning Constraint

TRUCHET Charlotte, ASSAYAG Gerard
Constraint Programming in Music

VICAT-BLANC PRIMET Pascale *et al.*
Computing Networks: From Cluster to Cloud Computing

2010

AUDIBERT Pierre
Mathematics for Informatics and Computer Science

BABAU Jean-Philippe *et al.*
Model Driven Engineering for Distributed Real-Time Embedded Systems 2009

BOULANGER Jean-Louis
Safety of Computer Architectures

MONMARCHE Nicolas *et al.*
Artificial Ants

PANETTO Hervé, BOUDJLIDA Nacer
Interoperability for Enterprise Software and Applications 2010

PASCHOS Vangelis Th
Combinatorial Optimization – 3-volume series
Concepts of Combinatorial Optimization – Volume 1
Problems and New Approaches – Volume 2
Applications of Combinatorial Optimization – Volume 3

SIGAUD Olivier *et al.*
Markov Decision Processes in Artificial Intelligence

SOLNON Christine
Ant Colony Optimization and Constraint Programming

AUBRUN Christophe, SIMON Daniel, SONG Ye-Qiong *et al.*
Co-design Approaches for Dependable Networked Control Systems

2009

FOURNIER Jean-Claude
Graph Theory and Applications

GUEDON Jeanpierre
The Mojette Transform / Theory and Applications

JARD Claude, ROUX Olivier
Communicating Embedded Systems / Software and Design

LECOUTRE Christophe
Constraint Networks / Targeting Simplicity for Techniques and Algorithms

2008

BANÂTRE Michel, MARRÓN Pedro José, OLLERO Hannibal, WOLITZ Adam
Cooperating Embedded Systems and Wireless Sensor Networks

MERZ Stephan, NAVET Nicolas
Modeling and Verification of Real-time Systems

PASCHOS Vangelis Th
Combinatorial Optimization and Theoretical Computer Science: Interfaces and Perspectives

WALDNER Jean-Baptiste
Nanocomputers and Swarm Intelligence

2007

BENHAMOU Frédéric, JUSSIEN Narendra, O'SULLIVAN Barry
Trends in Constraint Programming

JUSSIEN Narendra
A to Z of Sudoku

2006

BABAU Jean-Philippe *et al.*
From MDD Concepts to Experiments and Illustrations – DRES 2006

HABRIAS Henri, FRAPPIER Marc
Software Specification Methods

MURAT Cecile, PASCHOS Vangelis Th
Probabilistic Combinatorial Optimization on Graphs

PANETTO Hervé, BOUDJLIDA Nacer
Interoperability for Enterprise Software and Applications 2006 / IFAC-IFIP I-ESA'2006

2005

GÉRARD Sébastien *et al.*
Model Driven Engineering for Distributed Real Time Embedded Systems

PANETTO Hervé
Interoperability of Enterprise Software and Applications 2005

Printed and bound by CPI Group (UK) Ltd, Croydon, CR0 4YY

16/09/2026

14963059-0002